HANDBOOK of REGENERATIVE LANDSCAPE DESIGN

Integrative Studies in Water Management and Land Development

Series Editor
Robert L. France

Published Titles

Boreal Shield Watersheds: Lake Trout Ecosystems in a Changing Environment
Edited by J.M. Gunn, R.J. Steedman, and R.A. Ryder

The Economics of Groundwater Remediation and Protection
Paul E. Hardisty and Ece Özdemiroğlu

Forests at the Wildland–Urban Interface: Conservation and Management
Edited by Susan W. Vince, Mary L. Duryea, Edward A. Macie, and L. Annie Hermansen

Handbook of Water Sensitive Planning and Design
Edited by Robert L. France

Porous Pavements
Bruce K. Ferguson

Restoration of Boreal and Temperate Forests
Edited by John A. Stanturf and Palle Madsen

Wetland and Water Resource Modeling and Assessment: A Watershed Perspective
Edited by Wei Ji

HANDBOOK of REGENERATIVE LANDSCAPE DESIGN

Edited by
Robert L. France

CRC Press
Taylor & Francis Group
Boca Raton London New York

CRC Press is an imprint of the
Taylor & Francis Group, an **informa** business

CRC Press
Taylor & Francis Group
6000 Broken Sound Parkway NW, Suite 300
Boca Raton, FL 33487-2742

First issued in paperback 2019

© 2008 by Taylor & Francis Group, LLC
CRC Press is an imprint of Taylor & Francis Group, an Informa business

No claim to original U.S. Government works

ISBN-13: 978-0-8493-9188-0 (hbk)
ISBN-13: 978-0-367-38846-1 (pbk)

Visit the Taylor & Francis Web site at
http://www.taylorandfrancis.com

and the CRC Press Web site at
http://www.crcpress.com

"[It] is necessary ... to overcome the idea that just because Venice is unique ... its problems are equally so. Much can be learnt in Venice from other areas of the world, with regard to specific phenomena, research techniques and remedies."

—Da Mosto et al. *in* **Fletcher, C.A. and T. Spencer (Eds.)**
Flooding and Environmental Challenges for Venice and its Lagoon: State of Knowledge (2005)

Contents

PART 6 Regions

Series Statement: Integrative Studies in Water Management and Land Development

Ecological issues and environmental problems have become exceedingly complex. Today, it is hubris to suppose that any single discipline can provide all the solutions for protecting and restoring ecological integrity. We have entered an age where professional humility is the only operational means for approaching environmental understanding and prediction. As a result, socially acceptable and sustainable solutions must be both imaginative and integrative in scope; in other words, garnered through combining insights gleaned from various specialized disciplines, expressed and examined together.

The purpose of the CRC Press series Integrative Studies in Water Management and Land Development is to produce a set of books that transcends the disciplines of science and engineering alone. Instead, these efforts will be truly integrative in their incorporation of additional elements from landscape architecture, land-use planning, economics, education, environmental management, history, and art. The emphasis of the series will be on the breadth of study approach coupled with depth of intellectual vigor required for the investigations undertaken.

Robert L. France
Series Editor
Integrative Studies in Water Management
and Land Development
Associate Professor of Landscape Ecology
Science Director of the Center for
Technology and Environment
Harvard University
Principal, W.D.N.R.G. Limnetics
Founder, Green Frigate Books

Foreword

Visible Cities: A Meditation on Civic Engagement for Urban Sustainability and Landscape Regeneration

Robert M. Abbott

The past is not the past. It is the context. The past—memory—is one of the most powerful, practical tools available to a civilized democracy …. It reminds us of our successes and failures, of their context; it warns us, encourages us.

John Ralston Saul, LaFontaine–Baldwin Lecture, 2000

As regular viewers of the award-winning American television program, *The West Wing*, or keen students of political theory can attest, democracy isn't easy; you have to want it bad. Put another way, the bedrock of democracy is public participation in the functioning, planning, and decision-making of society. Those of us lucky enough to live in democratic countries have a great many rights and freedoms we should all hold dear—the freedom to think what we like, to voice those opinions, to worship or not to worship at all, and so on. However, these rights come with responsibilities such as obeying laws, paying taxes, and exercising the right to vote. The responsibility that can make the most lasting difference, however, is getting involved in the political and policy process. As Craig Rimmerman notes in his book *The New Citizenship: Unconventional Politics, Activism and Service*, "increased citizen participation in community and workplace decision-making is important if people are to recognize their roles and responsibilities as citizens within the larger community …. In a true participatory setting, citizens do not merely act as autonomous individuals pursuing their own interests, but instead, through a process of decision, debate, and compromise, they ultimately link their concerns with the needs of the community." This is how we can all best be citizens of our cities, of our countries. This is how we serve the common good, or to use that most vexing of words, how we might move towards *sustainability*. Regardless of what we call it, the fundamental challenge facing us—as neighborhoods, as cities, as societies—is to learn how to put aside our individual aspirations and get involved in shaping a truly democratic community. Each of us has a voice, but it is too often silent; we must learn to use that voice to demonstrate that the common good, that sustainability, lives in all of us and is, in fact, only truly made manifest when we speak as a collective.

This exciting new volume, *Handbook of Regenerative Landscape Design Lessons for Venice and Elsewhere*, does an inestimable service in highlighting the system conditions necessary for the repair of degraded landscapes, especially in the urban context. At the top of the list is an acknowledgment that "landscape reparation is both far too important and far too complex to remain the purview of any single group of practitioners," and it is "just as much about people as it is about the nonhuman environment." Robert France and his collaborators well understand that the long-range planning undertaken by many cities, while useful and important, is only part of something larger—the creation of community and the need to work with a divergent range of stakeholders to truly create a sense of community. Equally, France *et al.* recognize the need to shine a necessary light on stories of individuals who have contributed to something bigger than any one person—the repair of a degraded landscape and the restoration of community. But there is more. The chapters in this volume also offer hope and direction for one of the world's most precious cities, Venice, the Queen of the Adriatic. In his lyrical and poetic *Invisible Cities*, the novelist Italo Calvino places us beside Marco Polo, the young Venetian explorer, and Kublai Khan, ruler of an empire that encircled the known world. The stories of the cities that Polo has visited enthrall the emperor, and yet he can't escape the feeling that in telling him about these places, Polo is artfully describing, piece by wondrous piece, only one city, the city they both love—Venice. While it is intellectually playful to imagine Calvino positioning Venice as the "invisible city" that slowly emerges from the mists of Polo's stories, it is a far more serious matter to imagine the loss of this city if its challenges—of water and waste management, of transportation and urban form, of livability—are not addressed. Imagine it is the year 2100. What is Venice like? Is the city thriving, or is it in decline? Are the canals noxious, or have they been reborn on the strength of a galvanizing vision? Is there a sense of frustration, if not anger, on the part of residents who feel marginalized from urban planning decisions, or is there a genuine sense of access, of involvement? While the year 2100 may seem a long way off, it is within the lifespan of most urban infrastructure investments. It is also, more potently, within the expected life of our children and grandchildren. The chapters here speak to all decision makers and shapers in Venice, and elsewhere, and ask that we do things differently than before, better than before. They ask that we make choices that ensure Venice remains an enchanted city, a vital city, a city much like the one of color and light that beguiled the painter, John Singer Sargent. And so it is that in this collection Venice again becomes the connective tissue that binds together stories of cities and landscapes, stories of hope and repair, stories that reflect our better selves.

At the climax of Oscar Wilde's novella, *The Picture of Dorian Gray*, the title character, seeking distraction and refuge from his latest sin, reads a beautiful poem about Venice by Theophile Gautier:

> On a colorful scale,
> Her breast dripping with pearls,
> The Venus of the Adriatic
> Draws her pink and white body out of the water.
> The domes, on the azure of the waves
> Following the pure contour of the phrase,
> Swell like rounded breasts
> Lifted by a sign of love.
> The skiff lands and drops me off,
> Casting its rope to the pillar
> In front of a pink façade
> On the marble of a staircase.

The point–counterpoint between the beauty of Venice in the poem and the ugliness of Dorian Gray's soul is stark. He has made a Faustian bargain; his visage remains untouched by time and the ravages of his lifestyle, but his portrait, hidden from view, graphically marks his decay. In the vernacular of our age, Dorian Gray was not transparent about his choices and their consequences. And despite the artifice of a painting that changes to reflect moral disintegration, Wilde makes an important point about individual choice—a point that is relevant to discussions of where and how we live. Too often we mask our true selves; we don our professional trappings and make choices without regard for the cost to the community, or to our own soul. We toe the party line, we do what's easy, or what's been done before, or we do what will increase our individual welfare. We seldom think of what is possible, what could help people create lives of abundance, lives of access and opportunity, that replenish and restore nature rather than degrade it. We seldom think of how we are all impoverished if we don't correct infrastructure deficits, if we don't foster an ethic of personal, governmental, and corporate responsibility and service to others, if we don't provide opportunities rather than burdens for future generations. Too often we remain silent, our true selves hidden away. Nowhere is this sad drama more evident than in our cities (Abbott and Holland, 2008). Is it any wonder then, that so many of our cities are breeding grounds for despair and disillusionment?

Handbook of *Regenerative Landscape Design* gathers together brave voices that are not silent, that are sounding a clarion call to break with the conventional wisdom. To make the invisible choices and decisions that underlie the decay and degradation of our cities visible. To make an intentional choice to do things differently than before, better than before. As the authors make clear, this is not easy. Robin Nagle, for example, points to the importance of listening to differing views of what might be done to a degraded landscape, and crucially, being prepared to accommodate those views. Fritz Conradin and Reinhard Buchli note that true success in landscape regeneration is often dependent on local residents standing up to governments who are frequently resistant to change. Viewed in this light, Conradin and Buchli rightly put a premium on building trust with the public at a grass-roots level. Kit O'Neill and Peggy Gaynor echo this with a call to highlight the importance of local ownership of landscape regeneration projects. Thomas Goreau and Wolf Hilbertz name the metaphorical elephant in the room of landscape regeneration—namely, the tendency for it to be a top-down imposition by outside agencies or individuals that overrides or ignores real world local experience and expertise. And Steve Apfelbaum *et al.*, echoing a theme that informs much of my own work (Abbott, 2008a,b), point out that it is necessary to educate humans on the primacy of natural capital; in the absence of awareness and understanding in this regard, progress in overcoming socio-economic, political, and regulatory barriers to maintaining these essential services will be difficult, if not impossible, to achieve.

A story that is not included in the collection, but that aligns with many of the design and other principles at play, is unfolding where I live, in Calgary, Alberta, and merits a brief mention in this Foreword. Calgary has been a corporate center for the Canadian oil and gas industry since the 1914 discovery of the Turner Valley gas and oil field west of the city. Today, Calgary is one of Canada's fastest growing cities. The challenge facing virtually every city in the world, but especially ones like Calgary that are experiencing rapid growth, is the tendency to focus on individual symptoms of nonsustainability rather than taking a systems-based view of the whole. *Imagine Calgary* is a city-led, community-owned initiative to create a 100-year sustainability vision for the city and 30-year targets and strategies that align with the vision (www.imagineCALGARY.ca). The methodology that underpins this effort is a whole systems approach that uses the *Melbourne Principles* as a directional compass. The Melbourne Principles are a set of principles that apply the *Earth Charter* to municipalities. The work underway in Calgary is one of the most ambitious attempts to date to embed the ideas and ideals of the Earth Charter in community planning.[1]

In some respects, it is easy to see why some might arch an eyebrow at the very idea of something like Imagine Calgary. It is, after all, popularly (if improperly) viewed as little more than "another" planning exercise and what most people really want is action. Others might note that the majority of plans and strategies fail, or fail to have the impact they might because they can't be effectively implemented. Still others will carp that any effort to make Calgary sustainable will inevitably stifle the entrepreneurial verve that distinguishes the city.

While it is right to note that the proof of any planning effort lies in the extent to which vision and idea are translated into on-the-ground change (Abbott and Holland, 2008), it is equally right to note that not all approaches to planning are the same. And it is here that Imagine Calgary boldly stakes out new ground. Most strategic planning efforts lack an effective process to involve the front-line

---

[1] The Earth Charter is a set of principles for a just, peaceful, and sustainable future that was created by thousands of people from around the world. The principles clearly state that decisions about a safe and healthy environment must also address the disenfranchised, and those decisions should embrace democratic, nonviolent forms of interaction.

people ultimately responsible for implementation. Imagine Calgary breaks from this mold in at least four important ways:

- It is actively engaging members of the public in conversations about the future of the city over a 100-year time period.
- It has created a round table and series of working groups, comprising community leaders and organizations with a mandate to implement change, to backstop the broad community engagement effort.
- It has forged productive working relationships with a variety of experts and other interested people from Calgary and elsewhere to provide periodic "reality checks" on the process.
- Most crucially, it has created a mayor's panel comprising community leaders who can ensure the good ideas are translated into actions that resonate with and motivate members of the public; the panel is a tangible demonstration that there is a will to make the vision happen.

Collectively, these traits make for a powerful lens through which economic, social, and environmental challenges are viewed as interdependent parts of the same system. The decisions Calgary makes today about buildings and infrastructure, for example, will define the environmental impacts and cost profiles for energy, water, air quality, health, and many other social and economic issues within the city for the next 100 years. Further, local government decisions about cultural life, social connections, and governance systems will define how people from all walks of life interact with the city and each other. If we learned anything from watching the devastation of New Orleans by Hurricane Katrina, it was the need to ask searching questions about the resilience of municipal infrastructure and the thinking that underlies decisions about the shape and "nature" of a city.

Marcel Proust observed that the real act of discovery is not in finding new lands, but seeing with new eyes. Imagine Calgary represents a rare opportunity for the people of Calgary to do just this—to see their city as they perhaps haven't seen it before, to draw a picture of what they want. Can the city maintain or enhance environmental quality while remaining an economic engine, and if so, how? Can it provide meaningful work for the people who live there? Can it be a place that incubates and supports new ideas in business, the arts, health and wellness, and citizen engagement? These and other questions lie at the heart of any meaningful conversation about what might make Calgary (and other cities) more vibrant and livable now and in the future. Imagine Calgary provides a seat at the table of that conversation for every Calgarian. Better still, it provides an opportunity for individuals and groups to shape the conversation, own a piece of the action, and see themselves as profound culture-shifting agents. To date, an extraordinary total of 18,000 city residents have contributed to the conversation, making Imagine Calgary the largest citizen involvement in visioning of any city in the world—and a powerful model for other cities, including Venice.

Manual Castells, author of *Rise of the Network Society*, argues that technology and globalization are making networks of relationships a decisive business asset. In much the same way that the Ford Motor Company's assembly line was the icon of the industrial age, Castells argues that the globally networked business model is at the vanguard of the information age. The same can be said of the networks required to build a great city. It might start with a particular civic problem such as poverty, poor infrastructure, crime, or the need to repair a degraded landscape, but it must move to trust for the collaboration necessary to tackle these complex problems. Handbook of *Regenerative Landscape Design* understands this imperative. Its chapters, a few of which I cite here in something of a literary *amuse bouche*, are not afraid to acknowledge the failures of modern environmental management. Equally, they do not shy away from the hard work that lies at the heart of truly effective landscape regeneration and repair. In this regard, they call to mind the poet Joseph Brodsky's evocative description of Venice:

So you never know as you move through these labyrinths whether you are pursuing a goal or running from yourself, whether you are the hunter or his prey.

To be sure, recasting the urban form, especially in cities like Venice, is daunting work not unlike threading one's way through a labyrinth. What might sustain us, however, and what is ultimately most exciting about this collection is the message of hope carried in the papers, hope of something better, hope of making better decisions now that reduce burdens on future generations. That such a message is grounded in people, not technology, in practice, not abstract theory, is a most welcome addition to the canon of landscape literature.

Calgary, Alberta and Venice, Italy

2007

LITERATURE CITED

Abbott, R.M. 2008a *Uncommon Cents: Thoreau and the Nature of Business*. Green Frigate Books.
Abbott, R.M. 2008b. *Conscious Endeavors: Business, Society and the Journey to Sustainability*. Green Frigate Books.
Abbott, R.M. and Holland, M. 2008. *Where We Live: Chasing the Dream of Urban Sustainability*. Green Frigate Books.
Brodsky, J. 1992. *Watermark*. Farrar, Strauss & Giroux.
Calvino, I. 1978. *Invisible Cities*. Harvest Books.
Castells, M. 2000. *Rise of the Network Society*. 2nd ed. Blackwell Publishing, Incorporated.
Gautier, T. 1981. *Emaux et Camees*. Gallimard.
Ormond, R. and Adelson, W. 2006. *Sargent's Venice*. Yale University Press.
Rimmerman, C. 2001. *The New Citizenship: Unconventional Politics, Activism and Service*, 2nd ed. HarperCollins Canada.
Wilde, O. 1998. *The Picture of Dorian Gray*. Modern Library.

Preface

Environmental Reparation with People in Mind: Regenerative Landscape Design at the Interface of Nature and Culture

RESTORATION WRIT LARGE

Restoration has become one of the most economically important (Cunningham 2001), intellectually challenging (Gobster and Hull 2000; France 2007a), and newsworthy (Fletcher and Da Mosto 2004; France 2007b) of all forms of modern-day environmental management. However, it wasn't until I attended my first conference about environmental restoration that I realized the large gap existing between the mindset of ecologists preoccupied with fixing the broken bits and pieces of nature and that of landscape designers concerned with the process of place-making in order to enable people to reinhabit the repaired landscapes (see Ryan 2007). In short, this corresponds to the inherent differences between the established field of restoration ecology and the emerging field of what might be termed "restoration design" (*sensu* France 2007c).

A major reason for this professional disjuncture arises from the false dichotomy (Wilson 1991; Evernden 1992; Cronon 1996) held by many ecologists (among others) regarding what is referred to, in their minds, as *either* "nature" *or* "culture." This underlies why it has only been recently that ecologists have recognized that nature does in fact exist within our cities and is meritorious of both study and stewardship in its own right (e.g., Houck and Cody 2000). The fact that is rarely acknowledged (Higgs 2003), much less championed (France 2007a), is that environmental restoration *is* very much a design process, and one that that had its origin in the pioneering work of landscape architects and urban designers such as Frederick Law Olmsted as early as the 19th century (Spirn 1984).

The reclamation of previously abused and presently neglected landscapes for use by people is a very different aspiration then the quixotic attempt to replicate some imperfectly imagined previous state of nature, and it is an approach that the Harvard Design School has assumed a leading role in developing (Kirkwood 2001; Marshall 2001; Berger 2002, 2008; France 2007a, 2008a,b, this volume).

ORIGIN OF THE BOOK

A conference, an exhibition, a series of workshops, and a symposium were convened at the Harvard Design School titled "Brown Fields and Gray Waters." This conference and associated meetings were sponsored by the Center for Technology and Environment in the Department of Landscape Architecture and brought together more than 50 presenters from North America, Europe, and the Middle East. The purpose of the gathering was to focus on procedures for achieving environmental integrity through remediating and restoring degraded terrestrial and aquatic landscapes. There were three specific goals: first, to present promising options and limitations in the cleansing of polluted brownfield and industrial sites and the repair of wetlands and buried streams; second, to introduce new restoration and remediation approaches of particular interest to professionals, academics, and researchers working productively in the fields of design, ecology, engineering, and technology; and third, to examine technical studies and key information and the interdependence between innovative site technologies and novel planning and design strategies and processes.

The gathering was perhaps most successful and instrumental in demonstrating the extremely broad range of interests and professional approaches encompassed within regenerative landscape design. In addition to landscape architecture, the following disciplines were represented by the presenters: land-use planning, environmental engineering, conservation biology, real estate development, environmental law, public health, politics, hydrology, urban design, government regulation, watershed management, environmental education, soil science, and community management. The gathering was attended by over 200 practitioners, academics, students, and interested private citizens. Topics covered in the conference and the professional development workshops (the latter titled "Building a Restoration Toolbox") included the following techniques: waste disposal, end-use park design, urban renewal and planning, wetland restorative engineering, stream daylighting, mining reclamation, natural river channel design, industrial landscape restoration, phytoremediation, bioengineering, soil reconstruction, thermal remediation, groundwater monitoring, and restoration ecology; and the following nonstructural elements: community development, health and legal aspects, economic revitalization, environmental toxicity, international exchanges, government regulation, financing and real estate, education and ecotourism, environmental justice, public participation and technology transfer. Locations for these undertakings included: landfills, brownfields, postindustrial factories, abandoned mines, wetlands, former military bases, former gas manufacturing plants, rivers, buried streams, and degraded urban cores.

In addition, an exhibition titled "Reclaimed! Case Studies of Reclamation Processes and Design Practices," was opened at the conference and ran for two months afterward. Five of the most complex, challenging, educational, important, and inspirational projects from around the world were presented: Xochimilco in Mexico City (France, in prep.), Clark County Wetlands Park in Las Vegas (France 2008), Westergasfabrick in Amsterdam (Koekebakker 2004), Fresh Kills Landfill in New York City (Nagel, this volume), and the Wetland Centre in London (France, in prep.). These studies were specifically selected in order to promote the message that the most important regenerative landscape design projects cannot be undertaken without the cooperation and integration of many different disciplines. In other words, landscape reparation is both far too important and far too complex to remain the purview of any single group of practitioners, such as for example, ecologists.

Prior to the conference, workshops, and exhibition—most of which dealt with the more technical aspects of restoration, remediation, and regeneration—a half-day symposium, open to all conference attendees, was convened. Titled "Healing Natures, Repairing Relationships: Landscape Architecture and the Restoration of Ecological Spaces and Consciousness," the symposium addressed questions of "why" rather than "how," "what," or "when," and formed the basis of the first book to derive from the conference (France 2007a).

NATURE OF THE PRESENT BOOK

The present book is designed to introduce, describe, and demonstrate new interpretations to landscape design and development in a form to engage the broadest audience possible. The case studies presented in these pages were carefully selected to illustrate pioneering efforts and new directions in the rapidly evolving field of what I refer to here as—borrowing and adapting a phrase similar to one used by John Lyle (1994)—**regenerative landscape design** (this paradigm is fleshed out in the Conclusion at the end of the present book).

And as a volume in the ongoing series *Integrative Studies in Water Management and Land Development*, this book strongly advances the thesis that success in accomplishing these reparative endeavors will come about most easily through true interdisciplinary partnerships. Unlike other books about ecological restoration, therefore, regenerative landscape design, as described within these pages, is shown to be just as much about people as it is about the nonhuman environment. Importantly, the paradigm of regenerative landscape design as developed from the book's case studies is described in the Conclusion in terms of differentiating it from its sibling disciplines of environmental engineering, ecological restoration, low impact development, and sustainable development.

The chapters in this book have a strong and undisguised aquatic focus. This is a reflection of not only my own particular field of scholarship (e.g., France 2002, 2005, 2006, etc.) but also a result of the referential case study I use to frame the context and illustrate the emerging principles of the book. In this regard, sections dealing with landfill islands, canals and creeks, coasts, communities, heritage sites, and regions are not merely random topics but instead were chosen based on their adaptability to the regenerative landscape design of that most treasured, yet most threatened of cities: Venice (Fletcher and Spencer 2005). In this respect, the "Overture" to follow provides an overview of Venice's copious problems, the "Intermezzo" summarizes the techno-fix solutions that have been implemented in order to "save" Venice, and the "Finale" reviews the litany of concerns and criticisms raised about how Venice is managing to preserve and restore its heritage, as well as introducing several possible solutions.

ACKNOWLEDGMENTS

Chairman George Hargreaves, Dean Peter Rowe, and especially my fellow conference coordinator Niall Kirkwood are thanked for facilitating the Harvard reclamation conference from which this book originated. I would like to express my sincere thanks to the contributors to this publication from whose work, presented in these pages, I have been inspired: Steve Apfelbaum, Peter Becket, Sarah Clark, Fritz Conradin, Nathaniel Cormier, Phil Craul, Doug Eppich, Cheryl Foster, Peggy Gaynor, Thomas Goreau, John Gunn, Bradley Herrick, Wolf Hilbertz, R.J.A. Hoogeveen, Stephanie Hurley, Laura Kadlecik, William Launtenbach, Carol Mayer-Reed, Amy Middleton, Stephen Monet, Robin Nagle, Kit O'Neill, Mark Rasmussen, Clarissa Rowe, Rene Senos, Megan Wilson Stromberg, Neil Thomas, Joshka Wessels, Mike Wilson, and Stephen Vogel. Robert Abbott is especially thanked for his synthesizing Foreword, as is John Gunn for use of his compelling photograph on the back cover.

And to my contributors I would like to offer the following in gratitude: to participate in regenerative landscape design it is necessary to wish for and move toward an imagined better state as captured in the spirit of William Morris writing at a time when Blake's "dark Satanic mills" were coming to be recognized for their toll being exacted upon the health of both land and life: "Dreamer of dreams, born out of my own due time, why do I strive to set the crooked straight?" The case studies presented in this book can be regarded as inspirational dreams on their way to becoming pragmatic realities, and thus can be satisfying answers to Morris's formerly open-ended question.

Robert L. France
Cambridge, Massachusetts

LITERATURE CITED

Berger, A. 2002. *Reclaiming the American West*. Princeton Architectural Press.
Berger, A. 2008. *Designing to Reclaim the Landscape*. Routledge.
Cronon, W. (Ed.) 1996. *Uncommon Ground: Rethinking the Human Place in Nature*. W.W. Norton.
Cunningham, S. 2001. *The Restoration Economy: The Greatest New Growth Frontier*. Berrett-Koehler Publishers.
Evernden, N. 1992. *The Social Creation of Nature*. The John Hopkins University Press.
Fletcher, C. and J. Da Mosto. 2004. *The Science of Saving Venice*. Umberto Alledmandi & C.
Fletcher, C.A. and T. Spencer. (Eds.) 2005. *Flooding and Environmental Challenges for Venice and its Lagoon: State of Knowledge*. Cambridge University Press.
France, R.L. (Ed.) 2002. *Handbook of Water Sensitive Planning and Design*. CRC Press.

France, R.L. (Ed.) 2005. *Facilitating Watershed Management: Fostering Awareness and Stewardship.* Rowman and Littlefield.

France, R.L. 2006. *Introduction to Watershed Development: Understanding and Managing the Impacts of Sprawl.* Rowman & Littlefield.

France, R.L. (Ed.) 2007a. *Healing Natures, Repairing Relationships: New Perspectives in Restoring Ecological Spaces and Consciousness.* Green Frigate Books.

France, R.L. (Ed.) 2007b. *Wetlands of Mass Destruction: Ancient Presage for Contemporary Ecocide in Southern Iraq.* Green Frigate Books.

France, R.L. 2007c. Landscapes and Mindscapes of Restoration Design. *In* France, R.L. (Ed.) *Healing Natures, Repairing Relationships: New Perspectives in Restoring Ecological Spaces and Consciousness.* Green Frigate Books.

France, R.L. 2008a. *Restoring the Iraqi Marshlands: Potentials, Perspectives, Practices.* Sussex Academic Press.

France, R.L. 2008b. *Clark County Wetlands Park: Landscape Restoration and Park Design.* Island Press.

Gobster, P.H. and R.B. Hull. 2000. *Restoring Nature: Perspectives from the Social Sciences and Humanities.* Island Press.

Higgs, E. 2003. *Nature by Design: People, Natural Process, and Ecological Restoration.* MIT Press.

Houck, M. and M. Cody. 2000. *Wild in the City: A Guide to Portland's Natural Areas.* Oregon Historical Society Press.

Kirkwood, N. 2001. *Manufactured Sites: Rethinking the Post-Industrial Landscape.* Spon.

Koekebakker, O. 2004. *Westergasfabriek Culture Park: Transformation of a Former Industrial Site in Amsterdam.* NAi Publications.

Lyle, J.T. 1994. *Regenerative Design for Sustainable Development.* John Wiley & Sons.

Marshall, R. 2001. *Waterfronts in Post-Industrial Cities.* Spon.

Ryan, R. 2007. Understanding the role of environmental designers in environmental restoration and remediation. *In* France, R.L. (Ed.) *Healing Natures, Repairing Relationships: New Perspectives in Restoring Ecological Spaces and Consciousness.* Green Frigate Books.

Spirn, A.W. 1984. *The Granite Garden: Urban Nature and Human Design.* Bassic Books.

Wilson, A. 1991. *The Culture of Nature: North American Landscape from Disney to the Exxon Valdez. Between the Lines.* Harper Collins.

Overture

Acqua Alta: Venice, the New Atlantis?[1]

The city of the Venetians
With the aid of Divine Providence
Was founded on water
Enclosed by water
Defended by water instead of walls
Whoever in any way does damage to the public waters
Shall be declared an enemy of the State
And shall not deserve less punishment
Than he who breaches the sacred walls of the State
The edict is valid forevermore.

—*Magistrato alle Acque* **wall plaque, inscribed in 1553**

INTRODUCTION

Venice is likely the world's most beautiful, recognizable, beloved, and also threatened city, a place of dreams and imagination into which no one ever enters as a complete stranger free of expectations and culturally acquired "memories" (e.g., McCarthy 1963;[2] Plant 2002). The most recent and deservedly critically acclaimed James Bond film, *Casino Royale*, concludes in a very dramatic fashion, even for those who might otherwise be jaded after having watched four decades of 007 in action. In his latest exploit, the super spy narrowly makes his escape from a Venetian palace as it crumbles apart around him and rapidly sinks into the Grand Canal. By consulting a 13-m long panoramic photograph of the Grand Canal (Resini 2006) and pausing the DVD player, it is possible to determine that the special effects of the sinking palace are based on an unnamed 16th-century residence that is actually situated between the Palaszetto Dolfin and the Palazzo Remer near the Ponte Rialto. In a way, this current film merely follows a long established tradition wherein Venice, born of the waters, is prophesized to end her days, like that of Atlantis, by succumbing to the embrace of the ocean and disappearing forever beneath the surface of its watery grave. For example, in December 1966, a few weeks after the disastrous *aqua alta* or high water flooding event, the then-director general of UNESCO, stated "Venice is sinking into the waves; it is as if one of the most radiant stars of beauty were suddenly engulfed" (UNESCO 1979).

Lord Byron, a lover of the city, was one who was especially preoccupied with what he imagined to be Venice's imminent demise. In *Childe Harold's Pilgrimage* he writes:

"Venice, lost and won,
Her Thirteen hundred years of freedom done
Sinks, like a sea-weed into whence she rose!"
And in *Ode to Venice*, he continues with this theme of the inevitable:
"Oh Venice! Oh Venice! When they marble walls
Are level with the waters, there shall be
A cry of nation's o'er they sunken halls
A loud lament along the weeping sea!"

---

[1] An expanded version of this essay will appear in the book *Waterlogged: An Environmental Reflection on Venice*.
[2] See *Finale* section for combined reference listing.

Ruskin, too, begins his remarkably influential *The Stones of Venice* with the following acceptance of Venice's aquatic extinction:

> Her [the Garden of Eden] successor, like her in perfection of beauty, though less in endurance of dominion, is still left for our beholding in the final period of her decline: a ghost upon the sands of the sea, so weak—so quiet—so bereft of all but her loveliness, that we might well doubt, as we watched her faint reflection in the mirage of the lagoon, which was the City, and which was the Shadow. I would endeavor to trace the lines of this image before it be for ever lost, and to record, as far as I may, the warning which seems to me to be uttered by every one of the fast-gaining waves, that beat like passing bells, against the STONES OF VENICE.

The recurring threat of flooding has meant that Venice has become a metaphor for survival of the old, the delicate, and the exotic (Plant 2002), it's eternal struggle against the water an intrinsic part of the city's identity (Spencer et al., *in* Fletcher and Spencer 2005), the perfect allegory of decline, the price that must be paid for either egregious beauty or hubris. Venice has thus become a metaphor for the death of civilization and the triumph of elemental forces beyond human kin or control. Recently, however, has come recognition that Venice's possible demise may have as much or more to do with human mismanagement as any putative whim of Nature's retribution: "Venice is a city beset by misfortune, suffering from severe neglect, caught in a cycle that is destroying her" concluded Frey and Knightly in their sentinel 1976 *The Death of Venice*, playfully tweaking the title of Thomas Mann's iconic novella.

HISTORY

One million years ago the Adriatic spread north and west before retreating. After millennia of fluctuations in sea level, conditions began to stabilize into the coastline we now recognize about 6,000 to 3,000 years ago with the creation of the lagoon (Keahey 2002; Spencer et al., *in* Fletcher and Spencer 2005). The lagoon has a surface area of 550 km², of which 400 km² are open water with an average mean depth of about 1 meter.

A few Roman ruins are found in the lagoon but colonization really began in response to the Lombard invasions of the mainland during the sixth century. In this respect, mimicking Byzantines who had established nearby Ravenna many years before, the lagoon's first settlers picked out the location precisely just because it was so inconvenient (Nova 2002). The refugees found the lagoon ideal in that it was both too deep and too shallow for, respectively, either maritime or terrestrial invasions. In this regard, the defensive walls were the surrounding immense "liquid plains" (Crouzet-Pavan 2002). This lagoon security turned out to be justified when one realizes that Venice is the only Italian city which from its foundation to Napoleonic times has never been invaded (Lauritzen 1986).

Early colonists consolidated the mudflats with mats of woven reeds as do water dwellers in the Iraqi marshlands and on Lake Titicaca today (France 2007a, 2008a). With time, wood planks were sunk to shore up the spongy soil with mud obtained from early canal diggings and other lands "reclaimed" from the salt marshes. The first large city, Torcello, was eventually abandoned in the 12th century due to siltation from rampant deforestation and the disappearance of the salt marshes which had been replaced by freshwater swamps with consequent malaria outbreaks. Settlers then moved further out into the lagoon where they built the historic center of Venice on 117 small islands, originally mudlflats, by using cribs of wood pilings and flexible floors for differential settling (UNESCO 1979). This same building strategy continued into Medieval and later times; the Salute church today lies atop 100,000 such wood pilings (Lauritzen 1986). Venice eventually evolved into being the richest city in the Medieval world with a population of 200,000, many times that of either contemporaneous Rome, London, or Paris (Lauritzen 1986).

The history of Venice is, quite simply, like that of the Mesopotamian marshes (France 2007a, 2008b), the history of environmental (wetland) management: "In the broader history of the environment, Venice was perhaps a pioneer in a common search for an equilibrium between human life and the environment's time scale" (Crouzet-Pavan 2002). For example, the lagoon has needed continual maintenance since the 15th century. The alluvium deposited from the major inflowing rivers always threatened to fill in the lagoon, a situation made more serious by the longshore current which continued to seal in the open water mouths between the barrier islands. As the marshes threatened to spread so did fears that Venice would loose its protective moat and also be sealed off from shipping, its mercantile lifeblood.

The lagoon has always been in a fragile dynamic equilibrium (Crouzet-Pavan 2002) upon which humans have acted, no more dramatically so than the "remarkably bold series of modifications" of diverting the inflowing rivers in the 13th to 16th centuries (Fletcher and Da Mosto 2004). Fears that the lagoon would fill in remained a constant concern in Venetian life as, for example, in the mid-19th century when the Austrians diverted the Brenta River back into the lagoon in order to protect from flooding upstream Padua (by that time having emerged from the shadow of a moribund Venice and being deemed more valuable) (Pemble 1995).

The construction of the automobile causeway in 1933 (famously and jokingly referred to by Venetians as finally enabling the mainland of Europe to be linked to Venice) changed the city from a cluster of islands to a peninsula and resulted in a "devastation" of the western end of the historic city (Pertot 2004), providing a "bridgehead [allowing] the 20th century's assault on Venice to accelerate rapidly" (Lauritzen 1986).

PROBLEMS

SUBSIDENCE

In 1810, French engineers were surprised to find that the missing bases of the lower arcade of the Ducal Palace were actually 38 cm below ground level, having subsided over five centuries and been compensated for by a raising of the surrounding pavement (Pemble 1995). Subsequent archeological work has confirmed these early findings. Underneath the now abandoned island of Torcello can be seen evidence of the repeated buildup of land by the first residents in the lagoon. One floor was built atop another as these early lagoon dwellers struggled to stay ahead of the general subsidence. Indeed, a Roman walkway (one of the few such ruins found in the lagoon) now lies an amazing 2 m below the current sea level (Nova 2002; Ammerman *in* Fletcher and Spencer 2005). The crypt of the San Marco Basilica in the historic center of Venice is now actually 20 cm below sea level.

Venice lies atop a squashed salt marsh and mudflats which are themselves situated above a mile of river sediments that are slowly being compacted (Nova 2002). In other words, as Venice presses down, it squeezes out water and the sediments become thinner and thinner, the result being a general subsidence of about 2 mm per year (Carminati et al., *in* Fletcher and Spencer 2005). There is no doubt that the situation is made all the worse following the Medieval diversion of the major rivers inflowing into the lagoon, the result being that there has been no ongoing deposition of alluvium to counter the natural subsidence (Keahey 2002). And if that was not serious enough, humans have been found to have greatly exacerbated the problem.

Measurements made in Venice indicate a subsidence of 2.6 mm per year over the last century which is substantially higher than the global average of 0.9 mm per year (Frassetto *in* Fletcher and Spencer 2005). In particular, Venice has subsided by as much as 7 in. between 1908 and 1961 (Fay and Knightly 1976), the accelerated rate being linked to the industrial operations on the mainland. In contrast to almost all other decisions in Venice which are characterized by decades of debate before decisions are made, plans for constructing Porto Marghera were accepted a mere

5 days after submission due to being tied to the wishes of the nascent fascist party (Plant 2002). By 1969, the five tap wells in Porto Marghera were withdrawing an incredible 360,000 gallons an hour (Fay and Knightly 1976) to fuel the industries there. This withdrawal of groundwater (which was halted in the 1970s when the dangers became manifest) resulted in Venice losing about 10 cm against sea level in a single century (Frassetto *in* Fletcher and Spencer 2005). Had it not been for Porto Marghera it would have taken Venice until 2050 to reach its current elevation; in other words, humans moved ahead the natural subsidence by about 75 years. From the 1920s to 1970, Venice sank about 10 cm more than nearby Trieste and thus essentially lost a century in its battle against the sea (Fletcher and Da Mosto 2004).

Perhaps the most imaginative independent confirmation of Venice's sinking comes from Camuffo et al.'s (*in* Fletcher and Spencer 2005) study of the meticulously realistic 18th century paintings of Canaletto and his student Bellotto, which had been made using a camera obscura, a technique which records a reflected image of a scene in a dark box that is then copied as a painting and thus is like a real photograph, but one taken a century before photography was invented. This analysis showed that the everyday wetting of the same buildings and buildup of attached algae as a result subsidence (in addition to sea level rise, boat propeller wash, and increased tidal cycles as described below) was 60 cm higher since Canaletto's time, representing an average rate of increase of about 2 mm per year.

SEA LEVEL RISE

In a scene reminiscent of those in the movie *Water World*, the Nova documentary *Sinking City of Venice* contains a dramatic computer-simulation of a future flooded Venice with only the top third of the Campanile and the dome of San Marco peeking out among the waves. The sea level around the world is predicted to rise anywhere from 8 to 88 cm by the start of the next century (Fletcher and Da Mosto 2004; Keahey 2002). Venice is therefore the first city to face sea level rise as a consequence of climate warming due to being flush at sea level (Nova 2002). In this respect, the present conditions in Venice are a presage for what the future might bring elsewhere. Or as Tomasin (*in* Fletcher and Da Mosto 2004) dramatically worded it: "Sea level rise is like a rope that is tightening around the neck of Venice."

MORPHOLOGIC AND HYDROLOGIC ALTERATIONS

Venice's architecture has overshadowed the importance of its lagoon. In point of fact, it is the largest wetland in Italy—one of the most important in the entire Mediterranean—and would be famous today even had the city of Venice never existed. In addition to the 70 km^2 of wetlands, the lagoon is composed of 90 km^2 of closed-off fish farms, 50 islands, and 1,580 km of canals (MITVWACVN 2006a).

Lagoons are transitional landforms that represent a delicate balance between terrestrial and marine influences; they are therefore only temporary ecotones on their way to eventually evolving into either land or sea. To regard lagoons as static entities is to deny the inevitable processes of ecological succession and, in a way, this is exactly what Venice's environmental history is really all about: continually altering the lagoon by attempting to bend time's arrow back on itself to a desirable but ultimately unstable moment in time. It is important to recognize, therefore, that there is really nothing *natural* about the Venice Lagoon which would have disappeared long ago had it not been for human interventions (Keahey 2002).

The opposition of land and water has always been the key determinant for the city and its settlements (Plant 2002). An engraving from the 18th century of the lagoon ecosystem (which was aptly used as the cover image for Lasserre and Marzollo's 2000 book about the lagoon) shows a pitched battle fought between terrestrial and aquatic elements portrayed as two giant wrestlers towering over the city of Venice.

All lagoon ecosystems are vulnerable, and Venice's has through time lost its equilibrium. Since the 14th century when the major rivers were diverted from entering the lagoon, it has been an artificial, man-made body of water. And from the mid 19th century the lagoon has completely lost its ecological balance due to construction of jetties at the three inlets; compartmentalization of the lagoon into fish-rearing enclosures; conversion of marshes, mudflats, and shoals into dry land; dredging of two large shipping channels; and the aforementioned accelerated subsidence due to groundwater extraction (Spencer et al., *in* Fletcher and Spencer 2005).

Originally, the lagoon was a river delta, a swampy marsh into which three major rivers and many tributaries flowed (Lauritzen 1986). However, all the major rivers mouths were diverted to reduce sedimentation and the openings on the Lido were reduced in size in order to accelerate the tidal current and sweep the lagoon free of any accumulated sediment. Dredging of new deepwater canals, salt marsh destruction, and creation of fishing enclosures all contributed to altering tidal patterns in the lagoon (Pertot 2004). Specifically, the Venice Port Authority does about one billion dollars in business a year (about a third of the amount for New York City) and is thus a very important contributor to the regional economy. Openings between the barrier islands have consequently been enlarged and deepened in order to permit passage of increasingly larger ships into the lagoon. This has allowed for an increase of about 10% in the amount of water entering during high tides (Keahey 2002) and has meant that the total volume of water in the lagoon has doubled over the last century. The deepening and narrowing of the shipping entrances and canals has seriously interfered with sediment dynamics and longshore currents (Silvio, *in* Fletcher and Spencer 2005; D'Alpaos and Martini, *in* Fletcher and Spencer 2005). As a result, the lagoon is now depleted of sediment and contains vast areas of underwater desert (Fletcher and Da Mosto 2004).

Fish farming and industrial and urban infill have severely reduced the size of the lagoon and influenced water circulation. Problems with fishing weirs and holding pens inhibiting water circulation have existed since the 13th century but the need for self-sufficiency in food production always superseded any environmental concerns (Fay and Knightly 1976). Today, the situation is more serious in that the one-time nets have been replaced with earthen dams which have yet further reduced the lagoon's area and more severely interfered with the natural circulation patterns.

The city of Venice has not been immune to hydrodynamic alterations. Previously (and to a limited degree today) the city relied upon a twice daily tidal flushing to remove noxious substances discharged directly into its canals (i.e., the "dilution is the solution to pollution" strategy in waste management). However, the "pedestrianization" of Venice underway since the 19th century for tax benefits and to increase tourist access has meant that this natural cleansing system no longer operates. Of the original organizational network of canals and waterside *calli* pedestrian thoroughfares, 40 km were filled in between 1815 and 1889 (Pertot 2004). The *Rio Terra* conversion of Venice's original canals into pedestrian thoroughfares has significantly altered not only water flows but also substantially changed the city makeup; Venice is now, as it never was before, primarily a walking city.

Wetland Loss

The lagoon has the largest salt marsh on the Mediterranean and is especially regarded in naturalist circles for its beauty and as a bird-watching hotspot (the highest concentration of birds in all of Italy) and therefore would have world attention regardless of the presence of the city. Despite this, the salt marshes have been under attack since the time of the very first settlers in the lagoon. Plans in 1880 to fill in most of the lagoon for agriculture sparked an internal debate (Plant 2002) that has continued to today, particularly in relation to development of the mainland coast over the last 50 years.

A consortium of investors began Porto Marghera after WWI with the goal of restoring Venice's past glory as a center of maritime commerce and thus rescue the city from its post-Napoleonic poverty. The first and second industrial zones that developed into the port initially reclaimed 4,000 acres of wetlands followed by another zone of 10,000 acres in the 1960s in order to develop one of Europe's largest petrochemical complexes (Fay and Knightly 1976). The project has always had its critics who would wholeheartedly agree with Pertot's (2004) criticism that it was and is "an extraordinary, unbelievable intrusion on the Venetian landscape" developed under the guise of a harebrained pretension that the city could still be an economic capital.

Alongside grew the rapidly sprawling dormitory complex of Mestre, a jumble of apartment buildings and offices, many built on land formed from drained salt marsh. Mestre's population grew fivefold in the second and third decades of the 20th century due to its proximity to the industrial juggernaut of Porto Marghera and sprawled "like an oil slick as a consequence of the indiscriminate parceling out of land for building development and absence of any concern for urban and environmental values" (Pertot 2004).

As a result of these and other developments, from half to three-quarters of the surface area of the salt marshes has been lost from the lagoon since the start of the 19th century at the same time as the volume of water entering the lagoon has increased (Keahey 2002; Scotti, *in* Fletcher and Spencer 2005; Day et al., *in* Fletcher and Spencer 2005). Today, only 4000 ha of salt marsh remain. Not only has this meant a 20% loss in plant biodiversity and a 50% decrease in birds species in the lagoon (Keahey 2002), but the loss of wetlands, bottom scouring of sediments, and increasingly high waters means that the lagoon is functioning today less like a wetland and increasingly more like a marine environment.

The future looks bleak. Salt marshes continue to wither away due to increased wave action and tidal channel enlarging and also insufficient sediment deposition in relation to sea level rise (Day et al., *in* Fletcher and Spencer 2005). The lagoon is now starved of sediments to such a degree that the wetlands cannot establish themselves naturally. Estimates are that all the salt marshes could be gone within another century.

ENVIRONMENTAL QUALITY DETERIORATION

In *Casino Royale*, there are underwater shots of James Bond swimming in the crystal clear water of the Grand Canal as he struggles to escape from the rapidly sinking palace. This is perhaps the most inaccurate scene in the entire film, for in reality the water of Venice's canals is so filthy as to be almost opaque in color. Also, as any native Venetian knows and shakes her or his head in horror at the escapades of some egregiously naïve tourists, one simply does not swim in Venice's canals … ever. Historically, Venice's canals used to be regularly drained and dredged but this was stopped in the 1960s. As a result, the sediments in the canals are contaminated with accumulated heavy metals, mercury, and fecal coloform bacteria. By the mid-1990s, the Grand Canal was referred to as "a bath of poison" (Plant 2002) or "the only sewer in the world that gives the gawker the intoxication of an embarkation in the Marquesas" (Debray 1999). Perhaps if 007 had known that bioassys of sediment elutriates taken from inner city canals showed them to be highly toxic (McFadzen *in* Lassarre and Marzollo 2000; Livingstone and Nasci *in* Lassarre and Marzollo 2000; Widdows and Nasci *in* Lassarre and Marzollo 2000), he might have thought twice about plunging into the water in an heroic attempt to save his latest love interest. So, although the film ends before we know the ultimate implications of Bond's immersion, it is likely to have been repeated visits to a health clinic upon arrival back at home in the U.K. and many needles to counter the deleterious effects of his risky exposure to Venetian waters. Byron, after all, did not live all that long after his famous swim along the length of the Grand Canal at the start of the 19th century.

The most overt sign of pollution that the visiting tourist, poet, or spy to Venice notices today as they travel to or from the airport, is the dark clouds hovering over the myriad of smokestacks in the mainland industrial complex of Porto Marghera, a scene in glaring contrast to the extreme beauty that either awaits or is already missed on the nearby island of Venice. It is impossible to resist calling to mind Blake's still-powerful words of reference to the dark, Satanic mills of the English Midlands. Later, when one learns that these emissions of sulfur dioxide create clouds of polluted air that hang over the city and contribute to corroding Ruskin's "stones" (Pertot 2004; Buckley 2004)—seemingly immortal only in his own words—one cannot help but feel a little depressed.

Though more subtle, the degree of water pollution is possibly even more serious. The extent of the pollution in the lagoon only really became obvious in the aftermath of the 1966 flood which left tourists stunned as they stared at the scum that coated the entire city as the *aqua alta* receded. Lack of adequate sewage treatment in Venice, urban runoff from Mestre, industrial discharge from Porto Marghera and associated toxic waste sites, and intensive agricultural drainage from the catchment result in a contaminated cocktail of lagoon waters (Fletcher et al., *in* Fletcher and Spencer 2005).

Metal and organochlorine contamination is highest near Porto Marghera, being present in both sediment accumulations as well as suspended in the water column (Cleray et al., *in* Lassarre and Marzollo 2000). This should be no surprise given that the environmental damage wrought by Porto Marghera (thousands of tons of heavy metals dumped into the lagoon to create a toxic sludge) has been officially recognized since the 1950s as reflected in a technical planning document cited in Pertot (2004): "To be located in the industrial zone are mainly those plants that discharge smoke, dust, or emissions harmful to human life into the atmosphere, that discharge poisonous substances into the water, and that produce vibrations and noise." The plan was therefore to promote the area as a pollution haven, and this has succeeded, so that today the area contains factories, a thermo-electric power station, two plants for processing pyrite, steelworks, and a huge petrochemical complex—one of the biggest in Europe—all within a few km of the "Queen of the Adriatic." It truly boggles the mind.

The drainage basin of the Venice lagoon encompasses 1877 km² and is home to a million people. Nutrients in runoff from agricultural fields (which comprise 80% of the surface area) mixed with disposal of Venetian detergents directly into the canals, contributed to a serious eutrophication problem in the lagoon during in the 1980s (Bianchi et al., *in* Lassarre and Marzollo 2000). Noxious macroalgae proliferated (Riccardi and Foltran, *in* Lassarre and Marzollo 2000), which outcompeted the phytoplankton and, upon die-off, led to anoxia, resulting in the mortality of benthic organisms (Scholten et al., *in* Lassarre and Marzollo 2000) and deaths of thousands of fish.

TOURISM

Tourism is practically the sole point of reference for contemporary Venetian life. The city today has an inescapable Disney-like quality to it, "portrayed not as a present living city but as a phantom from the past" (Pertot 2004). This is not new. As far back as the turn of the 20th century Venice was referred to in derogatory terms as "pandering to tourists in her putrescence like an old whore" (Pertot 2004).

Venice is the most touristed city in the world. (Interestingly, given the comparative success of its own environmental restoration projects as touched upon elsewhere in the present book, Las Vegas, with its own Venetian casino, of course, is the second most visited city). Estimates suggest that up to 19 million tourists a year visit Venice, of which three-quarters are the "bite and run" daytrippers (Keahey 2002). The latter results from both the fact that hotel beds in nearby Padua, for example, are about a third the cost of those in Venice itself as well as the bad services characteristic of Venetian old hotels that leaves tourists disgruntled (Musu, *in* Musu 2001).

Rampant tourism has become one of the most pressing of all environmental and social concerns facing present-day Venice. For example, though tourists provide 35% of all economic activities on the island, they also account for 20% of all boat passengers and contribute 83% to the total amount of waste that is generated there (van der Borg and Russo, *in* Musu 2001). Venice's monoculture of tourism slows down innovation, displaces non-revenue-making activities, produces environmental impacts, and is in the long-run simply and completely nonsustainable in any sense of the word (van der Borg and Russo, *in* Musu 2001). What is left is a "tattered Disneyland" (Musu, *in* Musu 2001). This has led several to pessimistically state, like Lauritzen (1986), that "there is little or no future in Venice, save for tourism." For some knowledgeable observers (e.g., Kay 2006a,b) the obvious and unavoidable conclusion is the total abandonment of the city to its fate as an open air museum existing solely for tourists.

RESULTS

Venice's vulnerability to floods is due to its position at the northern closed end of the Adriatic (Keahey 2002) and also weather patterns and tides, the latter being the highest recorded in the Mediterranean (UNESCO 1979). Southeasterly *sirocco* winds, which have a fetch along the complete length of the Adriatic, pile up the water at the top end which bathymetrically gets shallower further north. The funnel-like shape of the coastline here serves to focus the wind and waves toward Venice. Then the *bora* winds from the east push the accumulated water into lagoon where a *seiche* develops that can produce water level differences of up 60 cm or more (Keahey 2002).

Flooding in Venice, just as the world also recently learned was the case for hurricane-ravaged New Orleans, has been made all the more likely due to years of cumulative destruction of surrounding wetlands. Wetlands operate like natural defenses against flooding (France 2003a). With dendritic creeks removed and the one-time rough bottom of the lagoon smoothed out, the ability of the wetlands to absorb and dampen tidal surges has been greatly reduced. Also, as the salt marsh was converted to terra firma, it has meant that there is simply less space for the incoming water to spread out, the result being, of course, deeper water (Keahey 2002).

The "perfect storm" of November 3 and 4, 1966, served as a major wakeup call for those concerned about Venice's future or in Pertot's (2004) words, "putting Venice's destiny on the world stage." The resulting terrible flood "revealed the negligence, lack of care and of skill from which Venice suffered and the scourge of industrial development" (UNESCO 1979). The storm which raged across northern Italy left over a hundred people dead and caused much damage to other heritage cities such as Florence. In Venice, water levels peaked at about 2 m over sea level and the entire city was flooded for only the fifth time in its history. The inundation lasted for over 15 hours and the city was without electricity or phone services for a week afterwards. Both Pellestrina and San Erasmo as well as other islands in the lagoon were also completely submerged. Although there had been several serious floods in Medieval times, the 1966 flood was the first time that the water had been high enough to inundate the interior of St. Mark's Basilica (Vio, *in* Fletcher and Spencer 2005).

Water was not the only problem. Just as recently occurred in New Orleans following Hurricane Katrina, there was a massive leaking of chemicals and other liquid and solid waste products that combined into a contaminated coating of every surface. Further, almost all buildings in Venice were in need of substantial structural repair once the waters receded and the grime was removed. However, despite the hysteria in the foreign press following the 1966 floods (Lauritzen 1986), the fact that no deaths occurred, nor that any art was damaged, led Venetians, once the shock of the magnitude of the event was over, to settle back into their characteristic lackadaisical attitude toward flooding. Residents in Venice have always been resigned to *aqua alta* events, regarding them as part of the life in the city in much the same way as Bostonians today regard occasional snowstorms. "What's the problem? The water goes up; it goes down. No one is hurt. This has been happening for centuries and we're still here," reported one observer (Keahey 2002). And even the mayor of Venice seemed unfazed, quipping that "Venetians have been getting their feet wet for centuries" (Fletcher and Da Mosto 2004).

The 1966 flood was, of course, not a "one off" affair, but was merely an extreme example of what has become an all to common occurrence in modern-day Venice. During the 1920s, for example, there were 385 recorded *aqua alta* events, the incidence of which have increased progressively throughout the rest of the century until the 1990s when there were 2464 such events (Pertot 2004). The situation is getting worse with 8 of the 10 extreme floods in the last century occurring only within the last four decades (Fletcher and Da Mosto 2004). St Mark's Square used to be flooded about 10 times a year a century ago, about 20 times a year in the middle of the 20th century, 40 times in 1989, and now floods up to 100 times per annum (Keahey 2002; Nova 2002; Fletcher and Spencer, *in* Fletcher and Spencer 2005). The situation has become so severe that water now invades the atrium of St. Mark's 150 to 180 times a year and often every day during the winter (Spencer et al. *in* Fletcher and Spencer 2005). Venice, it appears, is losing its long battle "against the sea" as the title of Keahey's (2002) book would phrase it.

In addition to the structural damages described below, flooding produces economic implications (CELI 1998): time lost because of reduced or interrupted mobility, interruption of revenue activities and services, and damage to goods stored in warehouses and shops.

Crumbling Structures

Venice has long been identified as the most overt example of the inevitability of Time's ruin. All buildings bear scars of 150 years of impoverishment and neglect, given that most have not received any regular maintenance since the fall of the Republic two centuries ago (Lauritzen 1986). Death is an essential element in Venice's appeal and many have visited the city to relish its slow and ravishing obliteration (Pemble 1995). Spurred on by the gloomy writings of Ruskin ("The rate at which Venice is going is about that of a lump of sugar in hot tea") and others, Victorians rushed to see the city before it disappeared forever beneath the waves. By linking the ruins wrought by time to the decadence of its human residents such as Casanova, many came to regard Venice as a sort of modern-day equivalent of Sodom and Gomorrah, its fate seemingly justified as retribution for its many sins. In this respect, Venice became the quintessence of stricken beauty at the same time as being the archetype of exquisite corruption (Pemble 1995). The "romance of decay" (Brown 1996) and ensuing cult of decadence has, however, been given more substance in recent years in response to the numerous floods that have increasingly ravaged the city.

Venice's tides have the highest salt and oxygen content of anywhere in the world (Lauritzen 1986), and given that the city is 90% composed of brick has always meant that it is a disaster in the waiting. And it appears that now that waiting is over. Because tidal exchange with the Adriatic has increased, the canals now have the same salinity as the sea. As a result, the corrosive damage to bricks has increased over time irrespective of anything else (Keahey 2002). Corrosive salt is deposited on bricks and because of the high relative humidity, crystallizes and bores its way into the structures. Salt expands when it dries, which then causes bricks to crumble and disintegrate. Further, during an *acqua alta*, the salt can be carried 2 to 5 m upward in the porous bricks due to capillary action (Fay and Knightly 1976; Biscontin et al., *in* Fletcher and Spencer 2005). The salt water and chemical mixture of contaminants essentially turns the bricks to pulp and the stucco begins to peel off which in turn exposes the ends of the wooden floor planks and iron rods to the corrosive atmosphere, adding to the overall instability. In this regard, exceptionally low tides (also a recurring feature of the hydrologically altered lagoon) are thought to be possibly even more harmful than *acqua alta* events due to the rotting of exposed heads of wood pilings, which can completely undermine building foundations (Lauritzen 1986).

The whole process of deterioration is exacerbated by boat propeller wash whose churning waves add another 40 to 50 cm of water exposure to buildings (Biscontin et al., *in* Fletcher and Spencer 2005). The motorized *vaporetto* service began in the 1880s and the danger of boat wash was first warned about in the late 1950s. The original building foundations were made of imperious Istrian marble with low porosity. Today boat wash added to general subsidence and sea level rise now expose porous bricks to water and salt (Biscontin et al., *in* Fletcher and Spencer 2005). The sucking action of the rebounding waves as they return to the canal can actually pull apart the salt-damaged bricks. Motorboat wash also resuspends sediments that can block drains and antiquated sewage outlets (Spencer et al., *in* Fletcher and Spencer 2005), causing them to rupture internally and damage canal walls (Carrera, *in* Fletcher and Spencer 2005). It is therefore possible that motorboat wakes may actually be just as or even more serious a problem to structural integrity than the flooding itself (Carrera, *in* Fletcher and Spencer 2005).

The overall damage due to high water levels is estimated to cost from 3 to 5 million Euros per year for on-site damages. And this does not cover indirect factors such as lost revenue and expenses necessary to clean out lower floors (Breil et al., *in* Fletcher and Spencer 2005).

DEMOGRAPHIC CHANGES

In one of the most moving scenes in Visconti's wonderful rendition of Mann's *Death in Venice*, Dirk Bogarde stumbles around an eerie piazza at night, completely alone except for several smoldering piles of disease-infected clothes admix splashes of lime disinfectant. With the tourists gone due to the plague, all that seems to be left is our unhappy protagonist alone as a voyeur engaging in a sort of melancholic cultural necrophilia. Today, the most moving part of the lagoon is the strangely haunting island of Torcello whose onetime population of many thousands have long since relocated to what is now the historic center of Venice. The ghost town of Torcello may be a portent of Venice's own fate given the massive exodus of 20,000 people every 5 years that has been occurring there since the end of WWII (Fay and Knightly 1976).

Since 1950, Venice has remarkably lost half of its population. In this regard, it is on its way to becoming another Detroit. The postwar total was 170,000 and today is about 60,000 with most of the young fleeing and leaving behind the oldest population in Europe with an average age of 48, of which one quarter are over the age of 64 (Lauritzen 1986; Keahey 2002). Since 1996, Venice has continued to hemorrhage 800 people a year, a rate that is predicted to increase (Hooper 2006). Should this trend continue, authentic Venetians could become extinct by 2030. "We've reached the point of collapse, the point where things could fall apart," recently stated one urban planner. Further, of the 85,000 living in the historic center in 1985, fully one quarter of these commuted to the mainland for work. And few tourists realize just how many service people don't actually live on the island on which they work: Mestre now sends 12,000 people into historic Venice every day. In the concluding chapter in the only book concerned with issues of the long-term sustainability of Venice, Villinga and Lasage (*in* Musu 2001) flagged these demographic changes and imbalances as being the most serious issue facing the city.

In 1986, half of Venice's houses had no bathroom and had toilets that drained directly into the canals which were swept clean twice a day by tides (Lauritzen 1986). Renovation costs are very high and strict laws make modernization extremely difficult. As a result, there is a great attraction in simply giving up on the decrepit place and relocating to the cheaper postwar housing in Mestre. For example, a recent article in a major newspaper was titled "Rising cost of living emptying Venice," and began with the story of a 31-year-old born and raised Venetian who described the process of moving to the mainland as a "trauma" she was forced into due to exorbitant property prices in historic Venice in the range of a million Euros for a tiny apartment of just over 300 sq feet. Venice is filled with hundreds of abandoned and dilapidated houses. Even palaces along the Grand Canal stand mostly vacant with distant millionaire owners who may visit only infrequently. Can one really blame them? After all, though it might be a nice place to visit every once in a while, who really wants to live in Disneyland? No wonder then that the noted Italian architect Aldo Rossi called for abandonment of the city and its conversion into a museum of monuments (Plant 2002). Fay and Knightly (1976) conclude that "Venice is a city beset by misfortune, suffering from severe neglect, caught in a cycle that is destroying her. Because she has been neglected, few of her citizens want to live there. Because fewer Venetians want to live in Venice there is less urgency to remedy the neglect."

As mentioned before, tourism has stripped the city of its indigenous life, culture, and needed facilities (Plant 2002). It *is* difficult to live in Venice; even in nontourised areas it often seems easier to find somewhere to buy a papier-mache carnival mask than a bottle of milk or a newspaper. And one can walk kilometers and kilometers without seeing a single hardware store in which to buy a hammer and nails, notwithstanding the fact that the oppressive restoration laws would probably prevent you from being able to do any simple home repairs in the first place (unless, of course, you are willing to pay the well known and customary bribes). At times, the actual Venice seems no more real than the replica one on the Las Vegas Strip. Again, Fay and Knightly (1976): "Venice has become a city without a role, colonized by tourists and by Milanese and Turin industrialists. They use Venetian labor in their Mestre factories but they ignore the place their workers come from." The crisis in Venice today therefore is really as much about people as it is about the nonhuman environment.

Contributors

Robert M. Abbott
Abbott Strategies
Calgary, Alberta

Steven I. Apfelbaum
Applied Ecological Services
Brodhead, WI

Peter J. Beckett
Department of Biology
Laurentian University
Sudbury, Ontario

Reinhard Buchli
Entsorgang + Recycling Zurich
Zurich, Switzerland

Sarah Clark
Phalen Creek Project
St. Paul, MN

Fritz Conradin
Oberrieden, Switzerland

Nathaniel S. Cormier
Jones & Jones Architects
    and Landscape Architects
Seattle, WA

Phillip J. Craul
Craul Land Scientists
State College, PA

Doug Eppich
Applied Ecological Services
Brodhead, WI

Cheryl Foster
University of Rhode Island
Kingston, RI

Robert L. France
Harvard Design School
Cambridge, MA

Peggy Gaynor
GAYNOR Inc.
Seattle, WA

Tom Goreau
Global Coral Reef Alliance
Cambridge, MA

John M. Gunn
Department of Biology
Laurentian University
Sudbury, Ontario

Bradley M. Herrick
Applied Ecological Services
Brodhead, WI

Wolf Hilbertz
Global Coral Reef Alliance
Cambridge, MA

Robert Hoogeveen
Sapiens Productions
Maastricht
The Netherlands

Stephanie Hurley
Harvard Design School
Cambridge, MA

Laura Kadlecik
Humbolt Water Resources
Arcata, CA

William E. Lautenbach
Department of Biology
Laurentian University
Sudbury, Ontario

Carol Mayer-Reed
Mayer/Reed
Portland, OR

Amy Middleton
Phalen Creek Project
St. Paul, MN

Stephen Monet
City of Sudbury
Sudbury, Onatrio

Robin Nagle
Draper Program
New York University
New York, NY

Kit O'Neill
Ravenna Creek Alliance
Seattle, WA

Mark Rasmussen
The Coalition for Buzzards Bay
New Bedford, MA

Clarissa L. Rowe
Brown, Richardson and Rowe
Boston, MA

René Senos
Jones & Jones Architects and Landscape Architects
Seattle, WA

Megan Wilson Stromberg
Rana Creek Living Architecture
Point Reyes Station, CA

Neil Thomas
Resource Data, Inc.
Asheville, NC

Stephen Vogel
School of Architecture
University of Detroit Mercy
Detroit, MI

Joshka Wessels
Sapiens Productions
Maastricht
The Netherlands

Mike Wilson
Humbolt Water Resources
Arcata, CA

About The Editor

Robert France has published several hundred papers on the ecology and conservation biology of organisms from bacteria and algae to birds and whales, on research topics from environmental pollution to theoretical biodiversity, and in locations ranging from the high Arctic to the tropics.

Dr. France teaches courses at the Harvard Design School on the influence of landscape processes and development on aquatic systems, and in the Harvard undergraduate program on the invention of nature. He also teaches a course on watershed restoration in the summer program in Venice jointly sponsored by Harvard University and Universita CaFoscari di Venezia. He has participated in many watershed development plans around the world.

Dr. France is senior editor at Green Frigate Books and also serves as series editor at CRC Press/Taylor & Francis for **Integrative Studies in Water Management and Land Development**. His most recent books include *Wetlands of Mass Destruction: Ancient Presage for Contemporary Ecocide in Southern Iraq*; *Introduction to Watershed Development: Understanding and Managing the Impacts of Sprawl*; and *Facilitating Watershed Management: Fostering Awareness and Stewardship*, in addition to the two forthcoming titles for Taylor & Francis: *Aquatic Responses to Watershed Clearcutting: Implications for Forestry and Fisheries Management* and *Restorative Redevelopment: Lessons for the Iraqi Marshlands and other Devastated Cultural Landscapes*. (See his Web page at www.gsd.harvard.edu/info/directory/faculty/france/cv.htm.)

Part 1

Landfill Islands

1 To Love a Landfill
The History and Future of Fresh Kills

Robin Nagle

CONTENTS

LANDFILLS AS COMMONS

Most people do not choose to spend time at active landfills. Livestock are not put out to pasture there, crops are not raised, game is not hunted, picnickers do not ward off ants, runners do not sweat, children do not gambol there. Cemeteries, schools, churches, hospitals, parks, housing developments, and libraries are not sited nearby (though landfills sometimes grow up near such facilities, much to the dismay of teachers, librarians, and homeowners, and many landfills are meant to become parks after they're capped).[1]

A landfill is an unfortunate, if common, answer to solid waste disposal problems. It is not understood as a commons, though its function as the repository of unwanted material goods is essential to the well-being of the metropolis that relies on it.[2] Often considered a blight, it is also a space to which all residents contribute, a "social sculpture" (Ukeles 2002). The artificial geography of a landfill is created by all, shared by all, and has the potential to be transformed after closing into something that all may use and enjoy—or not, depending on variables like what's buried there in the first place, how steeply the garbage has been mounded, how the closed faces are landscaped,

---

[1] Until the 1930s, landfills, as they were called, were little more than dumps, but for centuries they were seen as a solution to the problem of solid waste and to the problem of unusable land. In New York and in many other cities, landfills were created specifically to make taxable land from otherwise marshy, swampy areas thought to be useless—or worse, thought to be dangerous, since they served as breeding grounds for mosquitoes. One of the most infamous in New York was the expansion of Riker's Island in the East River. The project was started in the early twentieth century with the intention of eventually building a city jail and perhaps some hospitals on the resulting land. The hospitals aren't there, but today Riker's is synonymous with one of the largest penal institutions in the world.

[2] Not all communities despise landfills. Some towns can earn significant revenue by hosting landfills for larger cities or private waste haulers. Tullytown, Pennsylvania, for instance, saw a reversal of its economic despair when residents agreed to open a landfill for New York and Long Island trash in 1991. The town (population: just over 2,000) earned payments of between $2 million and $4 million a year for more than a decade, allowing a new Borough Hall, playgrounds, park, library, police station, fire trucks, and an annual "property improvement allocation" of $1,500 per homeowner per year. See Kilborn 2002.

whether or not and how methane is retrieved, and where the newly created land fits in larger local development plans.

Moreover, a landfill reveals unexpected details about the society that creates it. "The urban physiology of excretion," notes social historian Alain Corbin, "constitutes one of the privileged means of access to social mentalities" (1986). He was referring to sewage disposal, but the sentiment applies just as well to solid waste. That landfills are the disposal method of choice for much of the United States[3] reflects a particular set of relationships between citizens, municipalities, environmentalists, material culture, moral and aesthetic sensibilities, and the science of solid waste management.

Landfills let us get rid of our debris but also keep it indefinitely. Contrary to popular belief, much of our buried trash does not decompose.[4] When choosing between a landfill and an incinerator (or a waste-to-energy facility, as they're now called), a landfill is allegedly safer because incineration is thought to cause unacceptable air pollution. The off-gassings of methane and other volatile organic compounds at many landfills, however, are often greater threats to air quality. As landfills are usually sited far from crowded population centers, they allow the illusion that there is a distant, disconnected place to "throw" "away" rejectamenta. In the United States, with its vast open spaces, it seems impossible to run out of places in which to deposit refuse. In part because of this, source reduction—generating less trash in the first place—is not seriously explored.

Carefully designed and engineered landfills (as opposed to "spontaneous" dumps) became the prevalent means of refuse disposal in American cities when manufacture and consumption habits moved from home-based local handicrafts to mass-produced goods newly accessible to large numbers of people. This trend coincided with significant population shifts as people moved from the countryside into cities and as waves of immigrants arrived in urban centers. Because there were more things to be had (so to be discarded) and more people to have them (so to throw them out), urban garbage drew serious infrastructural attention from city governments almost for the first time. The Progressive Era of the late eighteenth and nineteenth centuries, inspired in part by concerns about sanitary conditions, focused on individual hygiene and public cleanliness as signs of a healthy citizenry and of civilized cities. Cleaning the streets was one of the movement's most urgent goals. It is no accident that many of today's municipal departments of sanitation were first established as branches of local boards of health.

Fresh Kills Landfill in New York City is an excellent example of these trends and assumptions. Historically significant and politically volatile, it has always been a commons, whether or not it is ever formally recognized as such. In this essay I will consider how a dump or landfill can serve as a commons, and explore the unique role Fresh Kills plays in New York's well-being.

First, to contextualize the challenge, it is helpful to illustrate how solid waste vexed even the ancients. After that I explore the parameters of commons generally, briefly review Fresh Kills' history, and finally investigate some of the social implications and cognitive difficulties of allowing a landfill the role of a commons. I argue that if we cannot appreciate a landfill as a commons, our understanding of our larger culture is incomplete. Though landfills can be considered abhorrent, they reflect an age-old human behavior and thus, perhaps, are not entirely despicable.

GARBAGE THROUGH TIME

Garbage, in the form of rejected and discarded material remains, has been part of human civilization since our first days as hominids and perhaps even before.[5] The transformation of garbage into a large-scale problem, however, had to wait for us to move from hunting/gathering groups and agrarian

---

[3] Landfilling, "one of the oldest and perhaps the simplest form of biotechnology," accounts for nearly 70 percent of municipal solid waste disposal in the United States. See Suflita et al., 1992.

[4] Ibid.

[5] Martin and Russell note that garbage generation is "a universal human activity …. Materials discarded are seen as refuse, things that, while they may have some residual value for reuse or recycling, are essentially a nuisance that needs to be removed from the places where people do things" and that thus can become useful artifacts for research (2000:57). Needham and Spence also point out that trash is "… rich in significance for many aspects of social organization" (1997:77).

communities into early urban sites. Although not uniquely urban, garbage presents particular challenges to city dwellers.

Rubbish in ancient Troy, for example, was simply dropped on the floors of homes or tossed in the streets. This was the custom, too, in some parts of Africa, where eventually street levels rose and new buildings were constructed atop the mounds of accumulated debris. In approximately 2500 B.C.E., the city of Mohenjo-Daro in the Indus River Valley had rubbish chutes, trash bins, a drainage system, and a scavenging service. Babylonians had cesspools, drains, and a sewage system. Evidence from China suggests that by about the second century B.C.E., some cities had "sanitation police" who removed animal carcasses from the streets. Israelites took a big step toward improving hygiene when Mosaic law directed Jews to remove their waste and bury it far from living quarters (Melosi 1981:3–20; Rathje 1989:1,2).

The ancient Mayans disposed of their organic wastes in open dumps on the edges of their settlements. The first municipal dumps known in the Western world were organized by the Athenians, who also enacted what may have been the world's first antilitter ordinance. Romans had more trouble coping with sanitation, and by the time the city's population reached its zenith of a million and a half inhabitants, there were unprecedented health and pollution problems. But at least the Romans had their baths and a version of a sewer. Europe forgot these niceties for nearly a millennium after Rome fell, despite Leonardo da Vinci's innovative proposals for indoor plumbing, flush toilets, and a below-ground sewage system.

Indeed, the filth of Europe in the Middle Ages and in the Renaissance is difficult to imagine. King Philip II of France ordered the streets of Paris paved in 1184 because he was sickened by the smells emanating from the garbage-soaked mud. It didn't help much. "This town is always dirty," wrote one visitor to the City of Light in the late 1500s, "and 'tis such a dirt, that by perpetual motion is beaten into such black unctuous oil that where it sticks no art can wash it off …. It also gives so strong a scent that it may be smelt many miles off if the wind be in one's face" (quoted in McLaughlin 1971:67).

In 1758, dumps were outlawed inside the borders of Paris, and by 1781, Montfaucon was the only one serving the city. It was already infamous; between the 13th and 17th centuries, it was the site of thousands of hangings. It was also where executed criminals or those killed by torture were strung up to rot; their remains were eventually tossed in with putrefying household trash and sewage. "Deep cultural associations of execrated criminals and society's excretions merged in Montfaucon," observed historian Donald Reid (1991:11).[6]

London was as rank as Paris. In 1347, "two men were prosecuted for piping ordure into a neighbour's cellar—it says a great deal about the general smell of London at the time that this economical device was not discovered until the cellar began to overflow" (Ibid.:27–8). The flow of the Thames was regularly impeded by the accumulation of trash and untreated sewage that bulged from the river's many locks and along the shoreline, slowing and sometimes completely stopping boat traffic.[7]

Sanitary conditions in Europe remained relatively awful until the Industrial Revolution, when they got worse. Besides transforming production, labor, trade, residence patterns, life expectancies, and family relations, among other variables, the era also intensified urbanization and thus the problem of garbage generation and accumulation. Industrialism "produced the most degraded urban environment the world had yet seen … " (Mumford 1961:447). England suffered the most devastating transformation. In 1801, about a twelfth of the population lived in cities, but by 1901, 77 percent of the nation was urban. There were few adequate measures for dealing with the tremendous pressure of

---

[6] Reid also notes that the practice of stringing up the bodies of dead criminals stopped with the Revolution, but a slaughterhouse built on the site maintained the association between waste and dead flesh.

[7] See Halliday 1999 for a detailed description of the Great Stink of 1858, when the Thames became so fetid that it stopped the city of London.

such an increase; sanitary conditions went from bad to unlivable. In 1843, for instance, one neighborhood of Manchester had a single privy for every 212 people.

The situation in New York wasn't as awful but only because the city wasn't yet as big a metropolis as some in Europe. Street-side trash collecting was legislated in the 1670s, but it happened only sporadically. Scavengers sold what they could salvage, but much was also dumped along the city's shore. This proved a popular solution to the problem of too much trash and too little space. Even in the 1600s, real estate was a hot commodity in New York, and much of urban life centered on the downtown waterfront. Merchants were eager to build on crowded spaces, and dumping helped create more land onto which they could expand. In Manhattan below City Hall, 33 percent of the land is built on "street sweepings, ashes, garbage, ballast from ships, dirt and rubble from excavated building sites, and other forms of solid waste dumped along the shore" (Corey 1994:72). Some parts of lower Manhattan have been filled three blocks out from the original shoreline (Rothschild 1990:16).

By the turn of the twentieth century, trash disposal in New York was shaped by the separation of various kinds of debris, which the 1905 Annual Report of the city's Department of Street Cleaning (precursor to the Department of Sanitation) took pains to elaborate. "Garbage" specified kitchen or table waste, vegetables, meat, fish, bones, fat, and fruit. These putrescibles yielded greases and fertilizers through a boiling process called reduction. Reduction was a sometimes profitable but always malodorous business, and early NIMBY (Not in My Back Yard) protests in New York centered on reduction plant locations.[8] "Ashes" included ashes, sawdust, floor sweepings, bottles, broken glass and broken crockery, tin cans, and oyster and clam shells from homes (but not from restaurants or fish dealers, who were responsible for discarding their own). These were most valuable for landfill projects and were often combined with street sweepings. Paper and rubbish, a single category, was made up of paper, pasteboard boxes, rags, mattresses, excelsior (wood shavings), straw, carpets, old furniture, old clothes, oil cloths, old shoes, leather and rubber scraps, tobacco stems, flower stems, and "house refuse generally." These were scavenged for salable items and then burned, "developing heat, power, and light," according to the DSC's optimistic commissioner, John McG. Woodbury; in fact, much was combined with ashes and used as fill.[9]

New fill opportunities were greeted enthusiastically. "The possibilities of this reclamation are almost boundless," crowed Woodbury. Of one site in particular, he noted, "The lowlands on Jamaica Bay afford an almost unlimited supply of dumping ground."[10] When fills reached capacity, or when the city's ability to carry away the trash was overwhelmed, the debris was often dumped at sea. In fact, ocean dumping was one of New York's preferred means for ridding itself of trash, though much of the refuse, "being light ..., easily floated in onto the beaches along Long Island and New Jersey shores, where its presence in years past has caused great complaint."[11] It was not until the U.S. Supreme Court intervened that the practice was finally halted in 1934.

By then the city was roaring ahead with incineration and with landfilling. City planners saw incinerators as a practical waste disposal technique, while accelerated landfilling allowed for some of the most ambitious public works projects ever attempted.

Regardless of solid waste management choices in New York or in the ancient world, neither the trash of antiquity nor of contemporary cities was acknowledged as a common good; rather, it has always been seen as a chronic problem, even a crisis. Attempts at source reduction—that is, decreasing the amount of stuff that becomes trash—have never received large-scale, serious attention except temporarily during war time.

---

[8] The longest-lived and most infamous was on the aptly named Barren Island in Jamaica Bay, Queens. See B. Miller 2000; K. Johnson 2000.

[9] One of the first waste-to-energy incinerators in the United States was built under the Williamsburg Bridge in lower Manhattan. It opened in 1905 and had a daily carrying capacity of 1,050 cubic yards of "light refuse or rubbish," according to the Department of Street Cleaning's 1905 Annual Report (p. 82). It provided electricity for some homes in the neighborhood, and for lighting the Bridge.

[10] New York City Department of Street Cleaning Annual Report, 1905; p. 74.

[11] Ibid., p. 82.

It is helpful to recall, however, that the trash of New York, and of many other American cities, has created thousands of acres of shared space that would not otherwise exist. The contours of New York differ irrevocably from their configurations before Europeans arrived, both inland and along the shore. About 20 percent of contemporary Manhattan, Brooklyn, Queens, and the Bronx is landfill. Archaeological evidence suggests that even before Europeans, indigenous residents dumped refuse along water's edges; discerning an "original" shoreline for the city would be virtually impossible.

There is little public memory of the source of so much land. Few people realize that both of New York's airports are built on fill, as are the foundations of the Triborough, George Washington, Verrazzano, Whitestone, and Throgs Neck Bridges. Numerous New York parks (Great Kills in Staten Island, Orchard Beach in the Bronx, Battery Park in Manhattan) were wetlands or water before they were filled.

Fresh Kills will someday be one of those spaces. A Department of Sanitation Annual Report from the 1950s bragged that Fresh Kills was the greatest land reclamation project ever attempted. This unlikely claim is being made real with current work to transform the formerly pungent geography into the largest green space within the borders of New York City.

INVENTED COMMONS

Its status as a future park is not the only reason Fresh Kills qualifies as a commons. Social arrangements that bring a commons into existence, or that recognize and protect certain resources as commons, are in continual flux. The idea of a commons challenges notions of private property, prosperity, and who has rights to define and control communal well-being. In England, the commons were sources of grazing pasture, game, fish, and fuel wood for hundreds of years, until enclosure laws written during the eighteenth century forbade access (a process that was repeated throughout much of Europe). Resulting hardships among the peasantry included starvation, and inspired violent reactions, which in turn provoked Draconian responses from the state. The punishment in Britain for removing the boundary markers of newly enclosed commons was death (Rykwert 2002:24–5).

A cheerier model of a commons is the grazing pasture sometimes pictured at the heart of colonial New England towns. Careful husbandry meant that it was available in perpetuity, or at least until advancing modernities made livestock a cumbersome possession for townsfolk. Public parks often replaced those older commons; now humans occupy space once dedicated to large ruminants. Grazing cows would be a rare sight in such places today, though the Sheep's Meadow in New York's Central Park hosted its namesake until 1934, when parks officials, fearing that the wooly mammals would end up on dinner plates made empty by the Great Depression, removed them.

More contemporary examples abound. The Clean Air and Clean Water Acts of the 1970s were explicit recognition that those basic elements constitute shared resources that must be safeguarded, transforming them from unmanaged (and thus exploitable) to managed (and thus at least potentially protected) commons.[12] Genetically altered food crops seem to many a threat to the safety of the food supply, endangering a resource that in practice is largely private but that in imagination is public and so is a kind of commons. The Georges Bank, once-plentiful fishing grounds off the northeast coast of the United States, was for centuries an unmanaged commons; today it suffers serious depletion after unbridled overuse. Biodiversity in some of the planet's last tropical rainforests inspires nations like Brazil to guard them from northern powers that would investigate. Such efforts impose a shelter on these wild commons but also place them at risk, since science that would strengthen the cause of preservation is rebuked with the same energy as attempts at exploitation.

---

[12] According to Garrett Hardin's famous explanation, an unmanaged commons will be used by its members for their individual benefit before any other consideration. This model of resource management dooms any commons, since unmitigated personal gain will always come at the expense of the larger good. See Hardin 1993 and 1998.

Next to these illustrations, a landfill seems a lowly and unlikely commons.[13] But it serves, in the sense of a resource set aside by the community for its shared use, to enhance the greater good. Without a functioning landfill or some other way of ridding itself of debris, no metropolis can survive.

A landfill commons is humble in part because of the stuff that makes it. Garbage imposes technical, environmental, social, and cognitive challenges that unite and commemorate the culture that creates it. Household rubbish in particular underscores the problem of trash as intimate, perpetual, and despised. It is intimate because there are few activities that occupy us in any given 24-hour period, except perhaps sleep, that do not generate garbage. Thus, our refuse reflects our simplest, most mundane behaviors as well as our more celebrated moments. It is perpetual because, if we partake of contemporary life at what I call average necessary quotidian velocities, there are few ways to stop its creation.[14] And trash is despised for many reasons, the simplest of which is its conglomerate power to disgust. A single moldy orange peel is not so gross, especially if it's my own moldy orange peel, but a bucket of rotting fruit from who knows what source can elicit strong negative reaction.[15]

Trash invites a willing ignorance that is nicely revealed in our vehemently vague language of discard. We don't "put" it away, which would imply that we save it for later use.[16] Rather, we "throw" it away, and the "away" is comfortably undefined. It is initially the kitchen trash bin or the bathroom waste basket. It becomes the trashcan in the garage, in the basement, on the curb, in the back alley. When it is dumped into a truck, the "away" becomes more real, since the refuse is no longer part of a home, but it is also more invisible, since neither the trash nor the "away" are in sight, though obviously both exist somewhere.

The "away" is often a landfill. As burial grounds for our discards, landfills force commonality on our material traces, whether or not such commonality existed before they were discarded. They hold startlingly accurate records of the people who form them, and unlike the people, they endure. "If I were a sociologist anxious to study in detail the life of any community," wrote Wallace Stegner, "I would go very early to its refuse piles …. For whole civilizations we have sometimes no more of the poetry and little more of the history than this" (1959:78). "The artifacts that will fully represent our lives are safely stored within mega-time-capsules, which we call landfills," concurs archaeologist William Rathje. "It is these anonymous, random remains that will tell our story to the future …. " (1999:88).

Landfills unite objects. They also sometimes unite citizens. In municipalities without garbage collection, they bring together residents who must travel to their landfill (or, in days gone by, to their dump) to discard trash. They often provide formal or informal recycling centers, where rejected but still useful possessions are claimed by new owners. Some landfills, or dumps, provide entertainment. When I was a child growing up in a small town, we went to ours on summer evenings to sit in the car with the windows rolled up and the headlights focused on the pits of trash, watching bears forage for food. We usually met neighbors who had come for the same attraction. But even more significant connections

---

[13] A dump may serve as a commons. The difference between a landfill and a dump, according to the Environmental Protection Agency, is the way in which they are constructed. A dump is established without any environmental controls, a polite way of saying that it can spring up nearly anywhere—in an empty lot, down a backcountry holler, along a riverbank. One of the most infamous dumps in the world is Smoky Valley outside Manila. A community of up to 80,000 people scavenged a livelihood from it when it collapsed after heavy rains on July 10, 2000, killing more than a thousand.

A landfill is not so haphazardly constructed. According to the EPA Website, a landfill is a repository for "nonhazardous solid wastes spread in layers, compacted to the smallest practical volume, and covered by material applied at the end of each operating day."

[14] It is possible to live without creating much trash, but such lifestyles are bracketed as "alternative" and thought to be impractical. They require at minimum a relationship between an individual and her notions of time that is different from what most Americans know or want.

[15] It's never simple, of course; that bucket of rotting fruit might be destined for a compost, which could alleviate some of the nausea that it could provoke. For a thorough discussion of many varieties and implications of disgust see W. Miller 1997.

[16] There can be a later use for trash, even landfilled trash. Landfill mining is an industry gaining strength in England and in some parts of the U.S. (see O'Brien 1999).

can occur. Nuptials were celebrated at the Bethel, Maine, transfer station—formerly known as the town dump—on September 1, 2003. The location made sense to the newlyweds because it was where Rockie Graham, a conscientious recycler, met her husband, Dave Hart, a new employee at the facility. Townsfolk contributed to their honeymoon fund with bottles and cans to be redeemed for nickels.[17]

FRESH KILLS: LANDFILL

Fresh Kills is a rich example of a landfill as commons, which its history—especially its recent and future history—suggests. In part because of its audacious scale, it is a commons not only for the city that created it but also for the larger civilization that the city represents.

Before it was a landfill, Fresh Kills was a series of inlet marshes, woods, and meadows nestled into the middle western edge of Staten Island, separated from New Jersey by a narrow strip of water called the Arthur Kill.[18] Indigenous artifacts estimated to be nearly 10,000 years old were discovered there. To the north, Linoleumville (now a neighborhood called Travis) was founded around the country's first linoleum factory, built in 1882 by British inventor Frederick Walton. In its heyday, the factory employed more than 200 workers. By the early twentieth century, nearby hamlets included Kreischerville to the south (now Charleston), where locally mined clay was made into bricks and drainpipes.

Fresh Kills provided treasures for locals. Old women roamed the marshes harvesting herbs, wildflowers, grapes for jelly, and watercress. Italians came for mushrooms and mud shrimp. In the fall, truck farmers harvested salt hay with scythes, while Jewish elders and rabbis cut carefully chosen willow twigs for Succoth.[19]

Early in the 1900s, through a breathtaking piece of political legerdemain, the city established a reduction plant at Fresh Kills. Dead horses, other offal, and garbage (the putrescibles described earlier) were to be boiled down into grease, fertilizer, and glycerin. The contractor who built the plant promised that odors would not be a problem, but it was regularly filled past capacity. Garbage and carcasses rotted in uncovered barges for months at a time. The odors—the very ones that city officials had promised wouldn't exist—were nauseating. Public outcry was immediate and loud, but the plant was not closed until the mayor who approved it lost his bid for re-election.[20]

Two decades later, in 1938, city planner and infamous autocrat Robert Moses wanted to build a bridge that would straddle the Arthur Kill and further his grand scheme to lace the New York region with highways. Fresh Kills' many bogs and swamps seemed the ideal place to fill for the bridge's foundations. As city parks commissioner and chairman of the Triborough Bridge Authority (among other titles), Moses already commanded the dumping of city trash to create the foundations for highways, bridges, and parks all over the city; Fresh Kills was merely one more place to fill in, to make "taxable."[21]

It took a while, and Staten Islanders did their best to thwart the plan,[22] but dumping started on Fresh Kills in 1948; soon complaints were sounding from every neighborhood. Moses assured irate residents that he only needed three years to fill the land. In 1951, however, he urged the mayor to allow more time. "The Fresh Kills project cannot fail to affect constructively a wide area around it," he reported that year. "It is at once practical and idealistic."[23]

---

[17] Associated Press, July 2003 and September 2003.

[18] "Kill" derives from the Dutch word for "creek" and place names that include it (like the Catskill Mountains) are traces of a long-gone colonial legacy.

[19] For a more eloquent elaboration, see J. Mitchell 1993.

[20] A thorough description is found in B. Miller 2000:127–35.

[21] See Rodgers 1939 for an admiring if naïve profile of Moses' influence on New York City, written while Moses was reaching the peak of his powers.

[22] Staten Islanders sued the Departments of Health and Sanitation to try to stop Fresh Kills from becoming a landfill, to no avail. See Fenton 1947.

[23] Quoted in Severo 1989.

By 1954, Fresh Kills covered 669 acres. Five years later, the city proposed extending its life by 15 years. In 1965, when pressed about a closing date, officials demurred, claiming none could be set. By 1966, the landfill consumed 1584 acres. A planning report in 1968 proposed a ski resort once the landfill's slopes were finally capped, but this surprisingly creative notion did not count on the perpetual presence of methane gas, a heat-generating by-product of decomposing organic matter.

By the early 1970s, other landfills in the city were closing, and Fresh Kills received almost half the city's refuse. In 1980, the state's Department of Environmental Conservation charged the city with environmental violations because Fresh Kills had been built and expanded without linings, gas retrieval systems, or leachate recovery plans, among other problems. By then the landfill was so far from compliance with existing regulations, most legislated after it was opened, that it was technically illegal. The same charges against the city were made again in 1985, when Fresh Kills was receiving almost 22,000 tons of garbage every day—nearly all of New York's trash. Tipping rose to an all-time high of 29,000 tons a day by 1987. By then the landfill employed 650 full-time workers.[24]

In 1990, for the third time, the state's DEC cited the city for violations at Fresh Kills but this time the charges included a tight schedule for bringing the landfill into compliance. In 1991, Edgemere landfill in Queens closed. Incinerators were already in decline as public protest against them grew, and the city's last three closed in the early 1990s.[25] Fresh Kills became the city's only option for disposing of household waste.

Staten Islanders had spent more than half a century protesting the landfill, to no avail. Always New York's outlier borough, Islanders resented Fresh Kills for its size and stench, but even more, they resented it as but the most visible sign of the larger city's contempt. The landfill topped a long list of complaints that fueled increasingly passionate talk of secession. When a referendum was allowed in the early 1990s, residents voted two-to-one to secede. Had the vote been only theirs, they would have won, but it was open to the entire city, and the other four boroughs refused to relinquish their suburban would-be secessionists.

In 1995, Staten Island officials sued. Fresh Kills, they argued, violated the city charter's "fair share" provision of the federal government's Clean Air Act. At the same time, the state legislature entertained bills mandating the landfill's closure by 2002. A few months later, the governor, the mayor, and the borough president announced that the landfill would close by the end of 2001, a goal signed into law in June of 1995. There was no alternative garbage management plan in place. The closing date was mostly arbitrary, but it let Mayor Rudolph Giuliani pay a debt. Staten Island, the least ethnically diverse and most conservative borough in New York, had awarded him huge margins of victory in both his mayoral bids. Without an overwhelming majority there, it is unlikely that he could have carried his first or second election.[26]

Because the decision to close the landfill was political, despite the sound arguments of the Staten Island lawsuits, knowledgeable sources scoffed that it would not be shut down. There was no other place to put the trash and no coherent plan to find one. Despite predictions, however, and months ahead of schedule, the last tugboat pushing the last barge of steaming trash arrived at Fresh Kills on a rain-drenched Thursday in late March, 2001. A few politicians and reporters gave the barge a modest send-off from the its departure point at the North Shore Marine Transfer Station

---

[24] The daily load at Fresh Kills was cut in half in 1988 when the city doubled tipping fees for private carters, forcing them to find cheaper disposal options.

[25] Incineration was a popular waste disposal method in New York for decades, starting in the late 1890s. In 1930, the city had nineteen incinerators, including the three largest in the world. By the 1960s, "refuse incinerators were deeply rooted in New York City's waste disposal infrastructure. Thirty-six furnaces in eleven municipal incinerators and some 17,000 apartment house incinerators—arguably the largest refuse incineration infrastructure ever assembled in a city—were burning about 40 percent of NYC's refuse and emitting about 35 percent of the city's airborne particulate matter" (Walsh 2002:321).

[26] In his 1997 bid for re-election, Giuliani won close to 90 percent of the vote in most of the borough.

in Flushing, Queens. Mayor Giuliani hailed it as it passed his home. A handful of photographers followed it down the East River. Fire boats blasted water cannons as it passed the Statue of Liberty. Five hours after it left Queens, Staten Island dignitaries met it at its destination, where the rain poured and the bunting dripped and after 53 years, the last building block of an extraordinary piece of architecture was loaded into payhaulers and sent up the hill.

Fresh Kills is one of the largest landwork structures ever built in the history of humankind and for years was the largest landfill in the United States.[27] It spans nearly 3,000 acres, about two and a half times the size of Central Park. Allotting two square feet of space per person, it could hold 33 million people. It comprises nine percent of the landmass of Staten Island, and until a methane retrieval system was initiated in 1998, generated six percent of the nation's and fully two percent of the world's methane. It is crisscrossed by fifteen miles of roads and bridges. It cannot be seen from its entirety on the ground, but only from the air; in fact, it is visible in space to the naked eye.[28] It holds approximately 108 million tons of trash and still has an estimated 80 million tons of remaining capacity.[29]

The city's municipal garbage is now exported upstate, out of state, out of the country. For the first time in its history, New York has no place for its own trash. Most of the garbage travels by diesel-fueled truck, severely stressing the city's highways and streets and adding significant quantities of pollution to the air. The cost of exporting has pushed the DSNY's annual budget above the one billion dollar mark. Rudy Giuliani's successor, New York mayor Michael Bloomberg, announced a plan to containerize the city's trash at existing marine transfer stations, which are to be retrofitted to accommodate the necessary technology. The containers will be loaded onto barges and shipped to rail transfer stations in New Jersey and elsewhere, and then sent by train to landfills. Recycling will be significantly enhanced, with new facilities planned in several locations.

The ambitious plan was fiercely debated for several years before it was approved in 2006; it still has big gaps.[30] No one is completely sure how long it will take or how much it will cost to reconfigure the marine transfer stations, but money and schedule estimates have already ballooned well beyond their original ceilings—and no construction has begun. The rail transfer stations don't exist yet, nor are there enough long-term contracts with enough landfills.

One thing is certain: the garbage will keep coming. And regardless of pressures in the future that might push toward reopening Fresh Kills, at least one section of the landfill will never again receive trash.

FRESH KILLS: BURIAL GROUND

One of the many arresting features of Fresh Kills is its view of lower Manhattan. Sketched delicately against the horizon, the city's skyline seems a hazy Oz from the landfill's austere hills fourteen miles distant. Workers on Fresh Kills watched both planes hit the World Trade Center on September 11, 2001, and watched the buildings fall. They knew what was coming. Even before official word arrived, they started readying Sections 1 and 9, the largest and last face closed, for the new loads. There was no other space in the city that was big enough and—just as important—that could be sealed off and secured for the ensuing criminal investigation and retrieval operations.

---

[27] A newer landfill outside Los Angeles can claim more square acres, but not yet more trash.

[28] Confirmed by NASA; see Schoofs 2001.

[29] According to Ben Miller, measuring just Fresh Kills' remaining capacity makes it the sixth largest landfill in the country.

[30] One marine transfer station scheduled to be retrofit for containerization is located on Manhattan's Upper East Side. The neighborhood includes some of the wealthiest ZIP codes in the world, and residents vow that they will continue to fight the siting of the facility. Residents of other neighborhoods that have hosted one or more transfer stations for many years, however, are relieved that each borough must now bear responsibility for its own trash. The policy, called "borough-based self-sufficiency," was inspired in part because of Fresh Kills.

The first wreckage arrived by truck in the early morning hours of September 12, but the Department of Sanitation soon had out-of-commission barges back in action and opened several marine transfer stations within days. At the height of the operation, more than a thousand people representing nearly 25 different city, state, and federal agencies worked at The Hill, as it was called, twenty-four/seven. A ceremony marking the end of the sorting process took place on July 15, 2002. By early August, the last piles of bent I-beams and tangled rebar were heading to recyclers.

The idea of purposely adding human remains to countless tons of trash compounded stunned public disbelief and incomprehension in the days immediately after the attacks. Initial reports said that the debris would be sifted at Fresh Kills and then taken elsewhere for burial, and city officials were careful to make no commitment about where the wreckage might finally end up. But it was gradually clear that it would be too costly to move the million-plus tons a second time, and Fresh Kills became the final resting place.

One of the workers sifting the rubble suggested that a memorial to the attack victims become part of Fresh Kills' future. He said he was glad that the remains of the buildings were staying inside the borders of the city.[31] Outrage was immediate. "I really do believe that for the sake of their souls and their families, to have the 'Dump' be a Memorial is a disgrace," wrote one woman on the local newspaper e-mail forum that evening. "… Please don't even consider such an idea of this sort; the dump is the 'DUMP.'"

"I agree with you," replied another. "Who in the world came up with that idea? A garbage dump as a memorial to the HEROS!!! [sic] My God, what were you thinking?"[32]

Mierle Laderman Ukeles had a similar but more nuanced reaction. Ukeles has been the artist-in-residence for New York's Department of Sanitation since 1977. A voluble, passionate woman in her 60s, Ukeles' work first brought her into contact with every sanitation worker on the force, and has more recently focused on Fresh Kills.[33] When she learned that the Trade Center debris was bound for the landfill, she was horrified. Many killed in the attacks had left behind physical traces and remains; thousands of these were reclaimed and given over to loved ones. But she was thinking of those who had been turned to what a Rosh Hoshanna prayer calls "flying dust." Their forced anonymity atop anonymous heaps of trash was too much. "That would collapse a taboo in our whole culture," she wrote.

> To call something "garbage" means stripping the materials of their inherent characteristics. So that even though differences are obvious, hard becomes the same as soft, wet as dry, heavy as light, moldy old sour cream as a shoe, wet leaves as old barbells—they become the same things. The entire culture colludes in this un-naming. Then we can call it all "garbage"—of no value whatsoever.

Because of this process of un-naming, she argued, the idea of a memorial at Fresh Kills required a very particular sensitivity. "This must become a place that returns identity to, not strips identity from, each perished person," she concluded. "This part of the overall Fresh Kills site must become a double place: the unnamed healed and the named re-named. Otherwise the doubling being done here tumbles necessity into obscenity."[34]

Another person intimate with Fresh Kills had a different reaction. "One can view with horror the decision to place what I consider sacred material on top of something profane. I do not," wrote Nick Dmytryszn, for more than thirteen years engineer for the Staten Island borough president's office and thus a long-time student of the landfill. He continued,

---

[31] Staten Island Advance, 19 September 2001; http://www.silive.com/forums/.
[32] Ibid.
[33] Ukeles is Fresh Kills' Percent-for-Art artist, sponsored by New York City's Departments of Cultural Affairs and Sanitation. In that capacity she is working in collaboration with James Connor's Field Operations, the design team charged with transforming Fresh Kills into its next iteration.
[34] Ukeles 2002.

This section of landfill is scheduled, as per the law, for final closure. In simple terms, closure involves first placing down *clean* soil, followed next by an impermeable barrier, to be then all covered with another thick layer of *clean* soil. Thus, what was enacted to protect the environment is now very relevant in protecting and respecting the final resting place of so many of our dead. Fresh Kills has now become a part of the landscapes of every American (his emphasis).[35]

FRESH KILLS: A NEW COMMONS

That Ukeles and Dmytryszn hold such divergent perspectives points to the volatile politics of memory that already marked Fresh Kills but that now promises to be even more controversial. Fresh Kills serves as an immense, inadvertent museum, with countless objects preserved until the future possibility that they are excavated, scrutinized, maybe even catalogued and displayed. It is a monument to what sociologist Wayne Brekhus calls the "unmarked" material relations of everyday life.[36] But now it also serves as a cemetery. It became both museum and cemetery by default. We had nowhere else to put our trash, and never worked much to diminish its quantity, so in our need for the "away" of throw-away society, we invented Fresh Kills. We had nowhere else to put the wreckage from the World Trade Center, so out of necessity for space and security, we transformed Fresh Kills—a name too horrific to say in the days immediately following the attacks—into The Hill (even as the Trade Center site was called Ground Zero and later, The Pit).

Before the attacks, Fresh Kills already memorialized many things. From its start until its closure, it represented the prevalent solid waste disposal technologies of the day. Now it will benefit from the best ecological science available as work is done to establish native plant species, insects, bird life, and mammals in its vast acreage.[37] It was located in a corner of New York that never received much serious respect or attention from the larger city, so complaints from locals, no matter how clamorous, could mostly be ignored. As it grew, its existence allowed the citizens of that larger city to continue living at break-neck speed, generating nearly five pounds of trash per person per day, without having to pull back or change consumption habits or consider that the "away" was always a real place.[38]

It is made of four huge heaps of objects we classed as untouchable, consigning them to uniformed workers who took them "away." Now it is also home to "flying dust" that was human beings. We reject the former, masses of things that we decided to separate from ourselves. We passionately embrace the latter, traces of victims who could have been any of us, killed randomly by a rage most of us never knew was so intense. The biography of Fresh Kills always pointed to lives lived fully, richly, even to excess, and now that biography includes some of the very beings whose cast-offs already resided in the landfill's quiet hills.

Anthropologist Mary Douglas reminds us that the sacred and the profane are both segregated from the larger society; both are marked by special places and require particular behaviors.[39] Landfills are designated locations for things we no longer want and that can therefore qualify as profane, especially when they are mixed together indiscriminately—as Mierle Ukeles notes. Often a landfill can be forgotten once it is covered over and turned into something else—a golf course, say, or a park. But landfills

---

[35] Dmytryszn 2001/2002.

[36] See Brekhus 1998. "Just as we visually highlight some physical contours and ignore others," he writes, "we mentally foreground certain contours of our social landscape while disattending others." Fresh Kills and other landfills hold the physical remains of objects once socially engaged and now purposefully "unmarked," or as Ukeles notes, unnamed.

[37] Steven Handel, an ecologist at Rutgers University, has been working on Fresh Kills since 1993 to determine which species might thrive on the closed landfill without compromising its underlying infrastructure. He has had good success with hackberry, crab apple, and mulberry trees, as well as with rose, beach plum, and other shrubs. See Carlton 2002; see also Young 2001.

[38] One of the earliest sanitary landfills in the country was established in Fresno, California, in the 1930s. It was closed in 1987 and recently proposed as a National Historic Landmark, arousing great derision. For the proposal, see Melosi 2002. For the disdain, see Dowdell and Thompson 2001, J. Martin 2002.

[39] Douglas 1970.

like Fresh Kills are too big to ignore, and so they pose a continual cognitive dissonance. They betray the lie of the "away." They confront us with part of the real physical cost of the way we organize our material lives. We dislike landfills because, among other unpleasantnesses, they stink. But we dislike them, too, because they make evident a cost of living that we would prefer to disregard.

Whether or not we acknowledge some of the more difficult lessons of a landfill, it is a geography with much to teach. Techno-artist and activist Natalie Jeremijenko has created robotic feral dogs fitted with chemical sensors designed to roam a landscape and sniff out volatile organic compounds and similar toxins.[40] "What if you could modify an ordinary robotic dog by adding an inexpensive, off-the-shelf gas sensor," she writes. "What if you could release packs of these feral robotic dogs at former Superfund sites whenever a school, housing development, market opens for business on or near the site?" Successful releases have already happened at two sites in Florida, where the dogs detected volatile organics, and along the Bronx River in New York City, where they found volatile organics as well as polycyclic aromatic hydrocarbons. Jeremijenko proposes releasing her dogs on Fresh Kills. Were the city to explore her idea, it could generate a cheap and accurate profile of just what stage of quiescence the landfill is in, and find trouble spots fast.

Such information could prove invaluable to the designers and planners at Field Operations, the New York-based team of landscape architects, ecologists, engineers, and artists chosen to take Fresh Kills into its next life. Awarded the project through an international competition that drew entries from around the world, Field Operations will work with the city, and particularly with Staten Islanders, on a parks project the size of which New York has not seen in more than a hundred years. The first phase focuses on community outreach and on technical training for the design team. Public use of some parts of the terrain is expected by the end of the decade; other elements will not be finished for nearly 40 years. The ultimate goal, in part, is to restore a corner of New York's suburban sprawl to what the Field Operations plan calls "nature sprawl," connecting Fresh Kills to a green belt that already graces other parts of the Island.[41]

The long-term design includes a memorial to the victims of the World Trade Center attacks, slated for Sections 1 and 9, where the categories "profane" and "sacred" are now and forever combined. Exactly that contradiction poses an insurmountable problem for some family members whose loved ones literally vanished without a trace when the Twin Towers collapsed. When they realized, near the end of the recovery process, that the material brought from Ground Zero to The Hill would not be taken to a different final resting place but would be left at Fresh Kills, they formed a group called WTC Families for a Proper Burial. They do not accept that the remains, however tiny—even microscopic—of their husbands and daughters and siblings and sons are to be left forever in a location that was one of the most execrable places in the world. It was a dump, they feel, and it will be a dump no matter how many feet of clean fill or what kind of shrub species cover the mounds, and it is not now nor will it ever be a fitting resting place for any human being.

The group sued to have the World Trade Center material unearthed and returned to Ground Zero. The court case has lingered for a few years now; as of this writing, a judge has ordered the Department of Sanitation and the Families for a Proper Burial to work out a compromise. An offer by the DSNY to move dozens of tons of earth from Sections 1 and 9 to a part of Fresh Kills that never held garbage was not accepted. In the mean time, while the lawsuit is unsettled, work on "Fresh Kills, the Park" is on pause.

The landfill before September 11, 2001, would seem to be nothing more than a massive profanity, and surely it is a dissonance of unfathomable pain to imagine one's son or wife or sister left there. But I believe that Fresh Kills was never a profanity, any more than the defunct landfills throughout New York or in other cities can be considered profanities. Perhaps our age-old tendency to create a category of object that we must reject is a profane act, but it has been part of our behavior since before we were *Homo sapiens*. Certainly we have often invented messy, dangerous means of disposal and have

---

[40] See Glassman 2003; also see http://xdesign.eng.yale.edu/feralrobots for a thorough description.
[41] http://www.nyc.gov/html/dcp/pdf/fkl/fien1.pdf.

imposed these unfairly on populations whose protests are ignored—Fresh Kills was one example. But we have needed and used landfills for too long to say that they are mere blights. We have even used them creatively to build our urban centers and to extend the reach of our municipalities.

No one can heal land that has been claimed for a landfill; Fresh Kills will never again be the salt marsh that it was before 1948. No one can heal a city, any city, wounded like New York was on September 11th, 2001, nor can families who lost loved ones heal to the wholeness they knew before the violence that ripped them asunder.

We can, however, acknowledge what landfills allow us and see them for the futures they help create, not just for the pasts in which they were difficult spaces. When they are closed, we have the chance to bring our most thoughtful efforts to their future, as the Field Operations team will demonstrate, by not forgetting what they are—what shapes the hills where children can gambol, runners can sweat, picnickers can ward off ants, aching citizens can mourn—and to make them a welcomed commons, not just a necessary one.

LITERATURE CITED

Associated Press. 2003. Couple Plans Wedding at Town Dump (July 15); Couple Weds at Dump Where They Met (September 3).

Brekhus, W. 1998. A Sociology of the Unmarked: Redirecting Our Focus. *Sociological Theory* 16(1):34–51.

Carlton, J. 2002. Where Trash Reigned, Trees Sprout. *Wall Street Journal*, January 23, p. B1.

Corbin, A. 1986. *The Foul and the Fragrant: Odor and the French Social Imagination*. Cambridge, MA: Harvard University Press.

Corey, S. 1994. King Garbage: A History of Solid Waste Management in New York City, 1881–1970. Unpublished Ph.D. dissertation, New York University.

Dmytryszn, N. 2002. Landscape and Memory. In Fresh Kills: Artists Respond to the Closing of the Landfill, exhibition catalogue, Snug Harbor Cultural Center, Staten Island, New York.

Douglas, M. 1970. *Purity and Danger: An Analysis of the Concept of Pollution and Taboo*. New York: Pantheon.

Dowdell, J. and B. Thompson. November 2001. Landfill or Landmark? *Landscape Architecture*, p. 16.

Fenton, R. 1947. An Analysis of the Problems of Sanitary Landfills in New York City. Report to the Department of Health, Bureau of Sanitary Engineering.

Glassman, M. 2003. Are Toxins Astir? Release the Hounds. *New York Times*, June 26, p. G3.

Halliday, S. 1991. *The Great Stink of London: Sir Joseph Bazalgette and the Cleansing of the Victorian Metropolis*. London: Sutton Publishing.

Hardin, G. 1998 [1968]. The Tragedy of the Commons. In *Managing the Commons*, 2nd ed. John Baden and Douglas Noonan, eds. Indianapolis: Indiana University Press.

Hardin, G. 1993. The Global Pillage: Consequences of Unmanaged Commons. In *Living Within Limits: Ecology, Economics, and Population Taboos*. New York: Oxford University Press.

Johnson, K. 2000. Bittersweet Memories on the City's Island of Garbage. *New York Times*, November 7, p. B1.

Kilborn, P. 2002. In Mount Trashmore's Shadow, the Gravy Train Slows Down. *New York Times*, April 1.

Martin, J. February 2002. Can a Landfill Become a Landmark? Letter to the Editor, Landscape Architecture, p. 9.

Martin, L. and N. Russell. 2000. Trashing Rubbish. In Ian Hodder, ed., *Towards Reflexive Method in Archaeology: The Example at Çatalhöyük*. Cambridge, U.K.: McDonald Institute Monographs, pp. 57–69.

McLaughlin, T. 1971. *Dirt: A Social History As Seen Through the Uses and Abuses of Dirt*. New York: Stein & Day.

Melosi, M. 2002. The Fresno Sanitary Landfill as a National Historic Landmark. *American Society for Environmental History News* 13(2): Summer, p. 1,3.

Melosi, M. 1981. *Garbage in the Cities: Refuse, Reform, and the Environment, 1880–1980*. College Station, TX: Texas A&M University Press.

Miller, B. 2000. *Fat of the Land: A History of Garbage in New York—The Last Two Hundred Years*. New York: Four Walls Eight Windows.

Miller, W. 1997. *The Anatomy of Disgust*. Cambridge, MA: Harvard University Press.

Mitchell, J. 1993. *Up in the Old Hotel*. New York: Vintage.

Mumford, L. 1961. *The City in History: Its Origins, Its Transformations, and Its Prospects.* New York: Harcourt, Brace and World.

Needham, S. and T. Spence. 1997. Refuse and the formation of middens. *Antiquity* 71, 77–90.

New York City Department of Street Cleaning Annual Report. 1905.

O'Brien, M. 1999. Rubbish-Power: Towards a Sociology of the Rubbish Society. In J. Hearn and S. Roseneil, eds., *Consuming Cultures: Power and Resistance.* New York: St. Martin's.

Rathje, W. November/December 1999. Time Capsules: The Future's Lost and Found. *Scientific American Discovering Archaeology*, pp. 86–88.

Rathje, W. December 1989 Rubbish! *The Atlantic Monthly.* pp. 1–10.

Reid, D. 1991. *Paris Sewers and Sewermen: Realities and Representations.* Cambridge, MA: Harvard University Press.

Rodgers, C. February 1939. Robert Moses. *The Atlantic Monthly*, pp. 225–234.

Rothschild, N. 1990. *New York City Neighborhoods: The 18th Century.* New York: Academic Press.

Rykwert, J. 2002. *The Seduction of Place: The History and Future of the City.* New York: Vintage.

Schoofs, M. 2001. Storied New York Landfill Live to Tell One More Tale. *Wall Street Journal*, September 28.

Severo, R. 1989. Monument to Modern Man: "Alp" of Trash is Rising. *New York Times*, April 13.

Staten Island Advance, online forum. September 2001. http://www.silive.com/.

Stegner, W. October 1959. The Town Dump. *The Atlantic Monthly*, pp. 78–80.

Sulfita, J., C. P. Gerba, R.K. Ham, A.C. Palmisano, W. L. Rathje, J. A. Robinson. 1992. The World's Largest Landfill: A Multidisciplinary Investigation. *Environmental Science & Technology*, Vol. 26, No. 8, pp. 1486–1495.

Ukeles, M.L. 2002. It's About Time for Fresh Kills. *Cabinet* 6(Spring), pp. 17–20.

Walsh, D. C. 2002. Incineration: What led to the rise and fall of incineration in New York City? *Environmental Science & Technology*, August 1.

Young, W. 2001. Fresh Kills Landfill: The Restoration of Landfills and Root Penetration. In Niall Kirkwood, ed., *Manufactured Sites: Rethinking the Post-Industrial Landscape.* London: Spon Press.

2 Restoration of Drastically Disturbed Sites
Spectacle Island, Boston Harbor

Phillip J. Craul and Clarissa L. Rowe, ASLA

CONTENTS

ABSTRACT

Drastically disturbed land is characterized by significant changes of topography, through large-scale cutting or filling or both, that changes relief and landscape shape, leads to disturbance or destruction of vegetation, interrupts or modifies hydrologic features, including both surface and ground water flow and drainage, and causes potential contamination by toxic substances to the site and adjacent areas. Thus, conditions are created in the disturbance process that pose serious problems for restoration of the site.

The resulting conditions created by the drastic disturbance of land are discussed to provide some important information that the restorer needs to consider in the planning and execution of the restoration process. Each disturbed site has its own characteristics of capabilities and limitations for restoration; no two sites are exactly alike. The objective of restoration must be carefully composed to form a best fit to these characteristics. Topographic stability and its reshaping to provide enhanced hydrologic and relief features are discussed. The importance of the soil in the restoration process is stressed. Different types of soil can be designed to meet sustainable needs in the restoration objectives.

A case study, the restoration of Spectacle Island, Boston Harbor, is presented in great detail. Originally timbered and used for hay production, later a quarantine hospital, gambling resort, horse and cattle slaughter house, and still later the City of Boston dump, became the major site for disposal of the excavated material for the I-93 tunnel through downtown Boston, i.e., "The Big Dig." The restoration objective is a recreational park area. Description is given of the soil materials on the island, how they required amendment, sourcing of the amendments, placement of the soil, the topographic design and the plant palette to meet the physical constraints of an island exposed to winds and saltwater spray and eventual public impact. Problems and lessons learned in the restoration process are fully covered.

The great connoisseurs of soil have been those who could assess at a glance the character of a piece of land.

—William Bryant Logan, 1995

INTRODUCTION

Drastically disturbed lands are those lands that have undergone severe topographic change due to excavation, filling, and leveling operations. A major characteristic of drastically disturbed lands is the large scale of the area involved, the large volumes of material disturbed and produced as waste, and the great changes in relief. These operations are usually, but not always, connected with mining. The degree of change is sufficient to have significant impact on the soil, hydrologic conditions, erosion and sedimentation, creation of dust, and other migrating substances which might be a source of contamination.

Bridges (1987, *in* Craul, 1992, 1999) lists in his book the characteristics of drastically disturbed sites. These include:

1. Unusual or complex topography, even holes in the ground next to piles of waste.
2. Varying degrees of unstable conditions caused by piled material, exposed, bare soil, etc., subsidence and erosion.
3. Presence of toxic substances.
4. Interrupted, disturbed drainage, and possible flooding.
5. Fragmented or absentee ownership.
6. May exist in areas with limited access by normal infrastructure.
7. Abandoned buildings and other structures in derelict condition creating hazards.
8. Unsightliness; ruined buildings, bare and erosing soil, or coarse vegetation.
9. Absence of infrastructure; probably removed from the site earlier.
10. Surrounding areas in depressed state; poor housing, etc.

It may be added to No. 10 that the surrounding areas may be contaminated from migration of residues, hydrologic contamination, or air pollution from the former industrial, mining, or commercial operations. In the particular case study presented below, the surrounding area is a body of water—Boston Harbor.

From this list it becomes obvious that a primary factor requiring attention is the soil or the material that covers the site. Attention will be given to the existing conditions of the materials covering the site, whether soil, overburden, or wastes, followed by the modifications required to create a satisfactory planting medium, or for more drastic conditions, the design and construction of a completely new soil. Modification of topography and vegetation planting choices and arrangements must be included here as they relate to the soil, but not discussed in detail from the engineering standpoint as that is beyond the scope of this chapter. However, the most important factor to remember about restoring a site is that the soil mantle must first be stabilized from the engineering standpoint before vegetation can stabilize the soil itself.

Premining or predisturbance planning, if possible, facilitates a more efficient and effective reclamation (Nieman and Merkin, 2000). There are situations where a progression of plantings with a planned succession of vegetation may be necessary to attain the final level of reclamation quality, suggesting an evolution of soil conditions similar to soil genesis in nature.

RECLAMATION SITE CHARACTERISTICS INFLUENCING SOIL DESIGN

TOPOGRAPHY

Bridges' first two items from his list given in the Introduction broadly describe the topographic condition of most drastically disturbed sites. They indicate that some modification to the topography is necessary to reclaim the area to a useful purpose. The type and degree of modification depends most strongly on the volume and relief of the material and of the site, the properties of the material (particularly its stability and whether or not it needs to be covered and protected from weathering), and the reclaimed use of the site. The requirement for protection depends on the soil characteristics, and the reclaimed use depends on all of these factors and the objectives of the reclamation.

Drastic disturbance to the land generates unconsolidated earth materials. Management of the materials is a key component to successful topographic reconstruction (Toy and Black, 2000). The material strength, consolidation characteristics, erosion potential, ability to support vegetation, and potential toxicity will greatly influence the planned reclamation. Evidence of erosion is indicative of hydrologic problems.

Topographic Position

The landform position of the disturbed site on the landscape is important to the assessment of the overall reclamation plan (Figure 2.1). A site located on a river terrace or lower floodplain will require a different reclamation plan than one on an upland slope. The former will require great

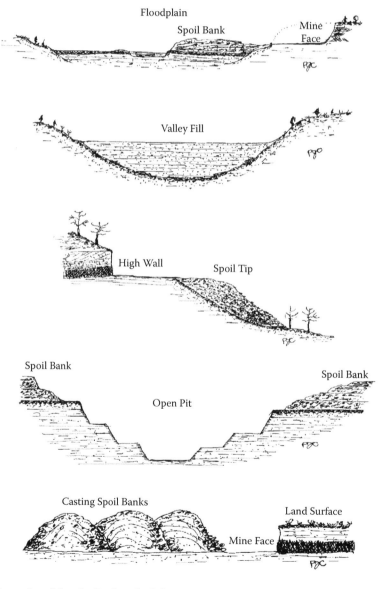

FIGURE 2.1 Examples of drastically disturbed sites on various landforms.

attention to the presence of water and its drainage together with the threat of flooding. The upslope site will require attention to slope stability, erosion potential, and drainage. A reclamation site situated on an upland plateau with relief only due to the disturbance itself may require less complex planning. The topographic position of the site will also influence its final planned use.

Surface Topography

The surface topography of disturbed sites range from simple to extremely complex (Figure 2.2). A simple surface may be a long slope of uniform gradient or perfectly level as if planed. The relief is small. These surfaces usually result from the tipping of materials on a natural slope or the deposition of waste slurry in a valley. A complex surface is one that is characterized by intermixed concave and convex surfaces exhibiting a definite pattern resulting from the method of excavation or deposition. Surface strip mining of coal and other deposits will have this pattern. The relief is dependent on the thickness (hence, the volume) of the overburden that must be removed to reach the desired vein. A very complex surface topography is one with a mixture of pits, piles, high walls, and flats. All of these features may be at the same average level or at quite different elevations, the latter rendering high relief to the site.

The simple surface topography may have relatively uniform material distributed over and along the surface, while the more complex surface topography will probably have widely differing materials varying in thickness across the surface. Detailed inventory of the area is required as a preplanning activity.

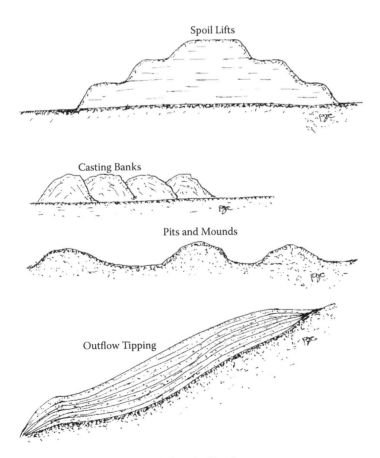

Spoil Lifts

Casting Banks

Pits and Mounds

Outflow Tipping

FIGURE 2.2 The surface topography of several disturbed lands.

SOIL CHARACTERISTICS

Inventory of the characteristics of the unconsolidated (soil?) materials on the disturbed site reveals their physical, chemical, and biological properties. Knowledge of these properties is necessary for a successful reclamation that also reduces any potential environmental impact. The inventoried properties and their importance to the reclamation process include the variability of the soil material over the site. The soil characteristics examined include: the physical properties of drainage, stoniness and texture, porosity and density, and organic matter content, with the chemical properties of reaction (pH), nutrient content, presence and concentration of contaminants, and soluble salts.

PRINCIPLES OF RECLAMATION SITE SOIL DESIGN

Reclamation of a drastically disturbed site requires a broad toolbox comprised of climatic, vegetative, hydrologic, and topographic and soil tools, in addition to design tools themselves. The emphasis of this section is on the soil and the closely associated factor of topography (Mays et al., 2000; United States Department of Agriculture, 1979). The other factors are not ignored but are given limited coverage owing to space considerations.

TOPOGRAPHIC RECONSTRUCTION: THEORY AND PRACTICE

The National Academy of Sciences (1974) in its report on the rehabilitation of Western coal mines suggested that in most cases geological material should be reshaped to achieve (1) the best ecological conditions for the landuse conditions, (2) proper hydrologic requirements, and (3) the most pleasing aesthetic experience for the viewer, admitting that the three criteria may not be compatible.

Some drastically disturbed sites may require severe topographic modification, whereas others, not as drastically disturbed or with an existing topographic condition that could fit the reclaimed use, may require little modification. Thus, the topographic modification required must be based on the individual site with no universal application. Topographic reconstruction becomes an opportunity for creativity and imagination. The final practical objective of topographic reconstruction is the minimization of long-term maintenance.

It should be noted in Figure 2.3 that the overall slope is 3:1, and the individual terraces have 2:1 slope. The shorter slope length allows for a steeper slope gradient and provides for the drainage interception benches necessary for slope stability. The compacted stepped subbase provides engineered stability to the entire system. One of the factors that must be considered in slope design is the angle of repose of the material to be shaped (Wiegle, 1966; Lyle, 1987). All material, when in loose condition,

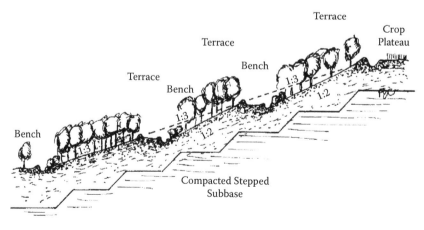

FIGURE 2.3 A reconstructed topographic slope with planting arrangement and engineered stabilization of the slope (from Darmer, 1992).

will assume a typical angle of slope with respect to the horizontal plane. This angle is the result of internal friction and packing among the particles composing the material. Thus, the particle size distribution greatly influences the angle of repose. Compaction will tend to increase the angle of repose but the resulting material may not be compatible with plant growth due to the high density.

In the process of reclamation, the influence of topography on factors affecting exposure to sunlight and winds, soil temperatures and moisture, and hydrologic factors can be employed to favor certain designs as well as creating problems for others (Mays et al., 2000). To illustrate these points, Figure 2.4 presents these factors on a hypothetical hill with assumed uniform soil. Exposure is greatest at the crest and varies on the slopes, depending on prevailing and local winds. Sunlight (or solar radiation) varies around the hill, being most intense on the southern aspect and least on the northern aspect (in the northern hemisphere), with the west and east aspects having intermediate sunlight. As the angle between the slope and sun varies, the amount of solar radiation received by a unit area of surface will vary. As the result of the differential distribution of solar radiation on the hill, soil temperatures and moisture will vary as well. As expected the greatest soil temperature occurs on the south aspect and is least on the north aspect. The soil temperatures of the west aspect are intermediate between the north and south aspects, but greater than the east aspect. The west slope is exposed to solar radiation in the afternoon when air temperatures are normally high. In the morning, air temperatures are cool from the night-time carry over, and soil temperatures are less than the west aspect but greater than the northern aspect. Average soil moisture is closely related to soil temperatures over a period of time. Thus, the soil tends to be driest on the south aspect, intermediate on the east and west aspects, and the northern aspect has the greatest average moisture content. On natural topography, the vegetation composition and growth response will vary on the hill according to the described conditions. Finally, as the relief of the hill is increased, the hill effects increase, and also increase with increase in the latitude of the site; however, the average solar radiation (thus, temperature) on a surface unit basis is reduced with increase in the latitude. The design process should take advantage of the attributes of these topographic effects.

Obviously, the reconstructed topography will be more complex than a simple hill. Possible landforms with their characteristics that might be constructed:

1. Plateau (level relief)—full sun and exposure, slight erosion potential, and may have slight soil variability.
2. Gentle slopes (width of concave surface or swale much greater than height of knoll)—less exposure with some shelter and shadow especially at low sun angles, low to moderate erosion potential, slight to moderate soil variability.

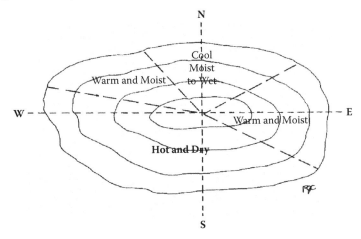

FIGURE 2.4 Slope aspect and relief effects on solar radiation, soil temperature, and moisture. The hill is segmented into cardinal direction sectors.

3. Hilly (valley width is equal to or greater than height of hill)—moderate exposure on the hills and shadows in the valleys, moderate to severe erosion depending on slope length and gradient (the LS factor of the universal soil loss equation employed by soil scientists) and the soil erodibility (K) factor, and probably a moderate to high degree of soil variability (Soil Survey Staff, 1993).
4. Ridge and valley (height of ridge is much greater than width of valley or no valley exists)—all factors increased: greater exposure on ridge, greater shadow and shelter in valley depending on orientation of the ridge and valley, severe erosion potential, high degree of soil variability.
5. Kettle (usually on level terrain)—intermingled mounds and hollows of relatively low relief with slight to moderate exposure, little shadow, slight erosion, but perhaps a high degree of soil variability.

Though these characteristics can be applied to natural topography, the same characteristics usually result during construction of the reclaimed site. The principles of design have their application in the planning at this point. A simple landform pattern is easy to construct, whereas a complex landform pattern is more difficult and requires extensive prior planning with management of the earth materials on the site. The complex landform provides more diverse habitat conditions and perhaps greater aesthetic values but also presents an increased intricate hydrologic pattern which increases soil variability and erosion potential. Figure 2.5 illustrates the contrast between the simple and complex landform patterns.

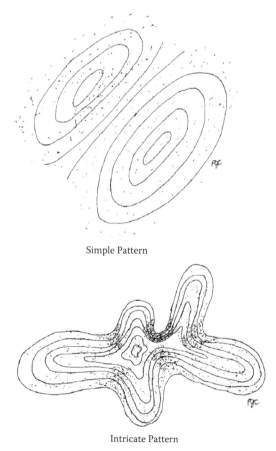

Simple Pattern

Intricate Pattern

FIGURE 2.5 The contrast between the simple and complex landform pattern.

There may be some question concerning moderate or high degree of soil variability on a reconstructed site for hilly or ridge and valley types. The placement process is usually by bulldozer or other front-end equipment or dumped from trucks. The material, even though relatively uniform in physical composition, may become segregated as the material tumbles down slope in the spreading process. Laying down lifts of material by pan equipment will provide soil uniformity but this equipment cannot be safely used on moderate to steep slopes except in unusual circumstances.

The conclusion then becomes that a well-designed reclamation plan takes advantage of the differences in topographic site conditions (and more than likely the soil differences as well) and creates a mode of diversity. It also suggests that components (vegetation, etc.) of the design cannot be confined to one simple alternative on all portions of the reclamation area. Uniform distribution of soils and plant species very seldom occur around a hill in nature. And, even if the soil is uniform around the hill as the result of disturbance (i.e., a waste heap), the topographic effect may preclude uniformity of plant growth response.

RECONSTRUCTION SOIL DESIGN

Reconstruction of the soil for a drastically disturbed site cannot be a hit-or-miss system ("let's use what's available!"), otherwise a mess may result. So-called "magic bullets" and "witches brews" should be avoided. Competent design and specifications drawn up by an appropriate team of professionals (engineers, soil scientists, chemists, hydrologists, and landscape architects) are required.

GOALS

Darmer (1992) presents several goals for soil development in reclamation:

1. Good texture and structure at depth
2. Increased humus and supply of nutrients
3. Good drainage and water retention
4. An active soil life

These goals apply to every vegetated segment of a reclamation site. There are three alternatives to achieving these goals:

1. Import natural topsoil from other large earth removal operations. Not really a good option as the topsoil on an off-site location should probably remain there for eventual restoration of that site unless it is to be totally occupied by structures.
2. Amend the existing on-site earth materials with various other materials to improve the properties so that vegetation may be supported and maintained. The amending materials may include sand, composted organic materials, and so on. This alternative is probably the most feasible for most drastically disturbed sites, but it must be accomplished with prior testing and analysis before the final specifications are formulated.
3. Design a specific soil profile for the site or separate portions of the site using off-site materials (not soil *per se* as in No. 1). These materials may be those of more favorable characteristics recycled from other sites or from a mixture of recycled byproducts such as glass, composted biosolids, and washings from stone and gravel operations. Testing and analysis is required before formulation of the final specifications.

As an example of the third alternative, the Commonwealth of Pennsylvania, the City of Philadelphia, the U.S. Army Corps of Engineers, and Pennsylvania State University, together with private consultants, are presently in the process of planning and executing a comprehensive research program to investigate the use of recycled construction demolition byproducts, municipal biosolids, short fiber paper waste, and river dredgings, among others, as variable ingredients for designed soil materials to be applied to a variety of restoration sites, including former mine sites and city green-scapes (T.A. Craul, 2006, personal communication).

The problem of scale and greater potential for very unfavorable materials as a planting medium or media, with greater emphasis on topographic planning differentiates the process from common urban areas (Sobek et al., 2000). Application of soil design and the specifications to a broader variety of land uses, rather than landscape architecture projects, is involved in soil reconstruction for drastically disturbed sites. Several applications follow.

SOIL RECONSTRUCTION APPLICATIONS TO RECLAMATION SITES

CROPLAND

Placement of suitable earth materials, as determined in the initial inventory area, is needed for cropland (Dunker and Barnhisel, 2000). The regular or periodic interference of the soil in agricultural use requires strong soil structure, a large nutrient reservoir, and good water relations (Darmer 1992). The material should be stable and compaction resistant, capable of being tilled (low stone and rock content), appropriate texture, and adequate depth. Intensive soil acidity adjustment (liming) and fertilization coupled with tilling is required. These conditions must be initially present to support crops and continuously maintained just as in cropland management on natural soil. If the conditions are not present initially and the expense and effort not possible, another land use alternative should be planned. A vegetative progression planned to improve soil conditions to the desired level for cropland may be instituted. However, this is a long-term effort and requires periodic attention and management (Bidwell and Hole, 1964).

RANGELAND

Reclamation to rangeland is usually considered under semiarid climatic regions where it is easier to maintain on a long-term basis. The highly productive tall grass prairies, once extensive on the moist eastern border of rangeland in the United States, have been converted to intensive agriculture. These are now the corn and soybean belt.

Rangeland does not require the same high level of quality of the physical and chemical properties that must be present for cropland (Williams and Schuman, 1987). Hence, the initial conditions existing in the earth materials of the site might better fit the requirements of rangeland than other vegetative types. Soil building is restrained under rangeland partly due to the climate under which it exists. The return of biomass to the soil is on an annual basis, thus sustaining the range productivity. Rangeland may fit a varied topography but should not be planned for very steep slopes unless the stand of grasses is heavy and maintained as such.

FORESTLAND

Soil building under forest cover is a slow process. However, it is relatively undisturbed because of the long life and rotation period of the forest, which may impart long-term stability required for a successful reclamation process (Torbert and Burger, 2000). The contribution of the forest biomass as organic matter to the soil surface and the lateral intermingling of the roots with their deep extension will aid the development of natural structure and maintain porosity and high water infiltration rates, as well as improve soil productivity over time (Buol et al., 1989). On extremely low fertility sites, initial fertilization may be required to establish an adequate forest cover.

Sustainable Forest Production

Operations supporting sustainable forest production come into play on long-term forest land management. They include the thinning of the forest by selective cutting and harvesting of products, or perhaps complete removal to provide space for the next vegetative stage in the succession. Access to the forest is required. Therefore, it is necessary to plan for the entry of forestry personnel and equipment to carry out the silvicultural operations. Access to the forest land is facilitated by appropriate road and trail layout and topographic arrangement. Silvicultural operations are modified according to the terrain and the forest type.

Windthrow Hazard

The influence of topography on exposure to wind, its velocity and gustiness should be considered in the shaping and configuration of the reclaimed landform. The location, arrangement, and spacing of the trees must be carefully planned. Further, the consideration of windfirmness in soil design for forest or individual trees is important to the planting success. The depth and density of the soil must be designed to prevent the blowdown and stem breakage of the trees.

WILDLAND

The term *wildland* takes on many connotations. The evolution of landuse allocation by society usually results in the most inaccessible land, or the land with little or no practical use being set aside or even neglected. Or, it may be land that has visual or aesthetic significance that is presently termed wilderness. The characteristics of wildland include: generally rugged or desolate terrain, rocky, thin soils or boggy or swamp conditions, and vegetation ranging from sparse or scrubby vegetation to large residual stands of timber. In these respects the reclamation of a highly disturbed site could easily become wildland by neglect, if it isn't already.

However, a more positive approach is to make a drastically disturbed site into useful wildland rather than being neglected. Thus, some planning and deliberate reclamation become appropriate. The objective may include creating use as recreation, wildlife habitat, or a combination of uses that includes both of these. The reclamation process should involve the reduction or elimination of the obvious features resulting from the cultural impact, leaving those that add to the aesthetic or reclamation value of the site.

RECREATION

Recreation requires accessibility. Corridors for roads and heavy trails must be developed on stable, compaction-resistant materials with gentle grades. These may need to be created in the reclamation process. More remote sections of the site will require less stringent standards for access. Areas where foot traffic will be heavy should be also located on stable, compaction-resistant materials. Actual surface hardening may be required.

WILDLIFE HABITAT

This option may be considered a modification or variation on wildland but with the deliberate objective of providing habitat for wildlife. As such, the application must provide food sources, water, proper shelter and protection, and adequate space appropriate for the wildlife. Soil quality becomes a major concern in the assessment and planning process. In the early stages of development the site may require protection from wildlife until the vegetation is completely established. Also, soil burrowing animals can be continued threats to vegetation, soil stability, or other deleterious conditions that negatively affect long-term reclamation success. The surface should be manipulated to create pools for water and provide a diversity of vegetation for feeding a variety of wildlife. These wet areas can be created on those materials that occur in low topography and/or exhibit poor drainage (low infiltration and permeability).

WATERSHED

Planning and developing a site as a watershed perhaps must be accomplished with greater care than any other land use (Mays et al., 2000). The soil must be particularly fitting to the use. There are several properties that the soil must contain to be totally appropriate. These include a high water infiltration rate, great volume of water storage without becoming saturated with a great degree of stability, and great resistance to erosion. The soil profile must be handled in such a manner so that no contaminated materials are exposed to erosion or the leaching of contaminants into surface water or the ground water. Isolation of contaminated materials may be required.

Even the topography should be subdivided and shaped in such a way to develop adequate drainage of the site with stable channels. Control of surface water is very important and standard diversion structures are necessary on steeper or longer slopes.

WETLANDS

Wetlands are a sensitive issue associated with disturbed land reclamation. Constructed wetlands tend to improve water quality, support populations of wetland flora and fauna, and require little maintenance once established. However, they are expensive to establish and become regulated under the same provisions as a natural wetland. The soil profile for a constructed wetland is relatively simple to construct but the essential materials must be present or available. Wetlands on a contaminated site may act as a bioremediation feature.

ALPINE AND SUBALPINE LANDS

The natural soils of these areas are mostly shallow, skeletal (stony and rocky), containing very little organic matter except in local wet or bog areas, cold, and usually of low fertility (Macyk, 2000). Since they occur mainly in mountainous areas, exposure to continuous wind, sometimes extremely high, is a major factor that subjects the bare soil to blowing and extreme drying conditions. Vegetation is sparse and stunted, increasing the area of potentially bare soil. The stoniness and rockiness does tend to reduce the effects of wind erosion in some areas. Soil weathering is slow and mainly of a physical nature. The low temperatures do not favor soil chemical processes which aid soil reclamation. Any reclamation plan for these areas must take a very long-term view and very low intensity of use. If a soil profile is desired initially or early in the reclamation plan, a designed soil is almost always required. The natural areas are fragile and very susceptible to damage by overuse or other abuses. Reclamation sites would have even greater potential for damage due to the nature of the disturbed conditions.

A CASE STUDY: SPECTACLE ISLAND, BOSTON HARBOR, MASSACHUSETTS

INTRODUCTION

Spectacle Island has been used as a dump for the City of Boston and is now to be covered and sealed to meet EPA requirements (Figure 2.6). To protect the integrity of the landfill it is necessary to construct stone walls in several locations to prevent breaching of the landfill and subsequent leaching into the harbor. Some foundations of former structures still exist along with the remnants of a wharf on the western shore and a sunken barge. The surface material on the island is highly disturbed with little original soil still existing. There is some glacial till material stockpiled on the island that is available for use as soil base material; otherwise, any covering material must be barged to the island.

FIGURE 2.6 An aerial of Spectacle Island looking toward Boston, near the beginning of the project. Spectacle Island is in the very center of the photo. (Photo by Alex McLean.)

It was proposed that the island be restored to a park-like condition for future use as part of the island chain of parks within Boston Harbor. Thus, the covering material and soil must be constructed to support a regional mixed plant palette and must withstand recreational usage and require minimum maintenance while being subjected to the weather typical of the northeastern coast, which is characterized by occurrences of high winds, intense rainfall, fog, wave action and its salty overwash and spray, freezing rain, snow and ice, with possible dry conditions during the late summer.

The stated objectives of the overall project are to:

1. Seal and cover the existing former landfill on Spectacle Island, Boston Harbor, Massachusetts (a separate phase);
2. Use the island as a disposal site for Central Artery Tunnel excavation material;
3. Cover the tunnel material with a substratum loam and then a topsoil in preparation for landscaping and planting as a future recreational park.

INITIAL CONDITIONS OF THE RESTORATION SITE

As mentioned in the Introduction, the island was covered with former glacial till subsoil, excavated material from an adjacent island undergoing construction, and other mixed debris from former activities on the island.

Historically, Spectacle Island has been home to an amazing variety of exploits. In the 17th century the island was used for pasture and timber cutting. In 1717, a quarantine hospital was built and later moved to another harbor island. Two resort hotels with (illegal) gambling activities were built in 1847 and closed ten years later following a police raid. The island was then taken over by a slaughter house/horse rendering plant until this business was abandoned in 1910. The site then became the city dump for Boston in 1921 and was closed in 1959 (Wallace, Floyd Associates, 1991). Over the years the leachate from the dump migrated into Boston Harbor, and eventually the federal government sued the City of Boston and the Commonwealth of Massachusetts for polluting Boston Harbor. Orders were given to clean up the island to stop the pollution. This work began in 1991 and was facilitated by the advances in the planning and initial construction of the I-93 Central Artery Tunnel through downtown Boston, now popularly known as "The Big Dig."

Essentially, there was very little, if any, soil material on the island useful for reclamation (Figure 2.7). The large volume of glacial till was not useable alone because of its natural susceptibility to self-compaction (Figure 2.8). It would require amendment.

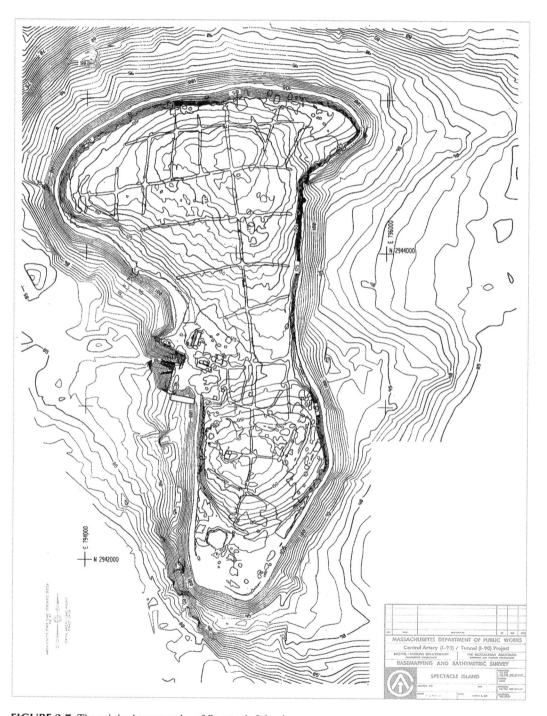

FIGURE 2.7 The original topography of Spectacle Island.

FIGURE 2.8 A view of a portion of Spectacle Island during placement of excavated fill and residual island subsoil.

RESTORATION LANDSCAPE DESIGN

The guiding factor in the grading design for Spectacle Island was placing 2.54 million cubic yards of excavated fill from the Big Dig. The Island was to double its size. Its height in 1991 was 190 feet above sea level; in 2002 its peak is 250 feet above the Boston Harbor. The "new" island mimics the earlier island form of "Spectacles." The fill placement on the island saved the Big Dig money because they did not have to place all of the fill on numerous surrounding landfills. This cost savings meant that there was money to restore the island and build a naturalistic park. About 580,000 cubic yards of loam and loam substrate were needed for the 105-acre park. Spectacle Island is like a park in the mountains. Its value derives from the views from two high hills to the surrounding Boston Harbor and Boston skyline. The challenge for the grading design was to provide stable 3:1 slopes and some relatively flat areas where most of the human activity would occur. Once the fill was placed, the whole 105 acres had to be planted. The last step of the grading design was the installation of 27,000 plants to stabilize the slopes. Planting of native plants will stabilize the slopes in the long term (Bidwell and Hole, 1964; Buol et al., 1989).

In the short term, the stability of the slopes relied on a series of "benches" in the slope. Large stone erosion control walls and winding pathways were placed along the slope to slow the flow of water (Mays et al., 2000). The other crucial design element was the composition of the newly manufactured soil itself. Controlling the surface runoff and the subsurface permeability were crucial design goals.

The concern about slope stability was most crucial on the north face of the North Drumlin where the harsh New England winter winds crossing the Boston Harbor are the worst. The height of the North Drumlin helps to buffer the rest of the Island. The North Drumlin has become a wildlife habitat. The areas of gentler slopes, where human activities are encouraged, are mostly on the southwest side of the Island. In the summer, the gentler southwest summer winds helps to create a cove that is hospitable to park users arriving by boat.

The planting design for the island derived from the grading stability concerns (Figure 2.9). Native plants were chosen for their root-holding capabilities, their self-seeding properties, and their ability to sustain periods of drought and neglect. Most of the plants were found on other Boston Harbor Islands. About 30 different tree species and 30 shrub species were chosen so that there was a greater chance of survival. There were four kinds of grass mixes chosen to cover the areas where the soil depth was the least. There were three kinds of clump grasses as well. A total of 2,389 trees and 25,600 shrubs and vines were planted. Plants were planted in small sizes so that they could more

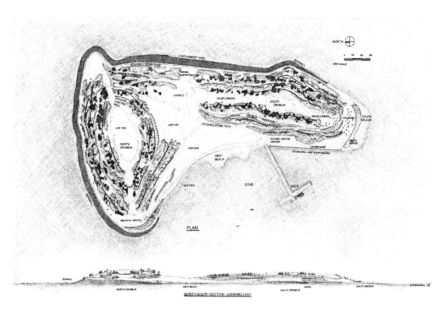

FIGURE 2.9 The restoration design plan for Spectacle Island.

readily adapt to the harsh environment. Plants needed to thrive in sandy, windy conditions and be able to survive on a New England waterfront.

THE SOIL DESIGN

The soil material installed on Spectacle Island needs to meet the requirements of several objectives. It must:

1. Cover and protect the landfill clay seal and the drainage layer;
2. Be relatively resistant to the potential of water and wave erosion, especially on moderately steep to steep slopes, and must be stable on these slopes;
3. Be an appropriate planting medium having sufficient depth for rooting and adequate volume to prevent windthrow of trees and shrubs, especially under wet to saturated soil conditions;
4. Have an adequate moisture retention capacity to prevent stress and mortality of the plant palette during droughty periods, coupled with at least minimum levels of plant nutrients;
5. Require minimum maintenance under various levels of recreational use and in varying weather conditions.

The Material on the Island

The large volume of residual glacial till material on the island was found, through testing (Sobek et al., 2000), not to be a suitable planting medium. Its very high silt content and well-graded particle size distribution (Figure 2.10) gave it great susceptibility to self-compaction and was a major limitation. It was also material that would have very low infiltration capacity and great erosion potential, both very undesirable properties for a planting medium. However, it was necessary to incorporate the residual glacial till into the designed soil in order to reduce the amount of material that needed to be barged to the island and to dispose of it on the island itself. It was not feasible to remove it from the island.

Various soil mix tests and calculations showed that a mix of a specified and locally available poorly graded medium to coarse sand (Soil Survey Staff, 1993) with the till in a ratio of two parts of

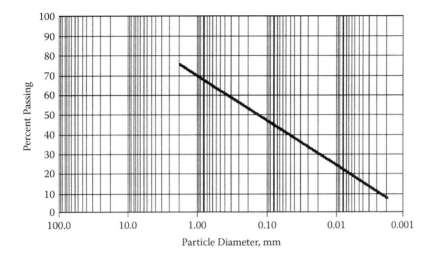

FIGURE 2.10 The particle size distribution of Spectacle Island glacial till. The near 45-degree, sloping straight line as the accumulation curve indicates susceptibility to self-compaction.

sand to one part of till (2:1) would provide a satisfactory planting soil mix. A ratio of 3:1 was found to be superior in compaction resistance and fitting the specified soil envelope, but the amount of sand required to be barged to the island did not make this choice feasible. Both of these mixes would provide a particle size distribution very close to the specified envelope for the planting loam and the planting loam substrate. Significantly, as will be seen later, it was determined that a mix of 1:1 was just as self-compacting as the till alone and would be unsatisfactory. The 2:1 mix and the source of the specified sand were approved.

The specification of a sandy loam or loamy sand as the planting loam requires that the plant palette contain those species and cultivars that exhibit moderate to low moisture and fertility requirements. They should have root systems that take advantage of the available soil depth (and volume) to increase water and nutrient uptake as well as to reduce windthrow hazard. Most native species would meet these requirements. Further, these species need to be resistant to salt spray and possible temporary accumulation of salts in the soil solution and on the soil cation exchange sites Brady, 1990).

The Planting Loam Substrate

A note needs to be made concerning the terminology used to describe the soil horizons of the designed soil profiles. The term *loam* is very specific in its reference to a texture class of the USDA soil texture classification (Soil Survey Staff, 1993) and should be used only in that context. Here, it is used to define the topsoil horizon as *planting loam* and the subsoil horizon as the *planting loam substrate,* even though the textures of the horizons are NOT loams—they are *loamy sands* or *sands*. These terms were used in earlier reports by engineers for preliminary planning of Spectacle Island soils, before the senior author was brought into the project. So that the terminology was consistent among the various reports and documents during the life of the project, the senior author agreed to continue the usage with the understanding by all parties that it is incorrect according to accepted International Soil Science Society and Soil Science Society of America standards.

The planting loam substrate is a mixture in a ratio of 2:1 of the specified sand to the island till. The resulting particle size distribution is given in Table 2.1.

The values given in Table 2.1 should not vary more than 5 percent finer or coarser in the coarser particle (medium sand and coarser) sizes while finer particle sizes should vary not more than 3 percent or less. The organic content should be 1 to 4 percent. The waterholding capacity of this material is approximately 1.5 to 2.1 inches per foot of soil and the infiltration capacity exceeds

TABLE 2.1

The Particle Size Distribution Range for the Spectacle Island Specified Planting Loam Substrate

USDA Particle Size Classes			ASTM Particle Size Classes		
Sieve Size	Percent Content	Percent Finer	Sieve Size	Percent Content	Percent Finer
3/8 inch	0	100	3/8 inch		100
#4	0	100	#4	0	100
		—	#8	6	94
#10	8	92			—
		—	#16	19	75
#18	23	69			—
		—	#30	26	49
#35	27	42			—
		—	#50	17	32
#60	14	28			—
		—	#100	10	22
#140	11	17			—
		—	#200	5.5	16.5
#300	2	15			—
silt (0.05–0.002)	10	—	silt (0.07–0.005)	8.5	—
clay (< 0.002)	5	—	clay (< 0.005)	8	—

10 inches per hour. Though the material appears to be very coarse in texture, it will drain easily with good aeration, and will be resistant to settlement and compaction. This enables its placement in relatively deep lifts.

The Planting Loam

The planting loam was the specified sand alone (Table 2.2), being the same sand as that used in the planting loam substrate mixture. The planting loam was not to contain any island glacial till. The sand was amended with organic matter to raise its content to 10 percent on a weight basis. The greater amount of organic matter increases the waterholding capacity and fertility level for this designed topsoil. The use of the specified sand as used in the substrate preserves a degree of compatibility between the two horizons (Figure 2.11).

The Specified Organic Matter

To meet the specification of 10 percent organic matter content for the planting loam, a very large volume of organic matter was required. And all of it had to be barged to the island (Figure 2.12). The search for a satisfactory material in sufficient volume became a major effort. The usual sources of organic matter, such as peat moss and so forth, were too cost prohibitive. An alternative considered early in the search was composted sewage sludge. Local sources were inadequate with respect to the volume and, more importantly, the quality available for use on a public park. For application to the island condition, the composted sewage sludge must meet the stringent requirements of EPA Part 503 specifications for Type A "exceptional quality" material. A source was finally located in nearby New Hampshire. To make certain that adequate volumes would be available when required, the material was stockpiled on a Boston wharf.

Less than halfway through the project, the approved New Hampshire source of brewers waste was depleted. The contractor found a new source of organic matter from a Holyoke, Massachusetts, wastewater plant. The new source of organic matter was used without approvals. While the organic

TABLE 2.2

The Particle Size Distribution Range for the Spectacle Island Specified Planting Loam. Ranges are Given in Brackets. Values Used for Calculations Represent the Midpoints of the Ranges

USDA Particle Size Classes			ASTM Particle Size Classes		
Sieve Size	Percent Content	Percent Finer	Sieve Size	Percent Content	Percent Finer
3/8 inch	0	100	3/8 inch	0	100
#4	0	100	#4	0	100
		—	#8	(6–17)	(83–94)
#10	11 (6–16)	89 (84–94)			—
		—	#16	(5–13)	(70–89)
#18	13 (7–18)	76 (66–87)			—
		—	#30	(5–13)	(57–83)
#35	9 (6–14)	67 (52–81)			—
		—	#50	(13–18)	(39–70)
#60	17 (17)	50 (35–64)			—
		—	#100	(20–21)	(19–49)
#140	29 (24–32)	21 (11–32)			—
		—	#200	(14–29)	(5–20)
#300	10 (6–14)	11 (5–18)			—
silt (0.05–0.002)	9 (5–13)	—	silt (0.07–0.005)	(3–11)	—
clay (< 0.002)	2 (0–5)	—	clay (< 0.005)	(2–9)	—

matter eventually met the EPA standards, the first shipments contained black plastic and small sticks which had probably been used as bulking agents for the wastewater product. Its use changed the quality of the manufactured soil. When this switch was noticed during a routine site visit, the work was stopped until the new product met project specifications. More testing of the organic matter was initiated and continued throughout the project. Often the organic matter was delivered to the site in an immature state and had to be stockpiled before use. Again, a compromise was made and the contractor was allowed to mix and spread the loam. No planting was allowed until the pH was in the allowable range for the plants. In order to speed the process of lowering the pH, aluminum sulfate pellets were mixed into the loam.

FIGURE 2.11 The specified sand component stored on the island. The foreground is the darker colored, mixed topsoil in place.

FIGURE 2.12 The composted sewage sludge as the organic component being unloaded from the barge.

THE SOIL PROFILES FOR PLANT PALETTE SEGMENTS

The diversity of the physical conditions across the site together with the variety in the proposed plant palette of the reclamation plan require that a series of soil profiles be designed (Figure 2.13). This process assures that the soil will closely match the needs of the plants for sustainability and

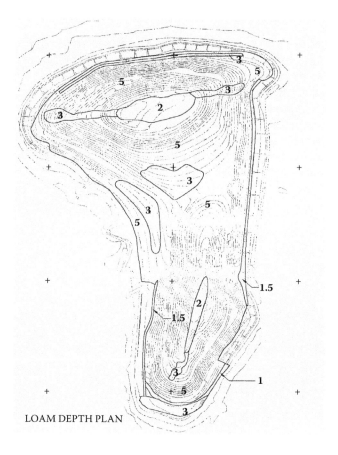

LOAM DEPTH PLAN

FIGURE 2.13 The distribution of soil profile depths across Spectacle Island.

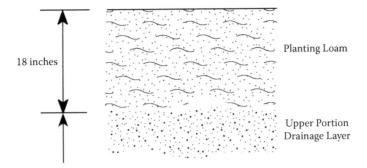

FIGURE 2.14 The 18-inch soil profile.

increases the efficiency of the reclamation process. Thus, there is no need to have deep soils over the entire reclamation area. To maintain simplicity across the profiles, the components and their mixtures comprising the planting loam and the planting loam substrate are similar in each case.

The 18-inch Profile

This is a shallow profile to be planted primarily to grasses (Figure 2.14). The relatively thick planting loam is necessary to provide adequate rooting depth for the grass sod. Its shallow depth precludes the planting of any other plant stock. It does not have a planting loam substrate horizon and contains no island till. The soil profile occurs at the foot of slopes on level areas behind a stone sea wall.

The 24-inch Profile

This profile has a 6-inch island till layer incorporated as the bottom layer with 6 inches of loam substrate (Figure 2.15). The island till covers the drainage layer. The planting loam is 12 inches thick. The 24-inch soil profile will have only low woody ornamentals and grasses as plantings. There are to be no trees planted on this profile; 18 inches of rooting depth does not provide sufficient anchorage for windthrow resistance.

The 36-inch Profile

This profile has a 12-inch layer of island till. This layer rests on the drainage layer (Figure 2.16). The loam substrate and the planting loam layer are both 12 inches thick. The profile provides 24 inches of

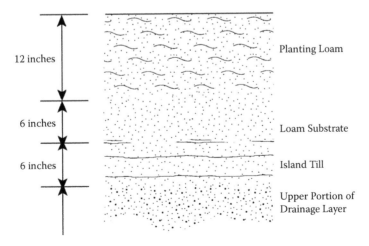

FIGURE 2.15 The 24-inch soil profile.

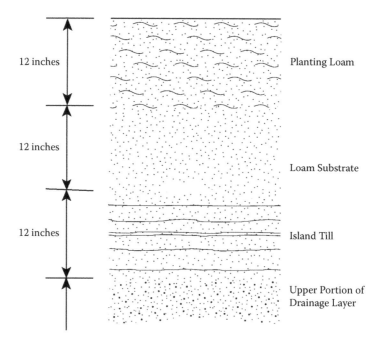

FIGURE 2.16 The 36-inch soil profile.

rooting depth that is adequate for anchorage of most trees against windthrow hazard. Also, since the soil is well drained, there is little likelihood that the soil would become saturated under extremely wet conditions resulting in windthrow damage.

The 60-inch Profile

This profile has a 24-inch layer of island till as the bottom layer resting on the drainage layer (Figure 2.17). The planting loam is 12 inches thick and the loam substrate is 24 inches thick. The profile provides 36 inches of rooting depth. This should be adequate for windthrow protection for the largest trees that may be planted on the island.

INSTALLATION OF SOIL

Traditional methods to apply the prepared mix upon the landscape and utilization of traditional landscaping equipment to blend the materials could not be employed for Spectacle Island. Large construction equipment would need to mix the material in place, particularly since the island glacial till was to be a component of the mix. The planting loam could have been mixed on-shore and barged to the island, but this would have involved handling the materials an additional time and the general contractor rejected the plan on grounds of costs.

Homogeneous mixing and placement of "topsoil" materials is critical to the success of the final landscaping. However, the contractor believed that competent bulldozer operators could mix the various components satisfactorily on-site by windrowing them and turning them over with the blade of the bulldozer. Because of the sheer volumes involved, the soil scientist and landscape architect had to concede to the contractor (Figure 2.18).

INSTALLATION AND POST-INSTALLATION INSPECTION

Continual inspections and analysis of in-place soil samples made during installation showed that the components were not well mixed and that too much of the till was mixed with the sand. Particle size

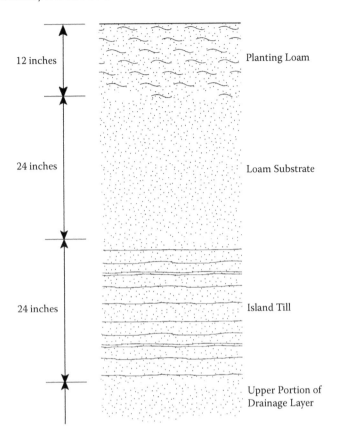

12 inches — Planting Loam

24 inches — Loam Substrate

24 inches — Island Till

Upper Portion of
Drainage Layer

FIGURE 2.17 The 60-inch soil profile.

analyses indicated that the mix for the planting loam substrate was closer to 1:1 rather than the specified 2:1 ratio. Excavation of soil in depth by augering, and using the knife-point test for relative bulk density observation indicated that the subsoil was self-compacting as compared to the original laboratory samples of the various mixes (Figure 2.19). Also, the organic matter content was quite variable across the final grade, in many cases being less than the specified 10 percent. Some glacial

FIGURE 2.18 The soil profile in place on the upper north slope of the north drumlin, Spectacle Island.

FIGURE 2.19 An auger hole through the topsoil and into the subsoil to examine apparent soil density, Spectacle Island.

till was observed in the planting loam. No island glacial till was to be included in the planting loam. Thus, the desired homogeneity of mixing was not obtained.

The original plan called for the slopes to be constructed from the bottom or foot upward to the top and the erosion control terraces constructed in place as elevation was gained on the slope. Unfortunately, the contractor proceeded to construct the slope to final grade completely to the top of the north drumlin without installing the boulder terraces. The contractor claimed that there was no barge space to haul the boulders when they were required. As a result of a severe "Northeaster" storm in December, 1995, several large gullies were eroded on the upper slopes (Figure 2.20) and sheet erosion was widespread at the foot slopes (Figure 2.21). Limited hydro-mulching on the slopes was inadequate to prevent the erosion. The contractor than constructed rip-rap swales (Figure 2.22) to channelize the runoff. These were not part of the initial plan and would not have been needed if the terraces had been installed as part of the slope-building sequence.

FIGURE 2.20 Gully erosion as the result of lack of erosion prevention terraces, Spectacle Island.

FIGURE 2.21 Erosion at the foot of the south slope of the north drumlin, Spectacle Island.

FIGURE 2.22 Rip-rap erosion control swale constructed by the contractor, Spectacle Island.

CONCLUSIONS AND RECOMMENDATIONS—LESSONS LEARNED FROM THE SPECTACLE ISLAND PROJECT

During the Project Design Phase

1. Educate the client about the importance of soil in the overall success of plant survival.
2. Make sure that you have enough fee to hire a soil scientist, with experience in construction, to work with the landscape architect from the beginning of the project.
3. If you are designing a public park, make sure that you use the public process to educate the public about the importance of the soil in a successful project. Explain the need for soil depth and quality. Explain the "scary" elements of any manufactured soil, like sludge pellets or wastewater compost, so that this science is understood. Explain the need for supervision of the process.
4. Develop site-specific soil specifications with the site's soil "recipe." Investigate sources for the soil elements as part of this process.

During the Testing and Pre-construction Phase

5. Test all the components of the soil recipe before making the recipe. Make sure that the testing includes samples of the soil components from all sources of the soil. Make sure that the samples meet the soil specifications. Keep the test samples and the results for comparing to the finished product. Make the test samples part of the discussion during the pre-construction meeting with the chosen contractor.
6. Develop a testing program for the soil components individually and for the mixed soil in place. A standard for testing the new soil is a test every 500–1,000 cubic yards.

During the Construction Phase

7. If possible, get the contractor to buy into the manufactured soil mix. A manufactured soil can substantially lower costs for loam. Be open to suggestions from the contractor, but hold the line on a "minimum" soil mix.
8. Be on site to supervise as much as possible. If that is not possible, make sure a really qualified, top-notch resident engineer is part of the construction team. Call for regular construction site meetings. It is important to have supervision of the soil recipe itself and how it is mixed. Check the required soil testing for changes in the test values, and then use all your senses when supervising the manufactured soil. Make sure that the test results that you are checking are for the soil that is being used on your project. Make sure that what you are looking at has the same look as the sample. Use your eyes, your hands, and your nose to make sure that the soil is what is specified.
9. Even after the recipe is mixed and approved, the testing should continue throughout the project to make sure that the original recipe is followed. There is a tendency to cut costs at the end of a project. Make sure that this testing is done by an independent entity, not the contractor building the soil and placing it.
10. Watch for material changes of the soil components as the job progresses. On large projects, supplies can run dry and alternative sources have to be found and tested all over again.
11. Once the plants are installed in the new soil, watch their health. If they fail, check their rootball, the soil compaction below the plant, and the pH of the soil.
12. Check the quality of any water used on the project. One of the failures on Spectacle Island was that a whole season of new plants were watered with salty water.

Keep reminding your audiences that a good soil will ensure that the final landscape is stable, durable, and sustaining (Figure 2.23).

FIGURE 2.23 The completed north drumlin with vegetation and terraces installed, Spectacle Island. (Photo by John Suter.)

CONCLUSIONS

The reconstruction of soil for a drastically disturbed site requires much testing and analysis, with forethought and planning, and thorough review of well-written, detailed specifications before a bulldozer blade is dropped into earth or the pencil is finished its movement on the drafting table. The process of designing a soil for a routine landscape architecture project in an urban area can be applied to the soil design process for a drastically disturbed site but on a much larger scale, with the attendant problems and with other special considerations as given in the first part of the paper. The planning of the soil reconstruction must be done in a cooperative manner with other parties on the project so that the finished soil fits the landscape, the plant palette and the final objectives of the reclamation plan. To do otherwise is to invite failure or even disaster. Or, according to what has happened in the past, leave the entire problem to benign neglect or to nature (Winterhalder, 2000)?

LITERATURE CITED

Bidwell, O.W., and F.D. Hole. 1964. Man as a factor in soil formation. *Soil Sci.* 98: 65–72.

Brady, N.C. 1990. *The Nature and Properties of Soils.* 10th ed. Macmillan, New York.

Bridges, E.M. 1987. *Surveying Derelict Land.* Clarendon Press, Oxford.

Buol, S.W., F.D. Hole, and R.J. McCracken. 1989. *Soil Genesis and Classification.* 3rd. ed. Iowa State University Press, Ames, IA.

Craul, P.J. 1992. *Urban Soil in Landscape Design.* John Wiley and Sons, New York.

Craul, P.J.. 1999. *Urban Soil: Applications and Practices.* John Wiley and Sons, New York.

Craul, T.A., and P.J. Craul. 2006. *Soil Design Protocols for Landscape Architects and Contractors.* John Wiley and Sons, New York.

Darmer, Gerhard. 1992. *Landscape and Surface Mining: Ecological Guidelines for Reclamation.* N. L. Dietrich, ed. English translation by Marianne Elflein-Capito. Van Nostrand Reinhold, New York.

Dunker, R.E., and R.I. Barnhisel. 2000. Cropland reclamation. In Barnhisel, R.I., R.G. Darmody, and W.L. Daniels. (ed.) *Reclamation of Drastically Disturbed Lands.* Agronomy Monographs 41. Amer. Soc. Agron., Crop Sci. Soc. Amer., Soil Sci. Soc. Amer., Madison, WI.

Logan, W.B. 1995. *Dirt: The Ecstatic Skin of the Earth.* Riverhead Books, The Berkley Publishing Group, New York.

Lyle, E.S. 1987. *Surface Mine Reclamation Manual.* Elsevier, New York.

Macyk, T.M. 2000. Reclamation of alpine and subalpine lands. In Barnhisel, R.I., R.G. Darmody, and W.L. Daniels. (ed.) *Reclamation of Drastically Disturbed Lands.* Agronomy Monographs 41. Amer. Soc. Agron., Crop Sci. Soc. Amer., Soil Sci. Soc. Amer., Madison, WI.

Mays, D.A., K.R. Sistani, and J.M. Soileau. 2000. Climatic and hydrologic factors associated with reclamation. In Barnhisel, R.I., R.G. Darmody, and W.L. Daniels. (ed.) *Reclamation of Drastically Disturbed Lands.* Agronomy Monographs 41. Amer. Soc. Agron., Crop Sci. Soc. Amer., Soil Sci. Soc. Amer., Madison, WI.

National Academy of Sciences. 1974. *Rehabilitation Potential of Western Coal Lands.* Ballinger Publishing Co., Cambridge, MA.

Nieman, T.J., and Z.R. Merkin. 2000. Pre-mining planning for post-mining land use: applying principles of comprehensive planning and landscape architecture to reclamation. In Barnhisel, R.I., R.G. Darmody, and W.L. Daniels. (ed.) *Reclamation of Drastically Disturbed Lands.* Agronomy Monographs 41. Amer. Soc. Agron., Crop Sci. Soc. Amer., Soil Sci. Soc. Amer., Madison, WI.

Sobek, A.A., J.G. Skousen, and S.E. Fisher, Jr. 2000. Chemical and physical properties of overburdens and minesoils. In Barnhisel, R.I., R.G. Darmody, and W.L. Daniels. (ed.) *Reclamation of Drastically Disturbed Lands.* Agronomy Monographs 41. Amer. Soc. Agron., Crop Sci. Soc. Amer., Soil Sci. Soc. Amer., Madison, WI.

Soil Survey Staff. 1993. *Soil Survey Manual.* Handbook No. 18. U.S. Department of Agriculture, U.S. Government Printing Office, Washington, D.C.

Torbert, J.L., and J.A. Burger. 2000. Forest land reclamation. In Barnhisel, R.I., R.G. Darmody, and W.L. Daniels. (ed.) *Reclamation of Drastically Disturbed Lands.* Agronomy Monographs 41. Amer. Soc. Agron., Crop Sci. Soc. Amer., Soil Sci. Soc. Amer., Madison, WI.

Toy, T.J., and J.P. Black. 2000. Topographic reconstruction: The theory and practice. In Barnhisel, R.I., R.G. Darmody, and W.L. Daniels. (ed.) *Reclamation of Drastically Disturbed Lands.* Agronomy Monographs 41. Amer. Soc. Agron., Crop Sci. Soc. Amer., Soil Sci. Soc. Amer., Madison, WI.

U.S. Department of Agriculture. 1979. User guide to soils: Mining and reclamation. USDA-FS Intermountain For. Range Exp. Station Gen. Tech. Rep. INT-68. USDA-FS, Ogden, UT.

Wallace, Floyd Associates Inc., 1991. The Development of Spectacle Island—Design Concept, Criteria and Guidelines, prepared for The Massachusetts Department of Public Works.

Wiegle, W.K. 1966. Spoil bank stability in Eastern Kentucky. *Mining Congr. J.* 52(4): 67–73.

Williams, R.D., and G.E. Schuman, (eds.) 1987. *Reclaiming Mine Soils and Overburden in the Western United States: Analytic Parameters and Procedures.* Soil Conservation Society of America, Ankeny, IA.

Winterhalder, K. 2000. Reclamation of smelter-damaged lands. In Barnhisel, R.I., R.G. Darmody, and W.L. Daniels. (ed.) *Reclamation of Drastically Disturbed Lands.* Agronomy Monographs 41. Amer. Soc. Agron., Crop Sci. Soc. Amer., Soil Sci. Soc. Amer., Madison, WI.

Part 2

Canals and Creeks

3 The Zurich Stream Daylighting Program

Fritz Conradin and Reinhard Buchli

CONTENTS

ABSTRACT

The Zurich Stream Daylighting Program is the concept of separating clear water streams from underground sewage pipes and surfacing stream water pipes by opening them up to simulate a natural environment. The concept is a solution to a problem that many cities face, as traditionally the way to cope with both stream water and wastewater was to amalgamate them into one waste system. As cities grow, the disposal of wastewater becomes a larger problem and the treatment of sewage an economic consideration. When a good proportion of the water being fed into the sewage system is clean, then the obvious solution is to separate the fresh water at source and thus the sewage treatment plants process only waste products. This is the ideal solution from an economic and from an ecological point of view. The concept also encompasses streams that have been channeled into pipes. Historically, this was the way to deal with water, however, from an ecological and aesthetic point of view, the resurfacing of streams is an attractive alternative in the urban landscape. The Zurich Stream Daylighting solution was to divert the stream water flows into a separate system that was custom built to accommodate stream water, clean storm water runoff, and groundwater. The concept of daylighted streams instead of pipes was strongly supported by a group of people from different city departments. The Zurich Stream Daylighting Program was set up and presented to the press and the public in 1988.

INTRODUCTION

The city of Zurich is surrounded by hills with numerous springs that form smaller or larger streams flowing into the city. During the last 130 years of the city development and construction, roughly 100 km of these previously open waters disappeared from the surface. Road construction, risk of flooding, and brook pollution due to the introduction of wastewater were the main reasons why these waters were channeled into large underground pipes. The consequence of this was the deterioration of the urban landscape, valuable public recreation zones were lost, and the natural habitat of plants and animals disappeared. Some of the channeled streams became sewers. When the sewage treatment plants were built, the streams that had become sewers were connected to the system. As a consequence of the introduction of stream water into the sewer system, about one third of the flow was clean extraneous water. As a result, the city's two treatment plants had to deal with large amounts of "clean" runoff, which increased operational costs and diminished the efficiency of the water treatment process.

As this was a nationwide problem, the revision of the Swiss Water Protection Law in 1991 stipulated that clean runoff and unpolluted rainwater should seep directly into the ground or, where this is not possible, should be diverted into a drainage system separate from the pipes carrying the wastewater. Due to the fact that too much clean water was flowing into the sewage treatment plant, Zurich started to modify the combined sewer system to accommodate a partially separate system. The existing combined sewers serve as sewers for polluted water, (i.e., wastewater from households and industries, rainwater from streets with heavy traffic flow and other polluted areas).

A new system will serve for the diversion of clean water. This clean water system has to be drafted in, in order to achieve the best results in the most cost effective manner. As a considerable part of the extraneous water in Zurich's sewer system originates from many streams, the separation of this water promises to be the most obvious solution. Even if the streams are small, the relatively continuous flow throughout the year adds up to a considerable amount of water. For this reason it is planned that all streams be diverted, except the less important ones, directly into the new system, which leads to the receiving water body. Obviously, where possible, clean water from springs, fountains, yard drainage, and cooling devices will be connected to this system (Figure 3.1).

This clean water system is generally designed to contain not more than two to five times the dry weather flow, so the dimensions are not very large. When it rains, the surplus overflows into the sewers for polluted water (formerly combined sewers), which normally have enough capacity.

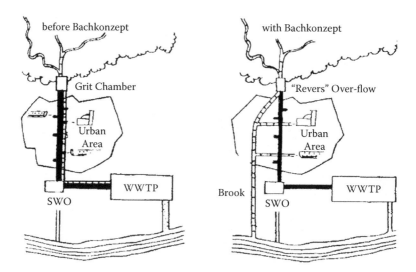

FIGURE 3.1 Clean water system.

THE ZURICH STREAM DAYLIGHTING PROGRAM

In our technical world we have become accustomed to the idea that urban drainage systems must be built with pipes. But this needn't be so, in that a daylighted stream is just as efficient as a pipe for diverting clean water. In Zurich this has been considered a more valid proposition in urban areas. That there is a need to bring a more natural environment into the cities and that the urban landscape needs to be more attractive is shown also by Cormier in Chapter 12.

For this reason, in the context of the Zurich urban drainage master planning, a so-called "concept of streams" has been worked out. On field investigations with an old map (Figure 3.2) as reference, the existence of the streams was examined. In some cases the water flow was measured. After that the need for space, legal and technical aspects, and other arguments were considered. The evaluation showed that over 40 km of streams could be daylighted or revitalized in the city of Zurich. In some instances there no physical stream existent but there was a lot of extraneous water in the sewer. In such cases a stream daylighting system was also considered.

In 1988, the stream daylighting and revitalization program (Conradin et al., 1988) was considered possible and reliable to the following extent:

- Stream daylighting in about 20 cases, when a stream had been channeled into a pipe (10–15 km)
- Daylighting of about 30 streams with the objective to separate extraneous water from a sewer system (total length 20–25 km)
- Stream revitalization in about 10 cases (10–15 km)

Figure 3.3 shows that in Zurich there are many opportunities to open streams, and that they are well distributed throughout the city. Streams instead of pipes have many advantages. A stream is a natural habitat for many plants and animals. The structure of the waterbed and the banks allow bacteria,

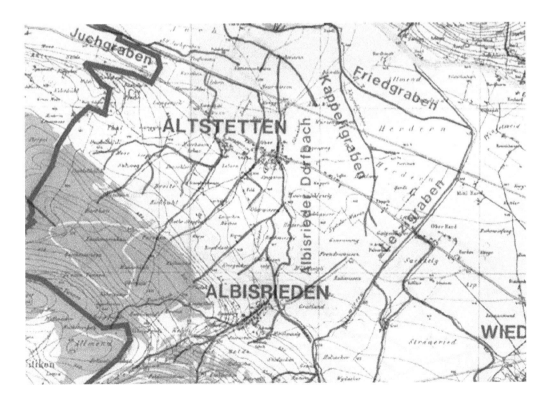

FIGURE 3.2 The *Map from Wild* (1855), the main reference to investigating disappeared streams.

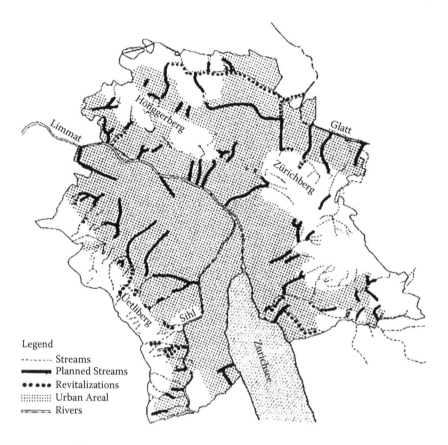

FIGURE 3.3 Opportunities to open streams.

insects, fish, and many other animals to develop healthy populations. A large spectrum of plants grow along a brook. The more natural a body of water is, including its bed, banks, and surrounding area, the more animals and plants will thrive.

The Zurich Stream Daylighting Concept is a typical result of integral thinking. The program approved by the Zurich City Council in 1988, aimed to achieve the following objectives:

1. To improve the recreational quality of urban neighborhoods and thus make them more habitable and attractive
2. To restore the lost habitat of plants, insects, and small animals, enhancing the relationship between city residents and their natural environment
3. To reduce the amount of "clean" water flowing through the waste water treatment plants and thus improve the quality and efficiency of the treatment process.

RESULTS

The success of the Zurich Stream Daylighting Concept is due to the support of all the relevant city departments, politicians, and the population in the following achievements:

- Since its conception, in more than 40 projects, 20,000 meters of streams were daylighted or revitalized.
- The daylighted streams represent in many ways the wishes and needs of city-dwellers for a more natural environment in the city. The streams are especially popular with children who often use them as playgrounds (Figure 3.4).

FIGURE 3.4 Children like to use streams as playgrounds.

- Brooks are an important factor in landscape and urban residential planning.
- To date, of the estimated 800 L per second total extraneous water in the sewer system, approximately 300 L per second are diverted into new streams. Additionally, the extraneous water has diminished due to the rebuilding of new watertight sewers. The result is that the sewage plant is less charged.

The following examples illustrate the transformation achieved by daylighting brooks.

DÖLTSCHIBACH

The Döltschibach is one of many streams originating from Uetliberg, a mountain that borders the western suburbs of Zurich city. The stream, upon leaving a woods, used to disappear underground where the stream water flowed with wastewater to the wastewater treatment plant. The Döltschibach was opened between 1993–1996 to a length of 2000 m. Along the In der Ey Street, despite width restrictions, it was possible to build the daylighted stream between the sidewalks and private properties (Figure 3.5). Further down, it traverses a private residential area where it forms a beautiful

FIGURE 3.5 Public sidewalks along the Döltschibach.

FIGURE 3.6 The Döltschibach traverses a private residential area.

landscape (Figure 3.6). Almost throughout its full length it is aligned with the sidewalk, enabling the local residents to appreciate the natural beauty of this stream. The daylighted stream has a capacity of 200 L per second. Rainwater from roofs and sidewalks is diverted to the Döltschibach. In the event of extreme storms, the overflow is channeled into the sewer as before.

ALBISRIEDER DORFBACH

The Albisrieder Dorfbach was daylighted to a length of 2,500 m and was built between 1989 and 1991. The stream flows through public and private properties. Figure 3.7 shows the situation before opening the Albisrieder Dorfbach in a private residential area. Figure 3.8 shows how it looks some years after the opening. The difference is evident; the landscape is now much more varied and interesting. Instead of the artificial look of monotone turf, an abundance of plants grow, and it is home to many small animals (Lubini, 2000). Stream daylighting is especially important in the city and brings a touch of nature into the lives of the city dwellers. Long sections of footpath accompany

FIGURE 3.7 Albisrieder Dorfbach: Situation before its opening.

FIGURE 3.8 Albisrieder Dorfbach: Some years after its opening.

the Albisrieder Dorfbach to enable the residents to enjoy the natural beauty of the water and lush greenery. The stream also flows through a children's playground and is much frequented by mothers and children. Its capacity is up to 200 L per second; the medium flow 10–20 L per second. The overflow goes through the existing sewer.

FRIESENBERGBACH

During the planning stage of the Gehrenholz residential area, the city civil engineers responsible for the concept and development of stream daylighting contacted the private investors. It was easy to convince them that an opened stream with a pond would be an attractive addition to their plans. The landscape architect developed a project with a daylighted stream (Figure 3.9), the result being a very pleasant recreation area with a pond (Figure 3.10) that can also be used for swimming in the summer. The maximum flow of the Friesenbergbach is 100 L per second; the surplus is deviated through the sewer system.

MANEGGBACH

In 1998 the Maneggbach was opened to a length of 100 m. The daylighted stream has been integrated into the landscaping of a new residential area (Figure 3.11). The dry weather flow is only 5 L per second, but in the case of storms the flow increases up to 800 L per second.

FIGURE 3.9 The result of the opening of the Friesenbergbach is a very pleasant recreation area.

NEBELBACH

The Nebelbach was built in 1991 to a length of 100 m. It represents an example of a daylighted stream featuring in a space restricted public street (Figure 3.12). To overcome the space limitation the stream was aligned to straightened walls, one of which can be used as seating. The stream bed was constructed to resemble nature as closely as possible. The result is that a profusion of plants grow there, and the water flow is so varied that even trout thrive in it. From the urban landscape perspective, the stream, with its greenery vastly improves the look of the street. From a social perspective it gives a focal point where children can get together and play. The Nebelbach deviates the dry weather flow up to 70 L per second. The surplus is deviated into a storm water sewer.

WOLFGRIMBACH

The Wolfgrimbach was daylighted in 1998 to a length of 800 m. The daylighting had to be done within a very narrow space between a public street and private houses (Figure 3.13). The initiative

FIGURE 3.10 The Friesenbergbach Pond can also be used for swimming.

FIGURE 3.11 The daylighted Maneggbach is part of the landscape.

of the neighborhood was very important and helped to raise the money for the project. It is built to deviate up to 100 L per second. Additionally, the rainwater from the roofs is introduced into the new system.

HORNBACH

The Hornbach is wide in relation to the other brooks in Zurich. After a flood in 1878 its banks were extended using hydraulic criteria only (Figure 3.14). In heavy storms, the flow is up to 50,000 L/s. This had to be taken into consideration in the revitalization project. To guarantee that the maximum flow would be deviated, and because of the narrow space, the stone walls on both sides had to be retained. In 2004 the brook bed was been revitalized for over 120 m with stones and gravel (Figure 3.15). Residents of the area were very pleased. Now it is planned to observe how the gravel and stone bed resists flooding in heavy storms. In the case of success, other sections of the Hornbach may be revitalized.

FIGURE 3.12 Nebelbach, an example of a daylighted stream in a street.

FIGURE 3.13 The Wolfgrimbach was opened due to the initiative of the neighborhood.

FIGURE 3.14 The Hornbach before revitalization.

FIGURE 3.15 The bed of the Hornbach after revitalization.

OTHER WATERFRONT PROJECTS

Traditionally, the waterfront of the lake of Zurich and of the Limmat River is well developed, and access for the public is widely guaranteed. New projects have also been realized, among them Gessnerallee at the Sihl River (Figure 3.16) and Witikoner Park at the Limmat River.

The Daylighting Program has developed in addition to these other waterfront projects.

The Sustainability of the Stream Daylighting Program

The sustainability of this program has been examined in a study that credits its viability to the implementation of six levels of systems (Hugentobler and Brändli-Ströh).

FIGURE 3.16 Gessneralle is one of various other waterfront projects in Zurich.

With respect to its *cultural system*, two important key factors in supporting the program are strengthening the relationship between humans and nature, and increasing an understanding of the natural world and the care required in dealing with nature.

The *social system* includes increased social interaction and networking in the neighborhoods, residents' participation in the planning and implementation process, and interdisciplinary cooperation across city government departments.

In the *human (individual) system,* opportunities for regeneration in the immediate living environment, aesthetically pleasing shapes and sensory stimulation, and opportunities (for children) to play in, experiment with, and creatively use the immediate living environment are all valuable.

In the *biological system,* the increased diversity of plants and animals in urban neighborhoods and more linkages between natural habitats lead to sustainability.

And in the *chemical/physical systems,* energy savings in the wastewater treatment process, and reduction in the amount of sealed urban land, are very positive. Risk of flooding of urban neighborhoods in periods of heavy rainfall, and the risk of water pollution with toxic substances/chemicals are the only factors that can move the program slightly away from sustainability.

ECONOMIC CONSIDERATIONS

Economic factors should be taken into consideration in the argument for the implementation of stream daylighting. The lower running costs of wastewater treatment plants and sewage systems are significant when clean water is taken out of the sewage system. Sewer systems that include clean water increase the hydraulic load and have to be larger, thus producing higher running costs. Additionally, the bigger flow causes more costs in maintenance and operation not only for the sewer system but also for the sewage treatment plants, e.g., for pumping energy. Estimations made by Entsorgung + Recycling Zurich, showed that annual costs could be as high as $5,000 per continuously flowing liter per second. That means for example, that when we have a medium flow of 20 liters per second of extraneous water in a sewer, the cost could potentially be as much as $100,000. This expenditure could be used to separate the clean water by utilizing the daylighting method. This calculation will, of course, vary, depending on the country and situation.

As the Swiss Water Legislation requires the separation of extraneous water from the sewer system, the question often is whether a new clean water pipeline is cheaper than an open stream. Often the daylighted stream is cheaper, and, even when it costs slightly more, this solution is chosen due to the other obvious benefits. In the case of the renewal of old stream pipelines, daylighting is a viable alternative that has to be considered, even if the costs are higher than the repair or the rebuilding, once these environmental advantages are taken into consideration. Finally, another consideration that is not necessarily financial but nonetheless should not be undervalued is the aforementioned benefits of bringing a natural environment and recreational facilities into the city. This is, of course, a political consideration and one that poses the question of what value the city council places on the attractiveness of the city and the well-being of its residents.

DEVELOPMENT OF A POLICY

Many projects such as stream daylighting cannot be realized because of resistance to change by the authorities when confronted with new ideas. Confrontations can occur with the engineers and politicians who were originally responsible for the streams being put into the pipelines in earlier phases. Bearing this in mind, it is very important to have a well thought out strategy before putting this alternative forward. Landscape architects and engineers are not used to "selling" their ideas and should be made aware that more emphasis should be placed on developing a sound policy, and, where necessary, a strategy. Just having a good project is not enough. Neighborhoods, city dwellers, city departments, authorities, and politicians have to be convinced.

The success of the Stream Daylighting Program is dependent on a group of strategic factors:

- The importance of presenting plans to the appropriate authorities at an early stage of the project. The Zurich City Council were approached at an early phase of the stream daylighting program. Their willingness to embrace and support new ideas has been very valuable, and a major factor in the success of the project over the last 15 years.
- The Stream Daylighting Group, an interdepartmental task force, was set up to coordinate the various field activities between city departments and to overcome resistance within the city administration.
- The argument for stream daylighting was strengthened by utilizing the planned revision of the Swiss Water Protection Law prescribing the separation of extraneous water runoff from the sewer system.
- The employment of private landscape architects and other experts, such as water engineers and biologists.
- The location of the program within the proper department. The choice of the sewer department in the case of Zurich was correct as it fulfilled all the requirements for the realization of such a project: vast experience in project management; financial capacity; and integrated water management expertise.
- Timely distribution of information to relevant departments. The listing and distribution of proposed stream daylighting projects, together with specific guidelines, enabled departments to recognize and incorporate stream projects into their activities.
- Early established guidelines for the technical design and the implementation of the stream opening projects helped to maintain a uniform philosophy.
- Direct communication with the public to inform and convince them of the project's importance. This was accomplished on many levels: through the local press, holding neighborhood meetings, discussions with affected property owners, and in school classrooms.

LITERATURE CITED

F. Conradin et al., *Das Bachkonzept der Stadt Zürich.*, Sonderdruck, Gas Wasser Abwasser, August 1988.

Bäche in der Stadt Zürich, Konzept, Erfahrungen und Beispiele, Entsorgung + Recycling Zürich, September 2000.

Sustainable Urban Development. A Conceptual Framework and its Application, M. Hugentobler and M. Brändli-Ströh, *Journal of Urban Technology*, August 1997.

4 A Multifaceted, Community-Driven Effort to Revitalize an Urban Watershed
The Lower Phalen Creek Project

Sarah Clark and Amy Middleton

CONTENTS

ABSTRACT

The Lower Phalen Creek Project is a community-driven effort to transform a strategically located urban watershed into an area rich with ecological, recreational, and social value. Situated along the Mississippi River in St. Paul, Minnesota, the project includes a range of activities to improve local water quality and restore and expand existing green spaces. Two key aspects of the project include:

- The successful community effort to daylight a section of Phalen Creek as it flows through Swede Hollow Park, an important local landmark that was once a settlement area for recent immigrants.
- The acquisition of a 27-acre brownfield site on the Mississippi River floodplain, and its transformation into the Bruce Vento Nature Sanctuary at Lower Phalen Creek.

INTRODUCTION

Across the county, urban communities are working to reclaim and improve green spaces, creeks, lakes, and other natural features. These efforts not only improve the local environment, they add to the quality of life and create new assets that can act as a springboard for broader social and economic revitalization efforts.

St. Paul's East Side and Lowertown neighborhoods are heavily-developed, core city communities. However, the neighborhoods are adjacent to several important natural features, including the Mississippi River, a forested ravine now known as Swede Hollow Park, and the nearby Phalen Chain of Lakes. Named for a degraded tributary creek that flows through the area, the Lower Phalen Creek Project is focused on making the most of these natural features and using them to enhance efforts to improve the overall health of the surrounding communities.

The fact that community members initiated—and continue to lead—restoration and redevelopment efforts is one of the significant aspects of the Lower Phalen Creek Project. Twenty-five partners, including the City of St. Paul and many government agencies, are now part of the project. A community steering committee coordinates the project's many partners and activities with assistance from two part-time staff people. By combining grassroots leadership and government teamwork, the project has succeeded in leveraging the funding, authority, and technical expertise needed to achieve a community vision for environmental improvement and revitalization.

SITE DESCRIPTION AND PROBLEM BACKGROUND

Lower Phalen Creek was once a spring-fed stream flowing from Lake Phalen through a deep ravine and into a low delta on the Mississippi River floodplain. Surrounded by forests and wetlands, the creek formed a natural corridor for migrating songbirds and other wildlife traveling between the Mississippi and Phalen Chain of Lakes. Native American tribes offered prayers, held councils, and formed a trading village in the area. As European immigrants arrived, the landscape was altered.

The changes to the land and the events that led to community restoration efforts are presented in two sections: work in the ravine now known as Swede Hollow Park, and the expansion of that work to include the Bruce Vento Nature Sanctuary, a large brownfield redevelopment project on the Mississippi River floodplain. The location of the two projects can be seen in Figure 4.1.

SWEDE HOLLOW PARK AND PHALEN CREEK

In the mid-1800s, Europeans began to settle along the river in what is now St. Paul's East Side. They built houses on the river bluffs and a railroad through the ravine near Phalen Creek. New immigrants with few resources settled along the creek in a shanty town that became known as Swede Hollow. For nearly a century, Swede Hollow was home to successive waves of immigrants from Sweden, Ireland, Italy, Eastern Europe, and, finally, Mexico. The houses in Swede Hollow did not have electricity or plumbing and outhouses were frequently built over the creek. (DuPaul and DuPaul, 1995).

As rail operations expanded, the creek was rerouted into the storm sewer section by section. By the late 1800s most of the creek had been buried (Rice's Ward Map, 1880). Despite the loss of most of the creek's open water, immigrants continued to live in Swede Hollow until the mid-1950s when it was condemned by the city Health Department and burned.

In the early 1970s, neighborhood residents and the St. Paul Garden Club joined with the St. Paul Parks Department to make Swede Hollow a city park. Neighbors and others formed Friends of Swede Hollow to provide a forum for ongoing citizen involvement in the park. The area has generated a

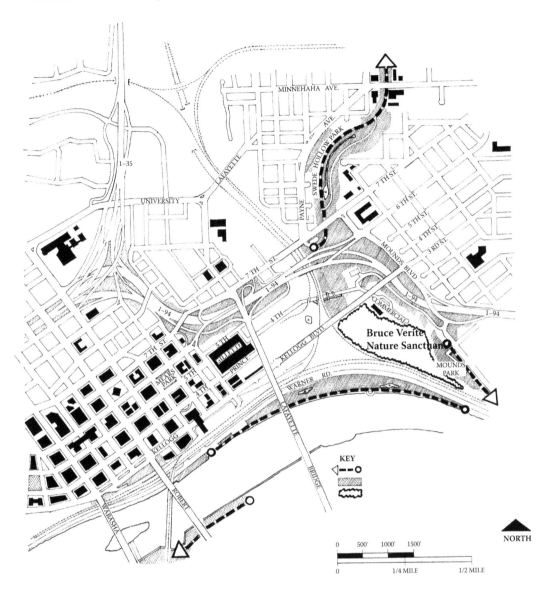

FIGURE 4.1 The Bruce Vento Nature Sanctuary and Swede Hollow Park are adjacent to one another in the project area just east of downtown St. Paul, Minnesota.

great deal of interest from people across the city because of its historic significance and its magical feel as an "enchanted forest within the city." In 1988 Friends of Swede Hollow worked with the city to bring Phalen Creek back above ground, a process commonly known as "daylighting." Open water was desired because it would be enjoyed by park visitors and attract migrating birds and other wildlife.

Today, Friends of Swede Hollow remains active in ongoing efforts to remove invasive exotic plants and restore native vegetation in the park. Swede Hollow Park was further enhanced in 1995 when a hard-surface bike and pedestrian trail was installed to connect the park to nearby Phalen Park and a network of other regional trails. This trail, the Bruce Vento Regional Trail, was further expanded in 2006 to connect Swede Hollow Park to the Bruce Vento Nature Sanctuary adjacent to the Mississippi River.

BRUCE VENTO NATURE SANCTUARY: EXPANDING RESTORATION TO THE MISSISSIPPI FLOODPLAIN

As improvements in Swede Hollow Park were made, community members developed a growing sense for the area's potential beyond the park limits. They determined that the whole 2.4-sq-mi Lower Phalen Creek watershed—which includes the East Side (Dayton's Bluff and Railroad Island Neighborhoods) as well as land right along the Mississippi River—was ripe for restoration. A stretch of Mississippi River floodplain just below the East Side's tall limestone and sandstone bluffs soon became the focus of the community's efforts. Reclaiming this brownfield site would not only provide much-needed greenspace to the community, it would help reconnect the East Side of St. Paul to the river and nearby downtown district.

A former rail yard, the 27-acre site was once a combination of floodplain forest and open wetlands. Altered many times during more than 100 years of industrial use, the site's soils were compacted and contaminated, and much of its landscape was taken over by buckthorn and other invasive plant and tree species. The land also became an illegal dumping ground for old appliances and other trash.

Despite its degraded state, community members saw the value of the land, which boasts a number of unique characteristics, including views of towering bluffs, caves, and seepage from natural springs. One cave is considered a sacred site to the Dakota Indians who call it "Wakan Tipi" or "House of the Spirits." In 1766 English colonist Jonathan Carver visited the cave and etched the British coat-of-arms into the cave wall. Following the publication of his journals, the cave was known as "Carver's Cave" and became a significant early landmark for European explorers.

The effort to acquire this brownfield site began in earnest in 1997 when Friends of Swede Hollow and other community members from the east side were joined by the Lowertown Redevelopment Corporation on the other side of the land. A historic downtown district, Lowertown, has achieved international recognition as a revitalized urban village with thriving arts, residential, and commercial components. The McKnight Foundation supported this connection and provided grant funds for project staffing.

In addition to its cultural significance and value for the surrounding communities, the land is critically situated along the Mississippi River flyway, a migration corridor used by more than 40% of North America's bird and waterfowl. The site is also located in a gap in the regional open space network and trail systems. For all of these reasons, and because of the community leadership from both the East Side and Lowertown neighborhoods, reclaiming the land and restoring its habitat were soon supported by city, county, and regional authorities who included it as a goal in numerous planning documents.

After nearly a decade of work to acquire and clean up the land, the Bruce Vento Nature Sanctuary opened to the public in May 2005. The actions involved in achieving this progress are detailed in the following section.

ACTIONS TAKEN AND RESULTS

The Lower Phalen Creek Project includes many initiatives designed to foster watershed-wide ecological improvements, increase public education, and create new natural amenities that will provide social and economic benefits to surrounding core city communities. As in the previous section, two aspects of the project are highlighted: the daylighting of Phalen Creek in Swede Hollow Park, and work to transform a floodplain brownfield into the Bruce Vento Nature Sanctuary at Lower Phalen Creek.

DAYLIGHTING PHALEN CREEK IN SWEDE HOLLOW PARK

Swede Hollow is a forested valley that has long been known throughout St. Paul for its colorful history and capacity to provide a glimpse of nature in the city. In the 1970s it became a park, and in the 1980s the St. Paul Public Works Department began making plans for storm and sewer line improvements in the area. When a major storm blew out much of the Swede Hollow's infrastructure in 1986, community members seized the opportunity to implement something unique.

With urging from residents, Friends of Swede Hollow and St. Paul Garden Club members, the city agreed to restore a portion of Phalen Creek's flow to the surface. Because of the very large storm volumes in Phalen Creek, the confined topography of Swede Hollow, and concern for other infrastructure running through the area, the city did not choose to daylight the entire flow of the creek.

In 1987, the city installed a 108-inch reinforced concrete culvert through Swede Hollow. At the culvert's entrance to the park, the city installed a pipe that allows water to flow out of the bottom of the main culvert into a 21-inch reinforced concrete culvert. This smaller culvert runs underground for about 100 feet, then opens to a small sediment settling pond. From there, a small creek channel runs a few hundred feet to a second, larger pond. This channel is surfaced with grouted limestone in places, due to its steep slope and the constraints posed by water and sewer pipes running underneath it. From the second pond, the creek meanders over 1,000 feet through Swede Hollow's flatter bottom land. It then enters a third, and the largest pond. During high water, a drain allows water to re-enter the main culvert for its remaining 0.75 mile trip to the Mississippi River beneath various streets, highways, and rail tracks.

The daylighted channel and ponds total about 0.4 miles in length. The diversion structure passes a constant flow of about 2 cfs into the recreated stream channel. Throughout the park, the creek and ponds are surrounding by planted grasslands, pre-existing woodland, and benches. The ecological benefits of the daylighted stream are clear, the ponds prevent some sediments from reaching the Mississippi River, invertebrates have returned to the creek, and some nutrients and other urban pollutants are captured. The stream's value as an amenity for park users is another important result of the project, and there are indications that it has improved the park's habitat value.

RECLAIMING A BROWNFIELD TO CREATE THE BRUCE VENTO NATURE SANCTUARY

Just south of Swede Hollow Park—on the south side of Interstate 94—lies a wedge of floodplain adjacent to the Mississippi River. A former rail yard, the land was owned by Burlington Northern–Santa Fe (BNSF), and was abandoned in the 1970s when the railroad moved switching and maintenance operations elsewhere.

Because of its strategic location, and the fact that it is primarily open space, community members around the site were very interested in this property and its potential to link the East Side to the Mississippi River and downtown St. Paul. In 1997, planning for the site began in earnest when interests from the Lowertown community, which is on the western side of the property, joined the effort to reclaim the site.

Over the years, community members faced many struggles that nearly ended the nature sanctuary effort. Crises over the sale of the BNSF land to industry occurred on several occasions, and community members spent hundreds of volunteer hours working to convince legislators and other city, county, and state decision-makers of the project's importance.

Community members worked hard to articulate the vision for the land as a park, and in July 2001 the Lower Phalen Creek Project published a concept plan (Martin & Pitz Associates, 2001). In October, the City of St. Paul adopted a summary of the community vision into their Comprehensive Plan. All parties agreed that the nature sanctuary would be operated as a city park, with a significant portion of the land preserved in perpetuity under a conservation easement agreement with the Minnesota Department of Natural Resources (DNR).

In November 2002 the land for the sanctuary was purchased with negotiations led by the non-profit Trust for Public Land and with key support from the National Park Service's Mississippi National River and Recreation Area (MNRRA) and DNR Metro Greenways Program. In 2003 the

Lower Phalen Creek Project organized volunteer cleanup events that included more than 100 people who removed 50 t of trash from the site. The Lower Phalen Creek Project worked with the city to leverage $400,000 in U.S. Environmental Protection Agency Brownfields Cleanup grants and in winter 2004–2005, soil remediation and site grading work took place.

Because of the land's many cultural resources, archeologists monitored the remediation process and identified resources that were uncovered. One of the most notable finds on the site was the intact foundation of the North Star Brewery, built in the 1850s. The foundation was documented and reburied, with specifications placed on land use for the area above the foundation.

By spring 2005, the land for the Bruce Vento Nature Sanctuary was safe for public use, and it was officially opened at a community celebration in May 2005. The strong volunteer and community leadership for the project has gained extensive recognition and received a 2005 "Take Pride in America" award in the public/private partnership category.

The community vision document was the first step in detailing plans for the site, and includes extensive history and other valuable information. Its preparation involved many public meetings and several rounds of comments from the Lower Phalen Creek Project's many partners. The next phase of planning and design involved development of plans for cleaning up the sanctuary's soils (Landmark Environmental, 2001), a ecological restoration plan (Emmons & Olivier Resources and MMC Associates, 2001) and a thorough review of the site history (106 Group, 2001). These documents were required before acquisition. These plans formed the basis for work on the sanctuary. Key components of the sanctuary project include wetland design, ecological restoration, walking paths, and interpretation and bicycle trail links. Many of these features can be seen in Figure 4.2.

WETLAND AND STREAM DESIGN

Natural springs on the sanctuary land once made the area a type of seepage swamp. Fill brought in by the railroads had altered the landscape, but spring water on the site was still apparent. In the soil remediation and grading process, springs at the toe of the bluff were channeled into a meandering stream and series of three wetlands. A clay liner was installed in all water features to keep the water above ground and to reduce pollution concerns. A wetland was placed in front of the mouth of Carver's Cave/Wakan Tipi to capture the abundant spring water that originates in the cave and create a natural barrier that reduces public access to this sacred site. The details of the wetland locations changed several times during the remediation and site grading process due to unexpected contamination.

Since the wetlands and stream were installed, wildlife has been attracted to the increased volume of water and shoreline of the pool. The wetland basins were designed with gentle side slopes that nourish more riparian plant species and improve wildlife habitat.

ECOLOGICAL RESTORATION

Lower Phalen Creek Project, the City of Saint Paul, and the Community Design Center's East Side Youth Conservation Corps have worked together to plan, implement, and maintain the restoration of the sanctuary's ecosystems. As noted in Figure 4.2, the primary ecosystems being recreated are floodplain forest, dry prairie, oak savanna, oak woodland, bluff prairie, and habitat around the wetland and stream channel. Through volunteer efforts, projects with local school children and work by the youth interns in the Conservation Corps, thousands of wetland forbs and hundreds of trees

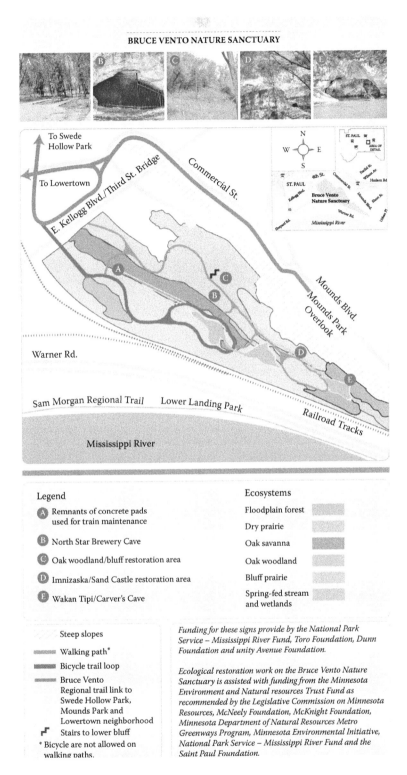

FIGURE 4.2 This map of the Bruce Vento Nature Sanctuary shows walking paths, the bicycle trail loop, ecosystem restoration areas, and cultural resources.

FIGURE 4.3 The wetlands in the Bruce Vento Nature Sanctuary in May 2005.

have been planted on the sanctuary. Figures 4.3 and 4.4 show the transformation of the sanctuary's wetland area. Figures 4.5 and 4.6 provide an overview of the site in May 2005 and a digitized rendering of how the area will appear once the oak woodland habitats are mature.

WALKING PATHS AND CULTURAL RESOURCE INTERPRETATION

A large and diverse group of interests is actively engaged in planning the interpretation of the cultural resources on the Bruce Vento Nature Sanctuary, including the National Park Service, Minnesota Department of Natural Resources, Dakota tribal interests, and the Audubon Society. The process is being coordinated by the Lower Phalen Creek Project and Historic Saint Paul. Interpretation of the sanctuary resources focuses on the following:

FIGURE 4.4 A digital rendering showing how the wetlands will look once restoration is complete. (Digital rendering by Mark Apfelbacher.)

FIGURE 4.5 A view of the Bruce Vento Nature Sanctuary and Mississippi River from the Kellogg Bridge.

- *Geologic history*—the evolution of the land over time, the carving of the valley, and the information shown in the makeup of the sanctuary bluffs
- *Ecological history*—the land's past as a confluence of tributaries and its connection to the larger Mississippi River and migratory bird flyway
- *Human history*—the Hopewell, Dakota, and European people on the land; the international and national significance of the site and how it fits into the formation of the City of Saint Paul
- *Park history*—how the land has been reclaimed and restored, and the partners involved, including Bruce Vento

FIGURE 4.6 A digital rendering of how the same view will appear once the oak woodland on the sanctuary is mature. (Digital rendering by Mark Apfelbacher.)

The site's walking paths, as noted in Figure 4.2, take visitors past key cultural resources, including remnants of the concrete railroad pads once used for train maintenance, the cave used to cool beer by the North Star Brewery, and Wakan Tipi/Carver's cave, a sacred Dakota site and landmark for early European explorers.

Although long-term plans call for having on-site signs that provide education on the topics above, the cultural history of the site is currently shared in detail in a kiosk-like structure along the walking path as it enters the sanctuary.

PEDESTRIAN/BICYCLE TRAIL DESIGN IN THE CONCEPT PLAN

The concept plan proposed two trail links, one linking East Side neighborhoods to Lowertown and downtown St. Paul, and the other providing a direct connection between the sanctuary and the Mississippi River.

The first trail connection, known as the Lower Phalen Creek Trail, is a mile and a half link that connects 85 miles of existing regional trails. The link between Swede Hollow Park and Indian Mounds Park will join a 14-mi and 15-mi segment together, and connect both to the 18-mi Gateway Trail to create 47 continuous miles of off-road urban trail. This link is slated for installation in summer 2006.

The second link is currently in the planning process as the Lower Phalen Creek Project, City of Saint Paul, National Park Service, and other partners work to determine the best method for overcoming the active rail lines and busy road that form a barrier between the site and the Mississippi River. More than $1 million in federal support has been leveraged for this connection.

"TAKE HOME" FINDINGS

The following have been key to the success of the Lower Phalen Creek Project and could potentially benefit other projects:

DEVELOP A VISION THAT ORIGINATES FROM THE COMMUNITY

Friends of Swede Hollow and the larger Lower Phalen Creek Project have involved a diversity of community interests from the very beginning. Together, we developed a vision for the watershed that defines our common aspirations. Throughout the visioning process we worked to reconcile different values and determine how we could simultaneously recreate natural settings, preserve history, and create new recreation opportunities. High engagement from the community helps ensure that the project goals and objectives will reflect what people really want—and the more people want the project, the more they will pursue it and maintain their passion through the often long and challenging process of implementation.

INVOLVE GOVERNMENT AND OTHERS NEEDED TO ACHIEVE IMPLEMENTATION OF THE PROJECT

No matter how strongly the community wants to achieve their vision for redevelopment, it will not succeed unless the agencies with relevant authority, technical skills, and funding are on board.

In the case of the Swede Hollow Park daylighting project, it was the city staff—who had been hearing from community members and the St. Paul Garden Club for many years—who ultimately led the daylighting of Phalen Creek in Swede Hollow Park. Community interests had convinced the city of the importance of having open water in the park. When an opportunity arose, the city staff pushed the project.

The Bruce Vento Nature Sanctuary effort has required more persistent and continual leadership from the community—and the community and nonprofit partners have had to take the lead in designing the park and fundraising for acquisition and redevelopment. However, the project would not be possible if the city and state Department of Natural Resources (DNR) and Pollution Control

Agencies were not involved. The city, which now owns of the site, has been steadily increasing its support of the project, as evidenced by the City Council's adoption of the community-developed concept plan for the site. Further, the legal details involved in the acquisition would have been difficult to navigate without the involvement of the national organization Trust for Public Land, which is also providing fundraising assistance to the project.

REMAIN PERSISTENT!

It is critical for community members and partners to work diligently to ensure that government changes and other factors do not alter the community vision. The near loss of the floodplain site to industry and other developers, for example, required extensive outreach to government decision-makers.

ACHIEVE FUNDS FOR PROJECT STAFFING AND COMMUNITY-CHOSEN CONSULTANTS

The Lower Phalen Creek Project steering committee is a primarily volunteer body. Maintaining a small paid staff has helped ensure that partners are coordinated, funds are leveraged, and the project continues to move forward. Project funds were also critical for achieving outside planning. For the Bruce Vento Nature Sanctuary project, is was necessary—and very valuable—to contract with a landscape architect who was able to articulate the details of the community's vision for the nature sanctuary in a professionally-designed concept plan. The community also needed to partially fund—and devote extensive staff time—to the development of the Natural Resources Management Plan and Remedial Action Plan. In choosing these consultants, it is vital that the individuals involved have experience in, and are comfortable with, community involvement and input.

LITERATURE CITED

Rice's Ward Map, City of Saint Paul, 1800.

A Swede Hollow Walking Tour. Angela DuPaul and Karin DuPaul, Friends of Swede Hollow 1995.

Bruce Vento Nature Sanctuary Natural Resources Management Plan, Emmons & Olivier Resources and MMC Associates, 2001.

Bruce Vento Nature Sanctuary Remedial Action Plan, Landmark Environmental, 2001.

Lower Phalen Creek Literature Search for Historical Archeological Potential, 106 Group, 2001.

A Community Vision for Lower Phalen Creek. Martin & Pitz Associates, 2001.

5 Retrieving Buried Creeks in Seattle

Political and Institutional Barriers to Urban Daylighting Projects

Kit O'Neill and Peggy Gaynor

CONTENTS

ABSTRACT

Through the last part of the 19th and the 20th centuries, urban and suburban streams nationwide were diverted into sewers in response to development pressures, in some cases despite the opposition of local residents. Efforts in Seattle during the past decade have attempted to reverse that sequence and bring creeks back to the surface as part of the broader national movement to restore nature in urban environments. This chapter examines the complex history of two grassroots daylighting projects within the context of development and planning in the city of Seattle during the past 10 years. The sagas of Ravenna and Thornton Creeks differ in sources of opportunity and in particular successes and defeats, but are driven by convictions held in common by their citizen proponents. They faced similar opposition from both developers and the city. The last section of the chapter addresses lessons learned, the larger question of daylighting when proposed by the community (bottom-up) rather than the government (top-down), and the effectiveness and value of these grassroots efforts in their communities.

INTRODUCTION

"They're going to *daylight* a river here—
That's what they call it, noun to verb. ..."

—Denise Levertov, "Salvation" (1996)

"To keep every cog and wheel is the first precaution of intelligent tinkering."

—Aldo Leopold, *Round River* **(1953)**

The act of exposing a stream and allowing it to flow on the earth's surface again, known as *daylighting*, is a compelling concept that has potential to change the urban context (Pinkham 2000). The prospect of daylighting means: (1) the stream has been buried, diverted to an underground conveyance, and (2) whatever has since occurred on the surface must be displaced as the stream is unburied and re-placed. The focus of this chapter is two daylighting projects located in dense urban areas of Seattle. Because both projects were initiated at the grassroots level, the changes required to return these creeks to the surface include public policy shifts as well as removal of the impervious combination of asphalt and concrete, of shopping center parking lots and roads, that now occupy the land surfaces where the creeks once flowed.

Natural water systems are frequently compromised as regions undergo development. Transportation impacts occur early in the process: roads or rails require bridges or culverts to provide for the spatial needs of wheel paths, disrupting the original continuity of stream or river beds for their own beds, and removing, at least at those fixed points, flexible floodplain response to temporal change. Development has traditionally meant loss of habitat; disruption of the local hydrologic cycle as increases in impervious surfaces lead to a decrease in infiltration; channelization, diversion and paving of waterways further restricting the length of drainage pathways; pollutants becoming concentrated; and stormwater runoff discharging abruptly into the abbreviated waterways that remain, with resultant flooding.

The history of streams in and near Seattle include not only these common hazards of development but some exceptional impacts due to the ability—and inclination—of city engineers in the early part of the 20th century to level topography using water and to re-channel water flows, for various civic purposes. The values of existing natural systems were accordingly discounted. Unanticipated repercussions are still being felt and dealt with, especially with newfound respect for Pacific Northwest salmon.

The primary players in these Seattle daylighting projects include the owners of the properties, dominated in each case by regional shopping centers and the City of Seattle; local citizens and the general public; and government entities and politicians at both the city and county levels. Among the issues are achieving a balance between protection of the last best places and "ecological literacy" (Mozingo 1997), a balance between saving any remaining pristine habitat in rural reaches and recreating natural systems as amenities in the cities and suburbs, where the people are. Lessons are learned from these two projects that may be valuable in approaching future similar projects, in exploring the larger issue of daylighting when proposed by the community (bottom-up) rather than the government (top-down), and in enhancing the effectiveness and value of grassroots efforts in the communities where these efforts are based.

CONTEXT

Daylighting combines many of the usual rationales for environmental restoration: reinstating hydrologic and/or ecological functions, wildlife habitat, aesthetics, open space, passive recreation, and even compliance with legal requirements. But daylighting projects also appeal on a fundamental level, arising from an intrinsic response to the dynamic attraction of water as an element, and maybe a desire for "environmental penance" in the sense of recognition of responsibility to and for the environment. Research has not yet been done in this area parallel to that elucidating other altruistic pursuits of the common good (Fehr and Gachter 2002). Nevertheless, proponents share a deep-seated conviction that daylighting is an imperative, that it is vital to return nature to the city in the form of freely flowing water, that development should be restorative, especially in correction of past errors.

The grassroots-driven agenda of these two daylighting projects is in contrast to the 20-step process of project development prescribed by Bays (2001), where the community is brought in by the planners at step 7 ("to ensure community support") and step 20 ("to inform and involve the public"). According to such a scenario, public support is sought through public relations rather than public involvement, a signature of top-down planning (Arnstein 1969). Grassroots projects on the other hand are democratic by definition: from, by, and for the people.

A bottom-up project may well, however, find it more difficult to garner acceptance, support, and approval from decision makers, policy makers, and politicians as the stages of development unfold. In some cases, bottom-up projects may succeed if they are successfully converted into top-down projects as a result of adoption by an elected advocate. Grassroots daylighting projects such as the two case studies described here incorporate strong local ownership, related to their origins within the community, which may prove difficult to hand off to a public entity.

SEATTLE: REGIONAL PHYSICAL CONTEXT

Although the impacts of dams as barriers to fish passage occupy national headlines these days, in Seattle early massive engineering projects lowered lake levels rather than raising them. Other engineered alterations to the local geography eliminated hills by sluicing them with water and redirected, in some cases reversing, river flow. Three of those projects in the early 20th century directly impacted the creeks that form the focus of this chapter. In 1911, the level of Green Lake in north Seattle was lowered. It had been an early focus of intensive residential development. Public open space was created in the lake's former shallows, but salmon-rearing habitat was removed from the ecosystem and the creek bed of Ravenna Creek that had drained the lake was left high and dry. In 1916, the level of Lake Washington, the eastern boundary of the city of Seattle, was also lowered (shown in process in Figure 5.1). The outflow of Lake Washington was redirected through the Montlake Cut, part of a ship canal created between Lake Washington and Puget Sound. The results were undoubtedly good for commerce at the time, but had long-term effects to the detriment of the biota, belatedly recognized and currently being addressed.

BACKGROUND FOR CASE STUDY 1—RAVENNA CREEK

The lowering of Lake Washington in 1916 impacted the outlets of both streams involved in the case studies. Union Bay, a deep indentation of Lake Washington into which Ravenna Creek flowed, was underlain by a hundred feet of peat in some locations. The change of lake level shrank the bay in size, transforming its shape into a shallower curve at the new mouth of the creek. The length of the creek was extended over nearly flat new land surface as mudflats were exposed where there had

FIGURE 5.1 This 1916 Seattle photograph shows the upper waters of Lake Washington flowing into the newly built Ship Canal through manmade Montlake Cut, observed by a few intrepid engineers.

been open water, and the former marsh, previously perforated by many waterways, became dry land. The mudflats were used as a municipal waste dump by the city between 1926 and 1965, later acquired by the University of Washington and allowed to become naturalized after the landfill was capped in 1971. The fine peat soils of the former marsh were used for decades as truck or nursery farms, irrigated by the creek. The land was purchased in the early 1950s for an open format shopping center named University Village. In 1948, prefatory to that development, the city decided to remove the waters of Ravenna Creek from the delta and divert the creek flow via sewage trunklines into Puget Sound.

The Ravenna Creek watershed at the turn of the 20th century drained 9 sq mi of land within the current city limits (Figure 5.2). It included the Green Lake basin that collected myriad mineral springs and a ravine, quite deep in sections, cut by Ravenna Creek between Green Lake and the open waters of Union Bay. Those waters and the adjacent marsh that formed the Ravenna delta had been fertile fishing grounds for the Muckleshoot tribe, who describe fishing traps laid in the "SLuwi/L," the Lushootseed name for Ravenna Creek (Waterman 1922).

When Green Lake was lowered in 1911, as recommended by the Olmsted Brothers in their 1903 plan for Seattle (Olmsted Brothers 1903), acres of public land were created by the reduction of the lake's surface. The creek bed that had connected the lake to the waters of Lake Washington was now six feet above the new water level however. For a time, an underground pipe carried the Green Lake drainage to the ravine that had become city parks.

One of those parks, Ravenna Park, founded privately in 1887, was a singular location where old growth cedar and Douglas fir trees could be found within the city limits. Those trees in a spring- and creek-fed ravine drew turn of the century tourists to the area. An entrance fee of 25 cents was paid by over 10,000 people in 1902 (Dorpat 1984). Figure 5.3 shows the creek and the park at the turn of the 20th century.

By 1949, Ravenna Creek existed only within the confines of Ravenna Park. It had been disconnected upstream from the Green Lake basin by the lowering of Green Lake. It had been disconnected downstream at the south end of Ravenna Park from its former Lake Washington receiving waters, and diverted into a sewer trunkline. The biodiversity of the Creek in the park remains an attraction for annual University of Washington entomology class fieldwork as well as for fourth graders from the local elementary school.

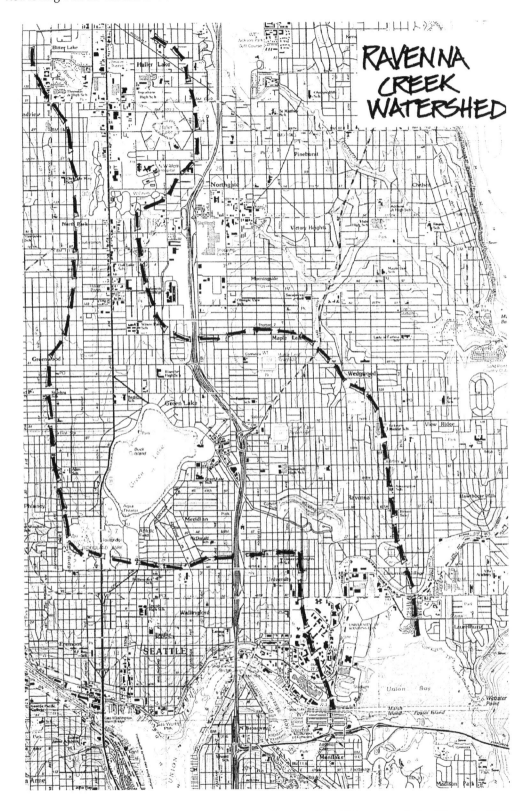

FIGURE 5.2 The Ravenna Creek watershed, which at the turn of the 20th century drained 9 sq mi within the current city limits.

(a)

(b)

FIGURE 5.3 Turn of the 20th century photographs and postcards of Ravenna Creek in Ravenna Park.

BACKGROUND FOR CASE STUDY 2—THORNTON CREEK AT NORTHGATE

The Northgate area of Seattle is the location of the headwaters of Thornton Creek's main south branch, draining one-third of the overall 11.6-square-mile drainage basin. Though not directly impacted by the lowering of lakes, these headwater stream channels and associated wetlands are nearly invisible today due to development impacts similar to those experienced by the areas where Ravenna Creek once flowed. The Thornton Creek watershed is shown in Figure 5.4.

Prior to settlement of the Northgate area by pioneers, a peat bog, springs, and stream channel filled a broad valley surrounded by ridges and hills, depicted in Figure 5.5 from a 1859 survey. Native Americans used the area, calling the wetland Bald Man's Bog. From the late 1890s through the 1950s, the area was actively farmed, including orchards, fields, and later nursery greenhouses. The peat bog was mined and, in an area south of the creek corridor, peat mining formed Square Lake. The majority of the original wetland was ditched and drained for farm fields. A series of changes in the creek channel location are indicated in Figure 5.6.

In 1948–1950, Northgate Shopping Mall was built on the uplands north of the stream corridor and wetlands. In 1961, a major north–south interstate highway construction project bisected the creek and wetland valley. Built on a filled strip of land, the highway was a boon for the shopping mall, but

ROOSEVELT

This monarch of the forest was christened "Roosevelt" November 31st, 1905, by Miss Ruth
Piles, under the auspices of the First Washington Conservation Congress and officers of the
Grand Army of the Republic. The Rainier Chapter of Daughters of American Revolution are
the sponsors for this tree.

(c)

FIGURE 5.3 (Continued)

terrible for the original environment. Initially the highway dike, 40 to 80 ft above the valley floor, was constructed as an impervious barrier. When the water table rose dramatically (and citizens complained), several culverts were installed to carry stream and wetland flow.

By 1969, commercial development had spread, and Thornton Creek was piped beneath an asphalt parking lot and a street, both built on fill. That pipe is currently 11 to 40 or more feet under the parking lot and street. The parking lot, now known as Northgate Mall's South Parking Lot, holds occasional car and RV shows but typically stretches as 12.8 acres of empty pavement (Figure 5.7a). Thornton Creek reemerges on the surface east of the South Parking Lot within a recently purchased open space, Thornton Creek Park 6 (Figure 5.7b). From that park, Thornton Creek flows as an open stream channel for 3½ miles to Lake Washington, supporting resident cutthroat trout, small runs of sockeye salmon, and occasional steelhead, coho, chum, and chinook salmon. River otters, beavers, muskrats, herons, and other waterfowl also use the creek corridor.

An adjacent expanse of asphalt west of the South Parking Lot functions as a county Park & Ride Lot and a transit center for buses and future light rail. Upstream from the highway dike, former wetlands have

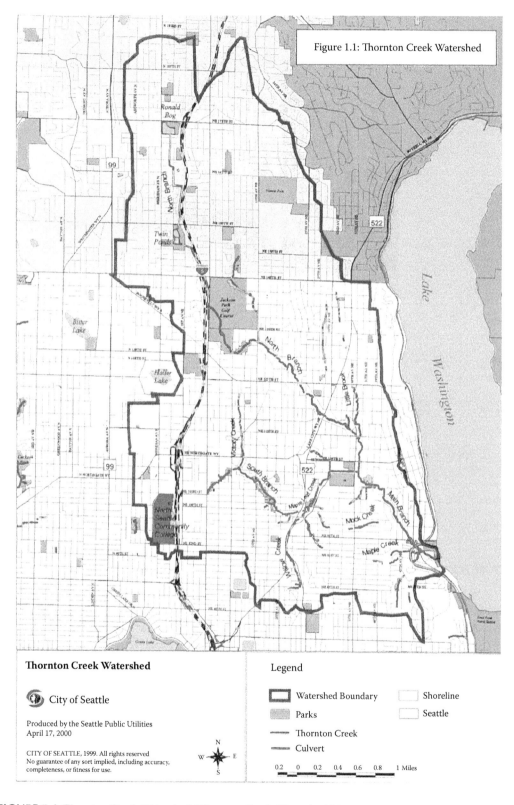

FIGURE 5.4 Thornton Creek Watershed (Thornton Creek Watershed Management Committee 2000).

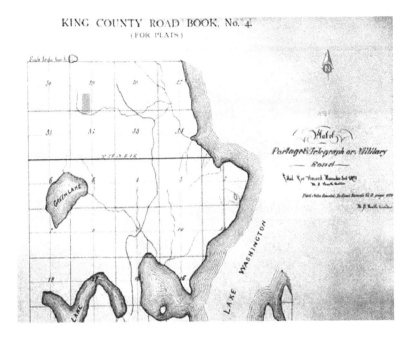

FIGURE 5.5 1859 survey of the Thornton Creek area.

become the campus for North Seattle Community College. At its northern end several wetland and woodland acres are being preserved and enhanced. New and old stormwater facilities are intended to mimic ponds and wetlands within the campus grounds. The region north of the college has been completely covered by commercial and institutional buildings and their parking lots. The spring

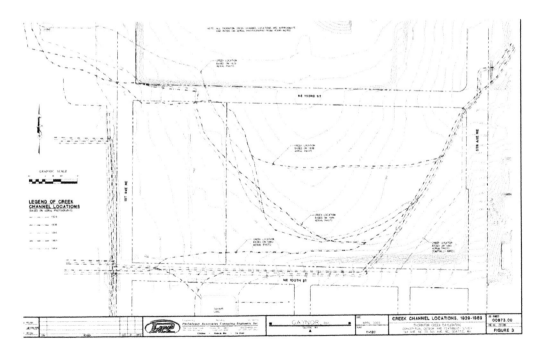

FIGURE 5.6 Thornton Creek channel locations, 1929–1969 (Gaynor/Penhallegon 2001).

(a)

(b)

FIGURE 5.7 (a) Photograph, Northgate Mall's South Parking Lot. (b) Thornton Creek Park 6. (Photographer, Shelly Baur.)

and creek waters on the western side of the highway are piped into the highway culverts, surfacing on the eastern side for short reaches, running north and south parallel to adjacent highway and streets, and then entering a pipe under the South Parking Lot to Thornton Creek Park 6.

So far, this development history is fairly typical in its uniformly adverse impacts on the creek system. The story becomes more complex beginning in 1998, when redevelopment of Northgate Mall and the South Parking Lot was proposed by their owner.

CASE STUDIES

The creek daylighting proposals described here have followed individual but similar paths, as the proposals responded to different sources of opportunity and to the involvement of other parties, including governmental entities, in the process. The accompanying table (Table 5.1) presents a comparison of

TABLE 5.1
Comparison of Two Seattle Creek Daylighting Projects

Waterway	Ravenna Creek	Thornton Creek
Watershed size:		
Original area	9 sq mi	11.5 sq mi, "700 backyards"
Drainage at issue	100 acres within Ravenna Park	1 sq mi upstream, including mall parking
Flow rates:		
Current base flow	1 cfs	0.7 cfs
Design flow, max	5 cfs	55 cfs
Reach last on surface	1948	1963
Length to be daylighted:		
Original vision	4675 ft	2535 ft
Current plan	560 ft	TBD
Scheduled for daylighting	2003	TBD
Costs:		
Preferred alternative	$7.5–9.8 m	$5.5 m (w/in 6 acres open space)
Current plan	$0.8m	TBD
Primary objectives	Natural urban amenity	Restored hydrology; improved community

Basic differences

Reasons for loss of surface stream: Extraordinary engineering projects impacted Ravenna; standard urban practices of past impacted both.

Hydrological connectivity: Head and tail with pipe body (Thornton) vs. head only (Ravenna). Thornton needs surface connection; Ravenna needs missing connection on surface.

Hydrological problem/flow source: Ravenna is groundwater fed, Thornton now stormwater dominated so that it flashes and floods. Of current Thornton watershed, 50% is paved, whereas current Ravenna watershed is a natural park.

Legal standing/grounds for legal action: For Thornton, existence of connection is protected by easement vs. no connection for Ravenna, so that state and local laws apply in Thornton's case and not in Ravenna's.

Players in Thornton legal actions: community vs. partnership of city and developers.

Maximum design flows: Ravenna—5 cfs; Thornton—55 cfs.

the two case studies including their basic statistics, such as length of design daylighting corridor, as well as some fundamental differences between the projects.

CASE STUDY 1—RAVENNA CREEK DAYLIGHTING PROJECT

Ravenna Creek daylighting has been called "one of the most studied daylighting projects in the country" (Pinkham 2000). In retrospect, its chances for success might have been improved by spending some of that study money on a couple of roundtrip airfares from Seattle to Zurich—or even to Berkeley—for site visits to observe the successful daylighting of Zurich's drainage system or of Strawberry Creek (Conradin Chapter 3; Mason 2001). Such site visits might have allowed the case for the benefits of daylighting to be made more forcefully both to the Seattle Drainage and Wastewater Utility, the municipal agency responsible for creeks in highly urbanized Seattle, and to the shopping center owner who acquired property where the Ravenna marsh had been, several years into the project.

The prospect of daylighting Ravenna Creek emerged in January of 1991, in response to a plan announced by the local sewer authority, Metro, to remove the creek waters from the combined sewer system as a way to decrease demands on the capacity of a new treatment plant. The Metro proposal was to remove the creek from the sewer trunkline into which it had been diverted in 1949 and to convey it via a new pipeline to an existing connection to Lake Washington. Metro had determined that such action would be cost effective regardless of demands on capacity, since the monies saved by not pumping and treating the creek waters over 20 years would more than pay for the pipe installation.

Local citizens saw the Metro plan for change as an opportunity to soften the local land use development trend with a return to nature, and suggested that Metro consider allowing the Creek to flow on the surface again rather than constructing additional pipe. Metro offered the community time to develop a plan—and to find additional funding for it. The community accepted the challenge, embarking on a radical undertaking, although not as "extreme" in scope or cost as some others presented at the Harvard Brown Fields, Grey Waters Symposium in 2001. Certainly no one anticipated in 1991 that it would be 2003 before construction of any aspect of the reconnection would take place. (For an update to 2006, see end of chapter.)

Studied, Ravenna certainly has been—first by a combined citizen-academic team in 1991 (O'Neill 1991), then by a professional-academic group in a 1992 design charette, followed by the community-driven professional pro bono development of a master plan in 1994 (Ravenna Creek Alliance 1994). More than a quarter of a million dollars was spent on the final county-commissioned feasibility study (SvR 1997b). That study, incorporating a parking garage demanded by the shopping center owner as compensation for lost parking spaces, came up with a price tag that effectively killed the original vision.

That original vision is drawn in plan view in Figure 5.8, one product of a 1992 charette held at the University of Washington. A surface creek, with occasional box culverts where bridges were impractical (under major roads), would flow out of Ravenna Park, alongside a redesigned arterial street, through a remnant wooded area and finally through the University Village shopping center to reach the arm of a bay on the western edge of Lake Washington (Figure 5.9). The resurrected creek would reconnect that part of the city, not only ecologically but socially, offering a pedestrian route between sites of interest and an outdoor "place to be," a meeting place (Figure 5.10). A fundamental

FIGURE 5.8 "Return of Ravenna Creek to Daylight," a product of the 1992 charette held at the University of Washington, with a perspective sketch illustrating the daylighted creek along a city street and a section illustrating the creek profiles within a shopping center.

FIGURE 5.9 Aerial overview, looking south, of the Ravenna Creek daylighting route. (Photographer, George White.)

(a)

(b)

FIGURE 5.10 Contemporary photographs: Ravenna Park trail; grate where Ravenna Creek terminates today; University Slough, which will become Ravenna Creek again.

(c)

FIGURE 5.10 (Continued)

requirement for that route was collaboration with the shopping center, but its owner never actually became engaged in negotiation.

Several formative actions followed the 1992 charette. A two-photo front-page article in one of the local dailies in mid-1992 attracted a critical mass to the original group. Citizens organized themselves into the Ravenna Creek Alliance (RCA) and incorporated as a 501(c)3 nonprofit, became qualified as a charitable organization under state law and obtained a city business license. The Alliance began collecting funding and support. By 1994, RCA had collected endorsements from a range of environmental and community organizations, from the local city and county councils and executives, and from the governor and the U.S. Congressional representative, along with several thousand signatures on petitions.

Other early efforts of the Alliance focused on fundraising, using the status developed as an organization. The first proposal to a local environmental foundation, the Bullitt Foundation, was turned down. The foundation thought Metro should fund the daylighting project as it was a governmental responsibility, and private funding was therefore inappropriate. They later funded RCA's organizational efforts for 2 years with a total award of $26,500. During that period a major effort was made to secure funding for the project itself from the Washington Wildlife and Recreation Program. Proposal development, including collecting legal opinions and a city council resolution, and a presentation were costly in time and materials, but proved ineffective. The selection jury judged Ravenna daylighting a "development" project and was favorably inclined in the opposite direction, toward projects they considered "preservation."

An interagency intergovernmental group was also formed in late 1992, with a charge to identify issues and their possible resolution. Attendees included politicians' aides and city and county staff, as well as representatives from the Alliance and the shopping center. The shopping center representatives were public relations people though, not members of the architecture firm hired by the new owner to guide development. A third group was formed in 1992, Creek People, a citywide collection of creek activists. Their objective was to pool resources and information toward the common aim of restoring creeks throughout the city.

The community's vision with regard to Ravenna Creek was refined and further defined in its 1994 Master Plan, the pro bono product of landscape architects and engineers based on a series of public workshops held by RCA (RCA 1994). The aim was to demonstrate both a feasible alignment and at least one way that each segment of the project could be accomplished. The Master Plan included an initial cost estimate of $2.3 million, exclusive of acquisition, part of which had been reallocated from the original pipeline funding. The master plan itself was funded by a matching grant for $5,000 (for a professional survey, matched by $8,100 in kind or cash) from the Seattle Department of Neighborhoods. A primary motivation for the master plan development was to demonstrate to the shopping center owner that the daylighting project could be done and would have value. The plan proved to be insufficient for that, although it added substance to and clarified the community's vision (Figure 5.11).

To take the project to the next level, the county agreed to fund a feasibility study. That study is the most significant document on the record in many ways, due to its influence on governmental reaction. The first major potential funding source for construction appeared during 1996 while the feasibility study was in progress, the "Fields and Streams" ball parks and green spaces levy

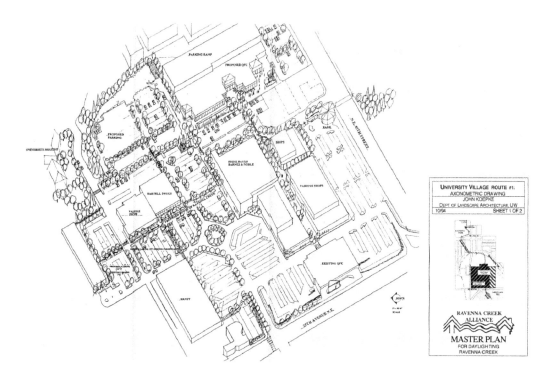

FIGURE 5.11 Bird's eye perspective of University Village shopping center, illustrating a route for a daylighted Ravenna Creek (RCA 1994).

at the county level. Although the issue of purchase of pristine versus restoration of urban ecology dominated the project selection discussions for the open space portion of the levy, the Ravenna Creek daylighting project was allocated $2 million in the levy, the balance required for the project according to contemporaneous estimates. The bond did not pass at the necessary 60% level countywide, but the Ravenna Creek project generated exceptionally high "yes" vote percentages in the precincts surrounding the current creek and the proposed route, as high as 85% and everywhere above 70%.

The elapsed time of the feasibility study was 3 years from consideration of scoping requirements to final draft. Bureaucratic gears grind slowly. The length of the process was important, as was the final product. The Alliance's view was that, instead of finding a feasible way to daylight the creek, the consultants had simply identified one of the many ways not to daylight the creek. Once under contract, it took the consultants a year to develop a first draft, which included construction of a parking garage. The cost estimate for the then-preferred route through the shopping center came to a total of $10.6 million, which was prohibitive (SvR 1997a). Abandoning hope for that route, the Alliance hired its own consultant to develop an alternate idea. The alternate route was located completely on public property, 95% of it city, the other 5% University of Washington property (Figure 5.12). That route was then also evaluated by the feasibility study consultants, who computed a new price tag (SvR 1997b).

The Alliance had objections throughout the study, from the selection process to the final report. The selection process was flawed in several ways. There were no stream experts on the

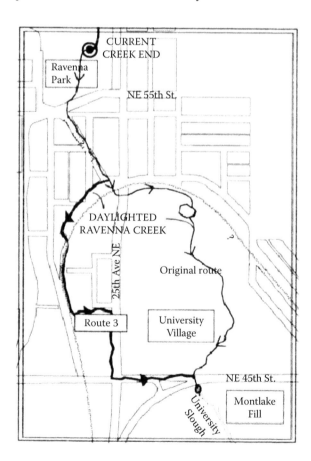

FIGURE 5.12 Schematic of Ravenna Creek Daylighting routes, through the shopping center (original route) and around it (Route 3).

panel choosing the consulting team, and the advisory experts from King County Surface Water Management, whose inclusion at the interviews had been promised, were in fact excluded. The Alliance had no role in selection and was only one of many parties interviewed during the study, with the result that the consultants chosen had no mission to figure out how to make the new route work, but merely to study. Due to various attrition factors, a junior engineer was left in charge of the study. The Alliance felt the emphasis was on bean counting, not problem solving, and that demands from the shopping center had been taken at face value rather than weighed. A parking garage may have been needed, but the creek was not the reason, since several years later a parking garage larger in size was approved by the city and is now under construction. The chronological entries in Table 5.2 indicate how feasibility study events relate to other milestones of the project.

ASSESSMENT OF THE RAVENNA CREEK CASE

As the costs of the project kept escalating and gradually occupying more of center stage, the benefits of the Ravenna daylighting project remained difficult to quantify. Instead of attempting to produce examples of comparable projects to demonstrate to the shopping center owner that the creek in fact would be a boon to commerce, the group tended to rely on signatures on petitions. Such support is easily dismissed. Although "nature" is part of the sales pitch of the shopping center where Ravenna Creek once flowed (a tree logo appears on banners and the advertising slogan is "Many shops under one sky"), what is actually occurring there is the commodification of nature. Artificial water features

TABLE 5.2
Ravenna Creek Daylighting—Major Milestones

Ravenna Creek Destruction

1911	First of two topography-altering municipal engineering projects impacts Ravenna Creek upper reaches. Following an Olmsted plan to create more open green space for the city, Green Lake level lowered 7 ft, leaving creek bed high and dry. Surface flow to Ravenna ravine replaced by pipe.
1916	Lake Washington level lowered 9 ft creating mudflats in place of open water in Union Bay and dry peat instead of marsh. Downstream from Ravenna Park, additional surface linkage now required for creek to reach former receiving waters.
1950s	City of Seattle disconnects flow downstream of Ravenna Park, routing Ravenna Creek into sewer trunkline instead of Union Bay.

Ravenna Creek Remaking

1991	Local wastewater treatment authority, Metro, recognizes economic benefits from reconnecting clean waters of Ravenna Creek to Union Bay via pipe. Community proposes daylighting as an alternative.
1992	University of Washington design charette demonstrates vision of surface reconnection; public support grows and nonprofit Ravenna Creek Alliance incorporates.
1994	RCA produces master plan with pro bono design and engineering expertise.
1997	Metro-funded engineering feasibility study results in project with a high price tag.
2000	City and county agree on limited daylighting, supplemented by public art, with piped conveyance downstream from Ravenna Park. City parks bond to fund design, County to fund pipe and creek construction.
2001	Gaynor, Inc., hired for design of daylighted creek within Ravenna Park.
2002–2003	Public process reveals resistance to change of status quo; ballfield will remain; designer produces innovative accommodation, creek will be daylighted within recreated ravine, surrounded by restored urban riparian forest.
2005	July 1 groundbreaking ceremony at Ravenna Park.
2006	May 14: Mother's Day grand opening to celebrate daylighting and reconnection of creek to Union Bay.

have been installed, and the shopping center owner has successfully blocked the creek not only from his own land but also from adjacent alternate routes. It turns out that a strong economic argument is difficult to make for the commercial value of ecological amenities to a shopping center such as the one involved in the Ravenna project, as naturalized systems for comparison are rare in highly urbanized areas. Recapturing is moot; high quality environment in the city only exists where it is already protected, as in a park. There is still no good template for the recreation of naturalized systems in urban environments.

The combination in 1990s Seattle of a municipal governmental structure that provides solely at-large representation with no locally responsible advocate and a climate favoring development without concurrent infrastructure, whether for transportation or for redressing past environmental damage, meant there was a critical absence of municipal clout to engage a shopping center owner who perceived his interests narrowly. The efforts to develop political clout for creek projects as a class in the early 1990s, by the formation of Creek People, had foundered after several years due to competitive jousting for individual project benefit. The resources available were too few to be successfully divided.

In contrast to the justifications for daylighting Thornton Creek, described below, daylighting Ravenna would not decrease flooding, because the creek is almost exclusively spring-fed and does not flash in its current configuration. Nor could proponents demonstrate conclusively that reconnection and increased reach length would absolutely increase usable salmon habitat, as the creek had been isolated within Ravenna Park for half a century and salmon runs were long gone although the creek does continue to support a resident population of a native salmonid, cutthroat trout.

Benefits from daylighting Ravenna tended to be cast in terms of reconstruction of a biologically diverse aquatic system, providing a pedestrian walkway along the course of the creek, creating a fish spawning and rearing habitat, and making improvements for bicycling at problem intersections in conjunction with the creek construction (priorities which are listed in the first RCA board meeting minutes, September 17, 1992). When the surface route linking Ravenna Park with University Village and Union Bay was blocked (by the high cost estimates presented in the draft feasibility report), the Alliance proposed an alternate route on public property. This route once beyond Ravenna Park would have been located almost completely within the public right of way of city streets or an urban bike trail.

When that route was rejected by the city, a compromise agreement was developed between the city and the county in 2000. Even though the city and county agreed not to daylight the creek except within the park, each body recognized the creek in a special resolution on Earth Day, 2000, with the city declaring that day as Ravenna Creek Alliance Appreciation Day in a proclamation signed by the Mayor and the City Council.

Rather than an ecological and surface reconnection, Ravenna Creek will be extended for 400–600 feet on the surface, within Ravenna Park, and from there will be piped in predominantly new line to a connection with its former receiving waters in Lake Washington. Construction of the daylighted reach in the Park and the pipeline, together with surface artwork, was scheduled to be completed in 2003 (but see update at end of chapter). This may be considered "partial success."

The word "daylighting" is now included in the local political vocabulary, and future proposed projects will perhaps face fewer hurdles. Extending Ravenna Creek as far as possible within Ravenna Park will both increase its visibility and improve the park. From the park's origins as a tourist attraction in the early 1900s, Ravenna has become a peaceful haven in the middle of a crowded city, where people can make a personal connection with nature and learn "to care enough for all the land" (Pyle 1993). The relationship between such personal connections and environmental stewardship is a theme developed by nature writer Robert Michael Pyle in his book *The Thunder Tree*. In its prologue, in fact, he describes how, for his mother, Ravenna Park itself was the place where that connection was made.

Case Study 2: Daylighting Thornton Creek at Northgate

Daylighting Thornton Creek at Northgate, although having its share of studies, has a shorter history but more current potential than Ravenna. The grassroots effort's success is not yet determined. Prospects for daylighting Thornton Creek in the Northgate area arose in 1998, during the later stages of the Ravenna process, when redevelopment was proposed at Northgate Mall and the adjacent South Parking Lot. The owner of both properties submitted a general development plan (Simon Properties 1998) for the two sites to the city for approval, incorporating commercial–residential development on the south lot, including a multiplex theater, office buildings, hotel and market-rate housing, but no creek. Table 5.3 contains a chronological list of milestones for the Thornton Creek Daylighting Project.

In 1993, a neighborhood planning process had produced the Northgate Area Comprehensive Plan (Seattle Planning Department 1993), containing goals for restoring Thornton Creek and for creating a linked system of open spaces and pedestrian connections, badly needed by the area. One proposed pedestrian route, the Urban Trail, would cross the south lot, generally following the expected path of a daylighted creek. Goals and benefits in the comprehensive plan for both daylighting and restoring Thornton Creek included a belief in its ability to breathe life back into a degraded neighborhood, revitalizing not only the creek's salmon habitat and reducing flooding, but enhancing a "town," with its increased business activity and desirable housing.

The wide differences between the comprehensive plan and the developer's proposal combined with a strongly-held local desire to daylight Thornton Creek during redevelopment led to the creation of two grassroots organizations:

1. Thornton Creek Legal Defense Fund (TCLDF), a nonprofit advocacy group made up of local citizens with the goal of daylighting Thornton Creek through the Northgate area, including Northgate Mall's South Parking Lot.
2. Citizens for a Livable Northgate (CFLN), a watchdog of development and promoter of open space, pedestrian-friendly amenities, reduced traffic congestion, and a daylighted Thornton Creek as components of a healthy, livable community.

TABLE 5.3
Thornton Creek Daylighting—Major Milestones

Thornton Creek Destruction

1948–50	Northgate Shopping Mall built just north of Thornton Creek.
1961	Major Interstate Highway I–5 construction blocks Thornton Creek and fills wetlands.
1963	Thornton Creek piped below ground; site becomes south parking lot of Northgate Mall.

Thornton Creek Remaking

Spring 1999	City approves Developer 1's major redevelopment of Northgate Mall and south parking lot without creek daylighting. Community groups start legal action against city.
Spring – Fall 2001	Developer 1 drops approved development plans and gives option to purchase south lot site to Developer 2. Collaborative design workshops held between Developer 2 and community to find solution that includes creek daylighting. Workshops fail shortly after September 11, 2001.
Fall 2003	City brings in local Developer 3. "Hybrid Design" concept involving both daylighting and water quality treatment developed during 1 month of intensive workshops between city, community, and Developer 3.
May 2004	Northgate Stakeholders Committee, representing diverse community interest groups, unanimously recommends approval of "Hybrid Design" as the best solution for Thornton Creek and Northgate. City proceeds with design development and construction documents.
June 2006	"Hybrid Design" now called Thornton Creek Water Quality Channel has first phase construction groundbreaking. Final phase of creek/water quality channel construction to begin in spring 2007.

After a series of developer-led (and city-sponsored) citizen advisory committee and design review meetings proved inadequate to address citizens' concerns, several appeals were filed against the plan with the city Hearing Examiner' Office. Limited by the hearing examiner to arguments based on city code and planning goals (environmental or drainage code elements were excluded), TCLDF and CLN nevertheless won their initial appeal against Northgate's General Development Plan, requiring revision by the developer (Simon Properties 1999). The Hearing Examiner's final ruling however led to a new controversy as it agreed with the developer's argument that the creek in the pipe was not a creek at all.

A long legal dispute continues over whether the creek is a creek or just a drainage conveyance, and over other environmental and drainage issues before the county and now the state court systems. When the county superior court ruled against the city and developer that the creek in a pipe is indeed a creek and thus requires a Supplemental EIS, it also ruled in favor of the developer and the city on drainage and storm water detention issues. Appeals continue by all parties (city, developer, and TCLDF/CFLN) to overturn those parts of the ruling that were lost.

As in the permitting process, the city and developer have joined forces against the grassroots organizations in this legal process. Basic issues are financial, the value of land, and maximizing urban development and tax base. The sequences of legal actions and appeals by all parties (city, developer, and community organizations) are expensive in time, energy, and money that might better be directed toward development of plans, actions, and future benefits for the affected community and environment. TCLDF, CFLN, and other community groups were convinced that was the case, and took their convictions to city and county politicians, developers, nonprofit foundations, and others. In hopes of gathering additional support, they produced a video, "Daylight Thornton Creek," and made a presentation model (Figure 5.13) to help people visualize the daylighted creek in place of an empty parking lot.

During the legal battle and political process, it became clear to creek advocates that major shifts in attitude, policy, and implementation are needed for a positive process that includes grassroots advocates and is effective for daylighting creeks (and other environmental restoration) within the city. Needed shifts may now be slowly occurring, as the community pushes for the public sector to catch up with its desires.

While the legal appeal process continued, the city and Thornton daylighting proponents engaged in a series of workshops and developed plans and reports that were sometimes at cross purposes. Although all related to either creek daylighting in particular or envisioning the future of Northgate in general, there were differences in scope, approach, and underlying attitude between elements sponsored by the city compared to those sponsored by TCLDF.

In August 2000, TCLDF used donations from private citizens to commission a conceptual design and feasibility study of Thornton Creek throughout the Northgate area, not just at the south lot. This proved difficult, as few details were known concerning the present creek in the pipe, the required depths for a daylighted creek alignment, and other basic facts.

FIGURE 5.13 Photograph, model of Thornton Creek daylighting project.

Simultaneously, the city sponsored a study of three daylighting alternatives for the South Parking Lot. That report, Northgate Daylighting Scenarios, Draft Report (R.W. Beck 2000), was released weeks prior to the TCLDF preliminary draft. The city's report was timed for use in city-organized Northgate "visioning workshops" (Strategic Planning Office 2000a; 2000b). Drafts of TCLDF's design and feasibility report (Gaynor/Penhallegon 2001) were also made available during the series of visioning workshops.

The limited scope of daylighting alternatives proposed in the city's report and their high cost estimates were construed to indicate a negative attitude within the city toward daylighting in the Northgate area. Although the city's purpose in convening the workshops was primarily a search for sites for a future community center and a library, the majority of design concepts developed during the workshops included some form of daylighting Thornton Creek on the South Parking Lot, usually in conjunction with the library and community center.

Then a new player was added to the mix, a developer who purchased an option to buy the South Parking Lot. In May 2001, the new developer and architect team presented a conceptual design for the south lot, which showed the creek daylighted through the entire site. Although this was an apparent victory for the advocates of creek daylighting, the design details became a source of conflict. The proposed stream corridor was stiff and narrow, with a right-angled layout that would be surrounded by parking garages, solid walls, and parking garage-filled bridges across the stream. For comparison, the Preferred Alternative C from the TCLDF feasibility study and the developer's conceptual plan are both shown in Figure 5.14.

During the following months the developer sought signatures on an agreement precluding lawsuits if the creek were daylighted. Daylighting proponents refused to sign, instead suggesting improvements to the developer's conceptual design proposal. The stalemate was broken following a behind-the-scenes intervention by creek designers and architects. The developer agreed to allow a limited design workshop process between its (by then) two architecture firms and what it called the "litigant" community, imposing a $5,000 budget limit for their architects. Three workshops were held during August, the first to list goals and essential elements for each side, the second to develop concepts focused on circulation system and creek corridor layout alternatives, and the last to present and improve the preferred alternative-collaborative plan. Creek designers and architects from both firms met to work on details and compromises.

The process produced a collaborative plan considerably improved in design, development mix, and creek considerations (Figure 5.15). The new creek corridor varied in width and included daylighting both branches along a larger park-like open space containing access and trails. It was surrounded by either residential or public uses and spaces. The total creek corridor acreage increased to 3 acres.

A joint presentation from the developer's architects and daylighting proponents was scheduled in early September to members of city council and representatives from the outgoing mayor's office. Proponents expected this meeting to be a positive turning point for the project. Instead the anticipated unified approach disintegrated as conflicts developed during the meeting between the two architectural firms. One architect and the developer defended the original plan and criticized the collaborative plan. In a surprise move, the developer presented the city with a request for $15 million for purchase of the property and construction of the creek corridor with its bridges. The parties have not met since then (but see update at end of chapter).

The community planned a "Thornton Creek Daylighting at Northgate" event for October 1, 2001, where they hoped to forge renewed community and political support for daylighting. Invitations were extended to the developer–architect team but they did not attend. Many politicians did attend, including a mayoral candidate who spoke in support of the project.

The mayoral candidate narrowly won the election, an apparent stroke of luck for the daylighting project as his opponent had gone on record in opposition. Early in 2002, the county purchased 4 acres of the South Parking Lot to expand its existing commuter parking lot and transit station. However, the county has since stated that the creek, i.e., the current pipe carrying creek flow, is not on their property. A daylighted corridor would nevertheless require relocation from beneath the street and onto adjacent land, including the county property. The collaborative design has become somewhat moot.

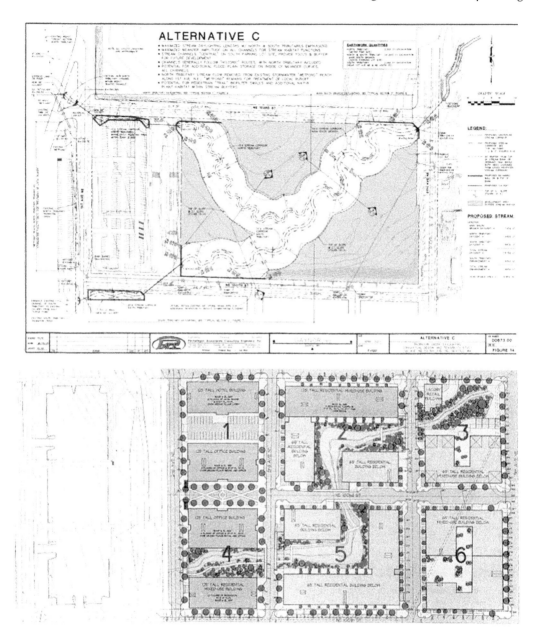

FIGURE 5.14 Thornton Creek preferred Alternative C (Gaynor/Penhallegon 2001) and developer's conceptual plan, showing flood plain terraces.

Creek advocates are now regrouping, forming new strategies to achieve Thornton Creek daylighting at Northgate. They seek to persuade the county to daylight Thornton Creek through its facility. They hope the new mayor will offer enough support from the city to make daylighting happen. They continue to raise funds and have begun to search for a land conservancy group that might purchase the creek corridor, as a possible mechanism for staging commercial development after the creek is in place.

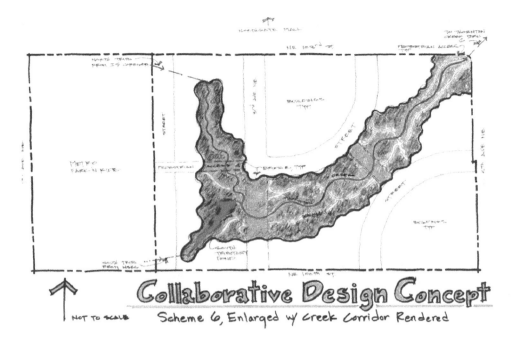

FIGURE 5.15 Version of collaborative design graphic prepared by Thornton Creek advocates.

FINDINGS AND RECOMMENDATIONS

Grassroots creek daylighting advocates need many elements for success, including luck. Major elements allowing progress in these two cases include: community credibility on the issue and persistence; financial resources available at the appropriate time; and partnerships/collaboration. Legal standing adds a significant and very useful advantage for Thornton, but legal action interferes with a cooperative and problem-solving spirit.

In the long term, facilitation of daylighting efforts will require both municipal environmental policy favoring nature (in a city where last year the parks department nearly replaced "nature" with "tourism" in its mission statement) and the identification of locations where streams now lay buried. Neither such an environmental policy nor any overview of daylighting opportunities is available in Seattle now. In Zurich, where a city engineer in the 1980s initiated a continuing effort to use surface waterways for solving flooding and treatment capacity problems to aesthetic advantage, city planners developed a map of opportunities that is implemented whenever development permits are granted (Conradin Chapter 3).

LESSONS LEARNED

These case studies suggest four required elements for urban creek daylighting to be propelled successfully from the grassroots community to often skeptical and dismissive developers, politicians, and municipalities. The elements are: (1) the grassroots organization's credibility and power, (2) a unified vision by all interested parties, (3) political will, and (4) funding.

Vision. Ravenna Creek Alliance formed with an active board and produced a strong master plan vision that, although it may have given impetus for a feasibility study by the county, never gained the city's concurrence or partnership. The process remains ongoing for the Thornton Creek proponents, but their vision, defined in the conceptual design and feasibility study and expanded during the design workshop process, has not been fully endorsed by all interested community groups, the developer, or the city.

Political will. Similarly, though both RCA and TCLDF have received support from individual politicians, the overall political will was not nor has it yet to be completely developed. In Ravenna's case, a county councilwoman recognized the value of the project and was instrumental in obtaining federal and county funding prior to city approval. However, the city, whose support as property owner was fundamental, came out in opposition to the daylighting of that creek (except for a small symbolic portion within an existing city park). Partial achievement of an element will not ensure the completion of the whole project.

The Thornton Creek project, on the other hand, has support from a successful political candidate who is now mayor of Seattle. A single high profile politician who has jurisdiction over the project can greatly improve the prospect of success. For Thornton Creek daylighting, the mayor's support may be persuasive in obtaining agreement from essential parties: the rest of the city, the county, and the developer.

Funding. RCA achieved a remarkable degree of financing (nearly $3.5 million) for its daylighting project. Lack of support from the city, which had jurisdiction over the project, to some extent voided these funding accomplishments. The federal funds had to be declined due to fiscal year expenditure restrictions. The county money will fund construction of the daylighting in the existing park, with city levy support for design.

A TCLDF member prepared a financing options report for distribution to all interested parties. One of the options identified is tax-increment financing, which relies heavily on state legislation to be feasible. After discovering that current state law authorizing such financing expires too soon to be of use for Thornton daylighting, continued promotion by some had a negative impact on county and city personnel regarding the project in general.

Currently TCLDF is building a larger coalition, a donor has offered a 2:1 match on donations, and other funding options are being pursued. Course corrections and even internal disagreements about direction are not uncommon for grassroots projects, although damage-control efforts may be required and some relationships may remain strained.

ROAD MAP

Thus, one case study project, Ravenna Creek Daylighting, had partial success, and for the other, Thornton Creek Daylighting at Northgate, it is too soon to tell. What is clear is that each grassroots organization, project, process, and politics is different. How, then, to proceed with the next grassroots daylighting or, for that matter, with any environmental project in an urban setting where broad support and large sums of money are required?

If only two or three of these elements are present, such as a unified vision, grassroots credibility and funding, they do not guarantee the project can proceed without grassroots power and the necessary political buy-in. Some suggested steps to achieve each element are summarized below.

Grassroots Power

1. Form partnerships with other grassroots groups, and/or large and/or well-established environmental organizations.
2. Amass widespread public support. (Figure 5.16 shows some development tools.) Consider qualifying a daylighting project initiative for public vote where such citizen-based governance is allowed.
3. Obtain legal representation to assist with (successful) appeals and suits, based on current environmental or other laws and regulations, against unfavorable permitting decisions, and establish legal precedents and standing for the grassroots project.

Vision

1. Keep a collective "eye on the prize" by staying focused on goals. Formulating a clear vision may require expenditure by the group for stream design, feasibility, cost/funding, and other studies.

FIGURE 5.16 Grassroots tools: 1994 t-shirt design; buttons; bumper sticker.

2. Determine which parts of the vision are open to potential compromise by confirming which elements are fundamental.
3. Involve all interested parties in an open and collaborative way. Include other community groups, environmental organizations, developers, politicians, and governing agencies. Identify and develop incentives for negotiation.
4. Consider a variety of construction and phasing options, while achieving a binding agreement that provides for the entire daylighting project.

Political Will

1. Cultivate political relationships.
2. Elect the supportive candidate to a position of power capable of persuading other politicians and entities.
3. Package a project that combines creek daylighting with other popular public facilities, such as public parks and open space, a community center, recreation facilities, library, public transportation facilities, public art.
4. Organize public pressure through communication campaigns, petitions and/or initiatives, and—last but not least—lawsuits.

Funding

1. Incorporate as a nonprofit 501(c)3 organization to allow tax-deductible donations to the organization.
2. Set up mechanisms for promoting and accepting donations, including pro bono.
3. Study all public and private funding options for design and construction.
4. Keep all funding options open, but discard those that become politically unpopular and/or unfeasible.

In addition, consider consulting professional fund-raisers. Effectiveness of professional fund-raisers may depend greatly on timing and communication to keep the goals and methods on track and costs under control. In RCA's case, a professional fund-raiser was ineffective, raising little more than fees paid.

Of course, a grassroots effort may follow all the right steps and still fail in its urban creek daylighting or restoration goal. Grassroots groups cannot form policy, build, or fund daylighting projects on their own. Help from political entities and others is basic. Success requires development of community-government partnerships that may be unusual or even unprecedented locally, the creation of political will, mobilization of resources, and, especially, a shared vision, as well as persistence from all parties to implement that vision.

Grassroots advocates have different perspectives from those of government and developers. Grassroots organizations and the general public may find themselves ahead of government philosophically and environmentally. Alternately, the government can be out of step with its public constituents. Developers may have different goals from those of governments, but governments at times can be more closely aligned in their goals and policies with developers than with grassroots organizations. Grassroots advocates may find themselves outsiders, trying to change large immovable bureaucracies, policies, goals and opinions without having the power, money, time, or connections to do so. From these case studies it is clear that among the things needed is development and implementation of laws and governmental policies that require correction of previous environmental damage, as laws do in some other areas of environmental management, notably wetlands and abandoned strip mines.

That any grassroots effort succeeds is generally a tribute to the individuals involved on all sides: the perseverance, flexibility, talent, and energy of grassroots advocates; some open minds and community consciousness from developers; and, especially, enlightened government individuals and politicians, powerful and courageous enough to change the ways of their bureaucracies. Even with all these ingredients going for a project, its timing may still be out of synch with successful levies or other funding options. In addition, if such citizen-initiated projects lose their sense of grassroots responsibility, they may not function well in the long run or even survive in the short run.

A sense of humor probably could not be counted among the vital ingredients, but it certainly helps to leaven the mix by refocusing attention, changing perspective, highlighting the absurd, and reexamining the situation. Two cartoons produced by a member of the RCA Board of Directors at various stages of the project are seen in Figure 5.17.

(a)

(b)

FIGURE 5.17 Cartoons by Thomas Whittemore, a member of the RCA Board of Directors, from various stages of the project.

CONCLUSION

If grassroots efforts to daylight creeks in cities are so difficult, time-consuming, emotionally draining, and often compromised, why attempt them in the first place? The individuals behind urban creek daylighting efforts may vary in their responses, but a common theme emerges. Putting nature back into the city is the right and imperative thing to do, not just for the obvious environmental benefits, but also for people. Nature brings life, literally and figuratively, back into neighborhoods.

Bringing streams and nature back into the fabric of the city is not only a compelling activity but it builds community as the vision is developed and the physical project proceeds. Although both top-down and bottom-up projects require the same resources in time and money and political will, bottom-up has concomitant benefits once completed. Whereas top-down projects must develop community support at various stages and sell the project to the community at completion, bottom-up projects have the long-term stakeholders engaged from the outset. Postconstruction success of such projects demands stewardship from the community to identify and respond to maintenance requirements.

Removal of institutional and governmental barriers and the alteration of policies to promote the restoration of nature in the city is an incremental effort. Every victory adds to the next, despite inevitable bureaucratic backsliding. The City of Seattle now has a stream biologist on staff, and the inauspiciously named (for creeks, at least) Drainage and Wastewater Utility has been revamped and is now located within the Seattle Public Utilities department rather than Engineering.

The Ravenna Creek Daylighting Project has been reduced in multiple ways from the original vision. Private property is no longer at issue; the number of city agencies has shrunk to two entities, the Parks Department and the Seattle Arts Commission. In a sense, Ravenna Creek will remain confined within Ravenna Park. Through art, a metaphorical connection will be made, though, and through the creek's reconnection with Union Bay, University Slough (the arm of the bay where it will flow again) will become Ravenna Creek once more.

With the current Mayor of Seattle's support, there is renewed hope that a creative daylighting solution for Thornton Creek at Northgate will emerge. Visions still need to be shared and reconciled. Can the community groups and the city come to agreement on the vision and on an approach that will make it happen and include the county and possibly developers? The time frame for such tasks could be several months—or years.

Ravenna Creek Alliance obtained permission in 1994 from the executors of Robert Frost's estate to use one of his poems along the route of the daylighted creek. The last lines, quoted below, give an inkling of the motivation to resurrect nature in the city:

No one would know except for ancient maps
That such a brook ran water. But I wonder
If from its being kept forever under
The thoughts may not have risen that so keep
This new-built city from both work and sleep.

—Robert Frost, "A Brook in the City" (1923)

UPDATE

When this paper was originally prepared for publication in early 2002, construction of the Ravenna Creek daylighting project was expected to begin in 2003 and the newly elected mayor of Seattle, Greg Nickels, was expected to advance the Thornton Creek project swiftly, as promised during his campaign. There were some substantial obstacles for each creek between then and now, however. For some details, see the milestones in Table 5.2 for Ravenna and Table 5.3 for Thornton.

As this update is written in mid-2006, 650 feet of Ravenna Creek have now been daylighted, adding an additional 20% of surface creek flow. Thornton daylighting is incorporated as part of a municipal–private partnership whose groundbreaking was celebrated on June 7, 2006, a month after the celebration of Ravenna Creek's daylighting and reconnection.

The Ravenna project includes a 4-acre urban forest restoration, incorporating 18,000 native plants. The final grading plan (Figure 5.18) illustrates how the original ravine was extended into flat fill. A pair of photographs shows the transformation of the ground plane in Ravenna Park (Figure 5.19) and several design details are illustrated in Figure 5.20. The water quality improvement produced by the creek's reconnection to its former receiving waters is dramatically visible in the photographs in Figure 5.21.

For Thornton, the third time's the charm. The community, city, and a third developer come together under a creative design vision with something key for each party: daylighting Thornton Creek, providing water quality treatment for over 600 acres of urban runoff, and creating a 3-acre natural park through a major new commercial and residential development. Sketches of this complex and innovative pubic–private project are shown in Figures 5.22 (concept) and 5.23 (related vignette sketches).

Persistence and determination are required in large measure. The pain is worth the gain, in our opinions.

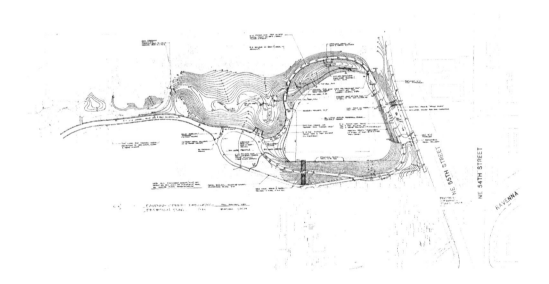

FIGURE 5.18 Ravenna Creek daylighting project final grading plan, by Gaynor, Inc.

(a)

(b)

FIGURE 5.19 Aerial views of southeast Ravenna Park before and after construction of the daylighted creek (a) June 2005, (b) June 2006. (Photography by Andrew Buchanan, Subtle Light Photography.)

(a)

(b)

FIGURE 5.20 Ravenna Park daylighting details (a) loop and bridge, March 2006, (b) close-up of log weir in (a) upper right, July 2006. (Photography by K. O'Neill.)

(a)

(b)

FIGURE 5.21 Water quality improvements visible in University Slough as a result of Ravenna Creek inflow (a) April 7, 2006, (b) April 30, 2006 (photos by K. O'Neill).

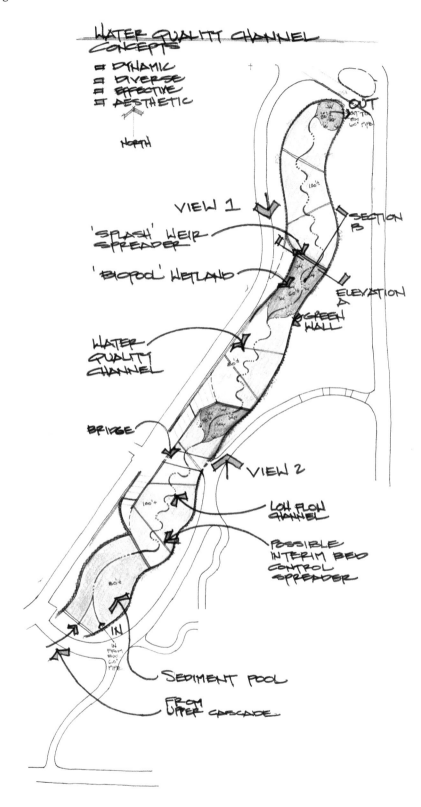

FIGURE 5.22 Sketch of water quality channel concept for Thornton Creek daylighted at Northgate, by P. Gaynor, Gaynor, Inc.

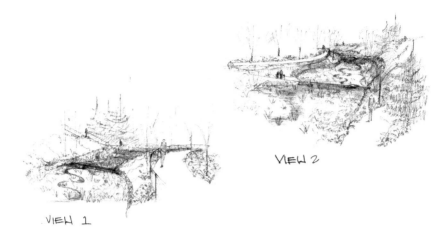

VIEW 2

VIEW 1

FIGURE 5.23 Thornton vignette sketches from viewpoints marked in concept drawing in Figure 5.22, by P. Gaynor, Gaynor, Inc.

LITERATURE CITED

Arnstein, S. R. 1969. A Ladder of Citizen Participation, *Journal of the American Planning Association* 35(4), pp. 216–224.

Bays, J. 2001. Wetland Engineering, Harvard Brown Fields, Grey Waters Symposium 2001 (http://www.gsd.harvard.edu/events/conferences/restoration).

Conradin, F. 2001. The Zurich Stream Daylighting Program, chapter 3, this volume.

Dorpat, P. 1984. *Seattle Now and Then*, Tartu Publications, Seattle, 240 pp.

Fehr, E. and S. Gachter. 2002. Altruistic Punishment in Humans, *Nature*, January 10, 2002.

Frost, R. 1923. A Brook in the City, from *The Poetry of Robert Frost*, (1969) ed. E. C. Lathem, Holt, Rinehart and Winston, New York, 607 pp.

GAYNOR, Inc. / Penhallegon Associates Consulting Engineers, Inc. 2001. Conceptual Design and Feasibility Study, Thornton Creek Daylighting: Final Report, April, 2001, 95 pp. + appendices.

Leopold, A. 1953. *Round River: from the Journals of Aldo Leopold*, ed. Luna B. Leopold, Oxford University Press, New York, 173 pp.

Levertov, D. 1996. Salvation, from *Sands of the Well*, New Directions: New York, 136 p.

Mason, G. 2001. Water Reborn: A Story about Stream "Daylighting" in the San Francisco Bay Area, Harvard Brown Fields, Grey Waters Symposium 2001 (http://www.gsd.harvard.edu/events/conferences/restoration/pages/program/abstracts/waterreborn.htm).

Mozingo, L. 1997. The Aesthetics of Ecological Design: Seeing Science as Culture, *Landscape Journal*, 16: 46–59.

Olmsted Brothers. 1903. Original Report of Olmsted Brothers, Landscape Architects, Adopted by the City Council October 19, 1903, reprinted in Parks, Playgrounds and Boulevards of Seattle, Washington, issued by the Board of Park Commissioners, 1909, pp. 72–154.

O'Neill, K. 1991. Ravenna Creek: Past, Present and Future: Report to the Working Group on the Daylighting of Ravenna Creek, July 1991, Preliminary Feasibility Study, 30 pp.

Pinkham, R. 2000. *Daylighting: New Life for Buried Streams*, Rocky Mountain Institute, Snow Mass, CO, Report for EPA, 63 pp.

Pyle, R. M. 1993. *The Thunder Tree: Lessons from an Urban Wildland*, Houghton Mifflin, Boston, MA, 220 pp.

Ravenna Creek Alliance. 1994. Ravenna Creek Daylighting Project: Master Plan. November, 1994, 1 v. (various pagings), folded plans.

R.W. Beck. 2000. Northgate Daylighting Scenarios: Draft Report, October 20, 2000, City of Seattle, Seattle, WA.

Simon Property Group. 1998. Northgate Mall, General Development Plan, September 2, 1998, full color, 11"×17", 23 pp.

Simon Property Group. 1999. Northgate Mall, General Development Plan, November 29, 1999, full color, 11"×17", 30 pp.

Seattle Planning Dept. 1993. Northgate Area Comprehensive Plan, September 1993, City of Seattle, Seattle, WA, 78 pp. + appendices.

Strategic Planning Office. 2000a. Workshop Findings: Understanding Northgate, October 26–28, 2000, City of Seattle, Seattle, WA, 11"×17", 22 pp.

Strategic Planning Office. 2000b. Northgate Community Workshops: Refining Our Choices, December 2000, City of Seattle, Seattle, WA, 50 pp. + appendix.

SvR Design Company et al. 1997a. Draft feasibility study: Ravenna Creek Daylighting Project, January 29, 1997, prepared for King County Department of Natural Resources, 2 v., folded plans.

SvR Design Company et al. 1997b. Final feasibility study: Ravenna Creek Daylighting Project, December 19, 1997, prepared for King County Department of Natural Resources, 2 v., folded plans.

Thornton Creek Watershed Management Committee. 2000. Thornton Creek Watershed: Characterization Report, November 2000, Seattle Public Utilities, City of Seattle, Seattle, WA.

Waterman, T. T. 1922. *The geographical Names Used by the Indians of the Pacific Coast*, American Geographical Society, New York, 194 pp.

Part 3

Coasts

6 Wherefore the Rhizome? Eelgrass Restoration in the Narragansett Bay

Cheryl Foster

CONTENTS

ABSTRACT

Using eelgrass restoration in the Narragansett Bay as the touchstone for discussion, this chapter clarifies some of the challenges faced in ecological restoration by analyzing those challenges under three distinct philosophical classifications: *epistemological, axiological,* and *normative.* Examining wide divergences among governmental, scientific, and popular articulations of estuarine matters, the chapter surveys how the fractured relationships governing reflection, value, and action has been addressed both within and beyond various principles governing restoration. It also recounts the efforts of the Rhode Island advocacy group Save the Bay to implement a community restoration initiative for children and then to redesign the initiative when original expectations about its effect did not come to fruition. Finally, the chapter addresses broader ambiguities about the relationships between fact and value, and between value and action, as they emerge from the restoration issues discussed. While the chapter avoids the prescription of particular solutions to the challenges facing eelgrass and other estuarine restoration, it strives to delineate and illuminate the tangled sources of value that can drive everyday decisions—sources that, left unexamined, may contribute collectively to the demise of eelgrass populations.

Prologue: August 20, 2003

The stench woke us up on Cedar Tree Point the morning of Wednesday, August 20. Something was wrong, dead wrong. The tide washed up not thousands but millions of bay residents. Menhaden, silversides, crabs, shrimp, eels, and other forms of bay life lay rotting at the water's edge (Early 2003).

The trail of tiny bodies cut across Greenwich Bay like a silver scar. Up to a million menhaden, their bellies swelling in the heat, floated on the surface … last week's sight of so many dead fish clumped onshore and drifting in oily slicks has prompted outrage, wonder, and even fear (Shea 2003).

The air was sour with the stench of rotten eggs yesterday morning as state biologist Arthur Ganz maneuvered his skiff through Apponaug Cove to view the aftermath of Wednesday's massive fish kill caused by polluted, oxygen-depleted water. "See that," Ganz said, pointing to an oily sheen that covered most of the cove. "Menhaden are a very oily fish and that is caused as they decompose." Although there was no wind yesterday morning, the cove water seemed to flutter, and Ganz said that the motion was caused by thousands of tiny silverside fish swarming just below the surface. "They're coming up to the top in an attempt to find oxygen." (Ganz said (Polichetti 2003b).

Torgan is Narragansett Baykeeper for Save the Bay, an environmental group dedicated to preserving Narragansett Bay and its surrounding watershed. About a week before the fish kill, Torgan and others observed heaps of dead steamer clams washed up along the banks of the Providence River. During one visit to Gaspee Point, Torgan scooped up handfuls of juvenile clams, lifeless as pebbles. He estimated that millions of them had suffocated, due to low oxygen levels in the river. "The kill was already happening," he said. "It just wasn't fish yet." Torgan and colleagues at Save the Bay shouted warnings to state officials, but there wasn't really anything that could be done. The dominoes were already falling. Several days later, dead fish clogged Greenwich Bay (Shea 2003).

Jack Early, a member of Defenders of Greenwich Bay, said that people in the Cedar Tree Point area are horrified by the occurrence, and that many gathered at the water's edge yesterday, where they cried at the sight of all the dead fish. "We know this is a significant [ecological] event," Early said. "This is something we've been screaming about for years. The people are weeping and the bay is weeping—she's weeping her dead." (Polichetti 2003a).

Unfortunately, it often takes a catastrophe to serve as the warning shot for the public to notice a problem (Fugate 2003).

INTRODUCTION

On August 20, 2003, an anoxic event of devastating proportions killed over one million fish in Narragansett Bay, Rhode Island. Less than two months before, on June 30, 2003, a page one photograph in *The Providence Journal* featured Rhode Island Governor Donald Carcieri helping restoration ecologists lower eelgrass shoots on frames as part of a Bay transplantation exercise. In the article accompanying the photograph the governor recalled scalloping on the Bay as a child but lamented the loss of that industry due to the erosion of eelgrass habitat. "Narragansett Bay is in my blood, and I feel strongly about it," Carcieri said. "I believe we have to do everything we can to preserve and protect the Bay (Emlock 2003)." Little did the Governor realize, on June 30, 2003, how very soon he would be called upon to fulfill the duty he prophetically cited that day.

Following the fish kill of August 20, Governor Carcieri announced the formation of a commission to function as a policy-making body and to develop a unified strategy to combat problems the Bay faces (Ortiz 2003a), including issues of how to restore ecosystem function on a broad scale. By January 2005 the Rhode Island House and Senate had worked together to pass landmark legislation that grew out of this commission. RI-GL Chapter 46-31 established a permanent Coordination Team for Rhode Island's Bays, Rivers, and Watersheds, one that would cut across existing agencies, councils, and boards to conceive, coordinate, and implement an ecosystem-based approach to managing the state's marine environment. In § 46-31-1 of the Legislative findings, the law states: "The general assembly hereby finds and declares as follows: (1) The bays, rivers, and associated watersheds of Rhode Island are unique and unparalleled natural resources that provide significant cultural, ecological, and economic benefit to the state (RI-GL 46-31-1)."

It seems a truism to proclaim that the marine environment in and surrounding Rhode Island provides significant cultural, ecological, and economic benefit to the state, but then again it took an event straight out of an Armageddon horror film—portentous waves of dead steamer clams, followed by a massive wave of dead menhaden (Figure 6.1), echoed by a third wave of silverside fish, all of it festering like rotten eggs—to push Rhode Island's legislative bodies into problem-solving cooperation with the state's executive office and nonprofit agencies. Until the August 2003 fish kill, the nation's smallest state had lacked an efficacious, elegant mechanism for managing those very resources that give the state its nickname: The Ocean State. Dead fish on the morning water certainly galvanized public and subsequent governmental awareness about significant marine resources, as did the pervasive stench of sulfur dioxide everywhere within reach of the shore.

Olfactory affront is an extreme example of what it takes to move elected representatives to action on behalf of our bays and estuaries, but political fractiousness and a congenital resistance to the anticipation of catastrophe often mitigate less dramatic (if concerted) efforts to form workable policy

FIGURE 6.1 August 2003 fishkill, Greenwich Bay, Rhode Island. Photo by Thomas Ardito, *Narragansett Bay Journal*, www.nbep.org.

about preservation and restoration across stakeholder groups. The 2003 fish kill in Narragansett Bay was an unprecedented environmental disaster. Nevertheless, it did focus public (and thus government) attention on the social behaviors and ecological conditions that combine to give rise to anoxic events. This in turn raised public consciousness about the degradation of native habitat in Narragansett Bay, including the dramatic decline of the ecologically-sensitive but little-known aquatic species known as *Zostera marina L.*, or eelgrass.

A close examination of restoration policies, practices, and principles in place prior to the 2003 fish kill in Rhode Island reveals an unsettling lack of consistency and clarity among various federal and national initiatives. One reason for this problem is the different professional or pragmatic communication styles used by parties to the restoration debate. Policy makers and elected officials examine restoration issues within specific contexts and publications, while professional scientists debate restoration in others, with the public turning to nontechnical journalism and sometimes lyrical popularizations in order to grasp the issues at stake. Ostensibly, this chapter considers a variety of philosophical and pragmatic ambiguities related to eelgrass restoration on the Narragansett Bay. More subtly, it attempts to think about the *value of* submerged species and aquatic communities in our lives—and the power *of values*, whether discrete or diffuse, whether explicit or unarticulated, to shape interactions between human beings and the natural environment.

Using eelgrass restoration in the Narragansett Bay (Figure 6.2) as the touchstone for discussion, I shall attempt to clarify some of the challenges restorationists face by analyzing those challenges under three distinct philosophical classifications.

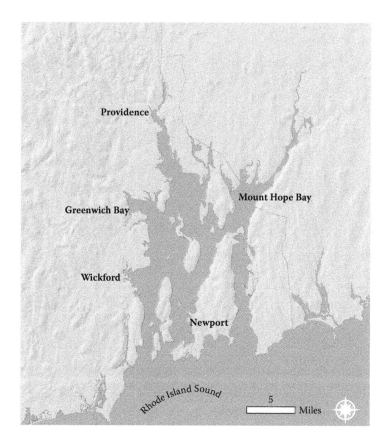

FIGURE 6.2 Map produced by the University of Rhode Island Environmental Data Center. Data sources are the Rhode Island and Massachusetts Geographic Information System databases and the United States Geological Survey (NED, GNIS).

Section I foregrounds the discussion by highlighting wide divergences among governmental, scientific and popular articulations of estuarine matters, followed by an introduction of the three classes of challenges—*epistemological, axiological*, and *normative*—that will govern the thinking about restoration difficulties undertaken here. Section II examines the Principles of Estuarine Habitat Restoration, generated jointly by Restore America's Estuaries and the Estuary Research Foundation in 1999, in order to identify the epistemological, axiological, and normative dimensions of those principles and to survey how the fractured relationship governing reflection, value, and action has been addressed both within and beyond those principles. Section III introduces the case of eelgrass restoration and discusses problems associated with it in the context of Narragansett Bay, Rhode Island, filtering those problems once again through epistemological, axiological, and normative lenses.

Section IV recounts the efforts of Rhode Island advocacy group Save the Bay to conceive, implement, and analyze a particular community restoration initiative for children—"Seagrasses in Classes"—and then to redesign the initiative when original expectations about its effect did not come to fruition.

Finally, Section V moves away from concrete discussion of eelgrass restoration as such to address broader ambiguities about the relationships between fact and value, and between value and action, as these emerge from the restoration issues thus far discussed. Although the chapter avoids the prescription of particular solutions to the challenges facing eelgrass and other estuarine restoration, it strives to delineate and illuminate the tangled sources of value that can drive everyday decisions—sources that, left unexamined, may contribute collectively to the demise of all eelgrass populations in Narragansett Bay.

CONGRESSIONAL MANDATE: PHILOSOPHICAL DIMENSIONS BENEATH ESTUARY RESTORATION

We begin with policy. In November of 2000, the United States Congress passed the Estuary Restoration Act of 2000 (Title I of Public Law 106–457) to establish a collaborative process for addressing the pressures facing our Nation's estuaries. As part of that Act, an interagency Estuary Habitat Restoration Council was appointed with this mandate: to develop and submit a strategy to Congress; to solicit, review, and evaluate project proposals; and to recommend projects to the Secretary of the Army (Ortiz 2002a,b). In December of 2002 the Estuary Restoration Strategy was put forward to govern the activities of the Council. The stated purpose of the Strategy is this: to ensure a comprehensive approach to the maximization of benefits derived from estuary habitat restoration projects; to provide incentives for the creation of new partnerships between public and private sectors; and to foster coordination of federal and nonfederal activities related to the restoration of estuary habitat. The Strategy's long-term goal is to restore one million acres of estuary habitat by 2010 (Ortiz 2002a,b).

The Act defines *estuary* as "a part of a river or stream or other body of water that has an unimpaired connection with the open sea and where the sea water is measurably diluted with fresh water from land drainage." For practical purposes the Act also includes near coastal waters and wetlands of the Great Lakes (Public Law 106–457). Within the strategy an estuary is limited to naturally occurring, non-man-made boundaries and "is considered to extend from the head of the tide to the boundary with the open sea." (Ortiz 2002a,b).

These policy-context depictions of estuarine boundaries and function, derived from the classic definition by Pritchard, have been criticized as being overly simplistic by marine scientist Daniel Alongi. Alongi argues that such depictions ignore the role of tides and other regional variations in salinity and dilution, and suggests that the following characterization (formulated by B. Kjerfve) is more comprehensive and hydrographically accurate:

An estuarine system is a coastal lagoon that has a restricted connection to the ocean and remains open at least intermittently. The estuarine system can be subdivided into three regions: (a) a *tidal river zone*—a fluvial zone characterized by the lack of ocean salinity but subject to tidal rise and fall of sea level; (b) a *mixing zone* (the estuary proper) characterized by water mass mixing and existence of strong gradients of physical, chemical, and biotic quantities reaching from the end of the tidal river zone to the seaward location of a river mouth or ebb-tidal delta; and (c) a *nearshore turbid zone* in the open ocean between the mixing zone and the seaward edge of the tidal plume at full ebb tide (Alongi 1998).

Such a characterization, however, while more scientifically layered, may not lend itself easily to the pragmatic purposes of a public strategy for restoration. As a contrast to the definition employed by the Estuary Restoration Act of 2000 and Kjerfve's more empirical and progressive account of estuarine zones, consider the following lyrical description offered by Tom Andersen in his history of Long Island Sound:

> An estuary is where salt water from the ocean meets and mixes with fresh water draining off the land. It is an arm of the sea coddled by the land. Estuaries are secure from the waves that pound the ocean's coast, but are renewed continually by the tides and currents and rivers—the word *estuary* comes from the Latin verb a*estuare*, to heave, boil, surge, be in commotion. The foundations of an estuary's biological richness—the array of life that starts with the marsh grasses and encompasses bacteria, diatoms, larval crabs, lobsters, striped bass—are the physical and chemical processes that stimulate life (Andersen 2002).

Andersen's image of an estuary as an "arm of the sea coddled by the land," coupled with his use of etymological definitions, offers a less objective but perhaps more memorable depiction of what an estuary might seem like on the affective surface of observation. More crucially, each characterization of an estuary thus far offered suggests a slightly different audience for its depiction: policy analyst, scientific professional, and general reader.

The gaps among pragmatic policy, detailed science, and affective attachment turn out to be crucial aspects of discussion in considering the particular topic of eelgrass (*Zostera marina L.*) and habitat restoration in Narragansett Bay, Rhode Island. This is because the gaps emerge between what scientists and, to a lesser extent, policy makers know about the challenges of eelgrass restoration (including a sense of what they do not actually know), and what the public in its ordinary attitudes and behaviors does not even consider in relation to that same set of challenges. For example, seagrass meadows *(estuarine aquatic beds)* are specifically cited as preferred features for restoration in the Strategy to implement the Estuary Restoration Act of 2000 (Ortiz 2002a), yet awareness of seagrass meadows (Figure 6.3) among citizens living adjacent to Narragansett Bay is at best minimal. According to a 1996 Bay Habitat Public Opinion Research Study underwritten by Rhode

FIGURE 6.3 Eelgrass flowering. Photo by Thomas Freeman, member, Save the Bay.

Island's Save the Bay and The Pew Charitable Trust, approximately 70% of Rhode Islanders and residents from southeastern Massachusetts do not know what eelgrass is (Save the Bay, 1). Public ignorance of both ecological processes and the consequences of human behaviors throws up a genuine impediment to the Strategy's goal of restoring a million acres over the next several years.

Public ignorance of what the scientific community has understood and what our elected representatives have mandated cuts just one way across the swath of challenges faced by restorationists. It can be argued that some members of the scientific and policy communities maintain a simplistic grasp of the role played by human values and personal reflection in the larger arena of ecological systems. Although the Strategy for implementing the Estuary Restoration Act of 2000 encourages social considerations such as land use, the recreation industry, and historical patterns in assessing estuarine trends, it also bases its claims about change primarily on faith with respect to public investment in the process of restoration and habitat change.

The restoration and maintenance of healthy coasts and estuaries will require the long-term support of a broad cross section of the public. Including local communities in planning and implementing restoration projects will build interest in protecting and maintaining restored habitat. Increased awareness of the attributes needed to sustain healthy habitat will increase local stewardship of the environment and will help to ensure the long-term success of restoration projects (Ortiz 2002a,b).

Intriguing philosophical implications lurk in the short passage quoted here. Firstly, it posits long-term public support as a necessary condition for the restoration and maintenance of healthy coasts. This raises the question of what kinds of initiatives gain and sustain public support, a troubling issue given the majority of citizens around Narragansett Bay who do not yet know what eelgrass is. How is such support to be cultivated?

Second, it makes two causal claims: one, that the inclusion of local communities in planning will build interest in the restoration of habitat; and two, that increased awareness of healthy habitat attributes will increase local stewardship. Although an allegedly causal relationship between public participation and the increased value of healthy ecosystems could perhaps be demonstrated over time, we are hard pressed to make such a case on any convincing scale, given the near-complete absence of direct involvement in community restoration in Rhode Island at present.

According to Wendy Norden, former restoration ecologist at Save the Bay in Rhode Island, a total roster of about 100 volunteers participates in eelgrass transplantation and harvesting throughout the course of a year—100 individuals in a statewide population of about 1 million. And, again, at a charrette to discuss coastal habitat restoration held at the University of Rhode Island's Coastal Institute in December of 1998, 100 individuals showed up to contribute to the conversation (Ardito 1999a). These are encouraging numbers (relative to civic participation in public events generally), especially as the charrette in particular has led to the creation of a Rhode Island Habitat Restoration Team to coordinate and communicate about separate restoration efforts across the region (Rhode Island Coastal Resources Management Council et al. 2003). Yet, in the shadow of sports events, rock concerts, and even antiwar protests, participation rates such as these do not appear to reflect widespread or active command of public attention on the issue of coastal habitat restoration.

Taken together, the implications of the passage quoted above reveal the presence of three broad classes of challenges facing those who wish to embark on eelgrass restoration as well as other forms of estuarine habitat restoration.[1] The first class of challenges is *epistemological*. This traditional area of philosophy examines theories of knowledge and the conditions under which a claim might pass from the status of a belief or hypothesis into knowledge on the basis of justified evidence. The epistemological dimensions of eelgrass restoration efforts on Narragansett Bay are circumscribed by both uncertainty about the exact location and density of past eelgrass beds in the estuary and

---

[1] I do not count among the classes discussed here the metaphysical—that is, issues concerning the status of a thing's being. Discussion surrounding the issue of natural vs. artificial status of restored habitats has been exhausted by attention to neat categories of genesis rather than to processual commitments to healthier environments. Such discussions have their uses in addressing broader issues of anthropocentricism and reductive attitudes towards natural "resources," but it is not the purpose of this essay to enter those debates here as they are well underway elsewhere.

the ambiguity of future success for decimated and degraded habitat given the benthic, atmospheric, and temperature changes over the last few decades. The epistemological dimension also includes more specific questions about matters such as: which techniques are best to employ in restoration; whether nonpoint sources of nitrate loading by fertilizer can be decoupled in their effects from those arising from atmospheric deposition; and whether or not restoration is anything more than a symbolic act, given the larger issue of water quality in the bay, which has the most tangible impact on eelgrass health and survival.

Alongside the epistemological class of challenges sits a second group of issues to be faced by advocates of restoration, a class I call *axiological*. In philosophy axiology stands for those areas of inquiry that take human values as their central subject, areas such as aesthetics, ethics, political philosophy, philosophy of religion, and sometimes the subject of the emotions. Although axiological issues necessarily intersect with other areas of philosophy such as epistemology and metaphysics, they nevertheless cohere as a distinct class by virtue of their emphasis on the nature of value rather than on that of fact. The axiological dimension of eelgrass restoration emerges most acutely in the public ignorance of, or perhaps indifference to, habitat health and maintenance. Ignorance and indifference manifest themselves in public practices such as the fertilization of lawns and the persistent use of septic systems in communities that directly abut the coast, which contribute to water quality problems in the bay. They also show up in the less certain but possibly connected contributions of emissions from cars and factories to atmospheric pollution, or to warming trends and rising temperatures in the bay. Axiological dimensions of eelgrass are additionally apparent in variations among individual preference judgments: eelgrass, because it is entirely submerged, is absent from aesthetic perception altogether; is seen as a nuisance to swimmers; but has value as a breeding ground for creatures attractive to fish caught for sport by recreational fishermen.

Finally, I carve out a subset of the axiological dimension to create a third class of challenges facing restorationists: the class of *normative* challenges. Normative issues in philosophy are those that involve rendering principles for right action or judgment, such that maxims or rules for ethical or legal behaviors are clarified. The entire Strategy for addressing the Estuary Restoration Act of 2000 functions as a mandate to create coalitions between federal and nonfederal agencies and groups, including academic, local, tribal, and civic entities committed to restoration. As a mandate it is a kind of normative claim, one that tells us what we must or ought to do, given the horizon of goals and values stated in the Act. On a more local level, Rhode Island faces a normative challenge in inconsistent standards, between federal and state policy, for activities pertaining to water quality and other eelgrass habitat health indicators (Save the Bay 2003). In this chapter, I suggest that although the normative dimension of eelgrass restoration has found expression in federal law, it also peeks out from under more epistemological disputes, such as whether habitat recovery is best served by broader water quality improvements instead of restoration of flora. The normative dimension lurks behind more axiological quagmires as well, quagmires such as how to raise public awareness of and attachment to habitat in an age when infotainment appears to dull the public's tolerance for complex information.

In other words, we find a counterpart for what we *do not know* scientifically about eelgrass restoration in what we *do not see* in relation to values—values that occlude restoration as a civic priority. The epistemological and axiological dimensions partake of the normative dimension insofar as they attempt to address each set of challenges through a set of legal mandates, scientifically recommended actions, or even social pressures. Here I argue that, science and policy notwithstanding, estuary restoration cannot succeed in the long term without a firm and lasting hold on the affective lives of those living on and around the water bodies themselves.

Opportunities for civic involvement of a conventional kind, such as those afforded by charrettes and public hearings, are certainly valuable as vehicles for representative involvement, but they are unlikely to *inspire* individuals, and then communities, towards a stronger sense of value with respect to eelgrass or estuarine habitat unless those individuals or communities are already committed to restoration. Community restoration, by contrast, while labor intensive and ultimately symbolic with respect to the scale of tangible goals for eelgrass and estuarine restoration, comes closer to instilling a sensual, active, focused involvement that engages the individual on the bodily and social levels. Affective engagement must be powerful, however, to overcome the allure of ecologically-questionable habits that remain deeply ingrained in popular life.

PRINCIPLES OF HABITAT RESTORATION AND EMBEDDED PHILOSOPHICAL KNOTS

Prior to the successful establishment of the Estuary Restoration Act of 2000, the scientifically-driven, nonprofit organizations Restore America's Estuaries and the Estuary Research Foundation collaborated to produce the 1999 monograph *Principles of Habitat Restoration: Working Together to Restore America's Estuaries* (RAE-ERF 1999). The document is comprehensive and ambitious, presenting 14 principles and accompanying illustrations for addressing those classes of problems I have called the epistemological, axiological and normative. The normative aspects of the principles appear throughout the document, as they should for a set of guiding maxims, but four points in particular highlight the urgency and complexity of right action on a variety of levels with respect to America's estuaries.

NORMATIVE DIRECTIVES

Principle 1: *Preservation of existing habitat is critical to the success of estuarine restoration* (RAE-ERF 1999). Two of the illustrative points attached to this principle have bearing here. One, that preservation of existing habitat must be the starting point for efforts to achieve a net gain in healthy, functioning habitat; and two, that preservation must be combined with aggressive restoration to meet national estuarine restoration goals. Within the first principle of the document we see an implied commitment to addressing the larger issues contributing to habitat degradation, such as water quality and pollution controls, and also to restoration as a necessary feature of habitat reclamation. A subtle emphasis falls here on the difference between halting the loss of present habitat (preservation) and creating new habitat (restoration). This is because at present the annual loss of healthy habitat outstrips the rate of current restoration. In other words: restoration efforts alone will not be sufficient to meet estuarine restoration goals for healthy habitat but must instead occur in tandem with efforts to control those factors that threaten existing healthy habitat. Restoration cannot occur in a vacuum. It is vitally and functionally linked to changing social patterns that degrade existing ecosystems.

   Related to this we see a mandate for change in Principle 11: *Ecological engineering practices should be applied in implementing restoration projects, using all available ecological knowledge and maximizing the use of natural resources to achieve goals* (RAE-ERF 1999). The illustrations given to justify this mandate make a sharp distinction between "ecological engineering practices" and "traditional engineering practices." Traditional practices do not consider natural and environmental factors in design but instead often ride the wave of public confidence in new technologies. In *Water Resources Management: In Search of an Environmental Ethic*, David Lewis Feldman notes that a "golden age" of engineering flourished from about 1850 to 1950, when some engineers achieved great technical feats and came to believe they were social revolutionaries, promoting technology as a panacea for the world's problems. Following this wave, however, environmental problems, the nuclear age, and the capacity for mass extinction soon cast a shadow on the optimism of unbridled technology. Feldman argues that "it has grown increasingly apparent that even small-scale engineering solutions to natural resources problems can generate irreversible impacts and may cause many new problems not amenable to engineering solutions" (Feldman 1991).

   Further implicating the effects of traditional engineering, Principle 11 also claims that "the best way to accomplish full restoration is to remove barriers to natural functioning, such as dams, ditches, and other man-made structures, and allow natural hydrology and drainage patterns to re-establish themselves (RAE-ERF 1999)." Not only must we abandon the isolated technologies of traditional engineering; we must actually dismantle the tradition's legacy whenever that legacy impedes natural systems. Historical data and appropriate references can be used to design ecologically appropriate interventions, but the mandate here is uncompromising: undo your mistakes or forfeit restoration goals.[2]

If these goals are indeed to be met, then another of the fourteen principles indicates the necessity of ratcheting up restoration activity in general. Principle 3: *The size, scale, and amount of restoration activity must increase substantially to have a significant effect on overall estuarine functioning and health* (RAE-ERF 1999). Essentially this principle mandates making a transition away from "demonstration projects," such as those used by advocacy groups to promote volunteer and civic involvement, and toward larger scale projects with more efficient yield in acreage. This, of course, means collaboration between scientists and restoration practitioners, as well as investment in technologies that will make such larger yields possible. In Rhode Island, for example, this would imply the allocation of fewer resources for the labor intensive method of transplanting eelgrass remotely with frames (the TERF method) and, conversely, the allocation of more resources for the development and use of mechanical seeding devices such as those that use boat-pulled sleds to deposit gelatin-encased seedlings below the sediment surface (Rhode Island Coastal Resources Management Council 2003).

Thus far, the highlighted normative aspects of estuary restoration stand in some tension to each other. On the one hand we have an emphasis on economy of scale and more effective methods of restoration, which can mean an increased use of technology (Principle 3), yet on the other we have an inherent cautionary note about having too much faith in traditional engineering (Principle 11). Exacerbating this tension is the idea of moving away from demonstration projects (Principle 3), which normally exist for the benefit of public education and participation, yet in another principle we see a re-emphasis on that very public participation.

Principle 6: *Estuarine restoration plans should be developed through open regional processes that incorporate all key stakeholders and the best scientific thinking available* (RAE-ERF 1999). This principle dictates that restoration planning should involve as many stakeholders as possible in order to promote a sense of ownership in the plan, while also suggesting the establishment of core planning teams made up of scientists, agency representatives, policy makers, and representatives of organizations with an interest in the future of the estuary. Although worthy in intent and admirable in scope, this principle remains silent on the biggest obstacle to Principle 1, which mandates attention to preservation of existing habitat first, as well as on the maxim within Principle 11 that we must remove barriers to natural functioning. If we conceive of barriers as literal, physical impediments, then no real tension emerges among the Principles. But if by "barrier" we also mean the more subtle and sometimes insidious set of human practices and attitudes that contributes, individually and collectively, to the ill-health of the estuary—septic leaching, lawn fertilization, automobile emissions, dumping of toxins in sewers—then it is unclear how we might begin to define "key stakeholders" in the restoration planning process. Too often the "key stakeholders" are already among the converted —those who have a pre-existing commitment to preservation and restoration—whereas the more widespread but less identifiable "stakeholders" among all coastal residents remain uninformed, unconvinced, and ultimately unchanged in habit and attitude. Normative claims for action in the absence of an axiological context for their implementation will be futile. Fortunately, the RAE-ERF principles give explicit attention to some aspects of this axiological context. I examine four of their points of emphasis here.

---

[2] A case where this might apply in Rhode Island, for example, would be among the artificial breachways plowed from the ocean into five of the south coast's nine salt pond lagoons. Such breachways, constructed when the fragile ecology of the ponds was valued far less than their potential for marinas and waterfront dockage, are now being reconsidered in ecological engineering terms with respect to dredging, maintenance, and the possibility of allowing them to close off naturally to boat traffic. Many thanks to Michael Traber, marine research specialist at the URI Graduate School of Oceanography, for information on the latest developments concerning the lagoons.

Axiological Context

One principle especially gives voice to precisely those barriers I have indicated within the discussion of normative directives. Principle 4: *Greater public awareness, understanding, and involvement in estuarine habitat restoration are necessary to the success of individual projects and to achieve national restoration goals* (RAE-ERF 1999). In particular, successful restoration requires an informed public ready to support the policies, funding, and changes in lifestyle needed to restore and maintain healthy estuaries. Considering once again the 70% of residents around Narragansett Bay who do not know what eelgrass is, the barrier to restoration posed by public ignorance seems substantial, as does that implied by the requisite changes in lifestyle and public spending to meet the demands of Principle 1 for halting present habitat degradation. To address this, the authors of the Principles elaborate on their vision for a supportive public. Members of the public must be involved in *direct and meaningful* ways (emphasis mine) at all levels of the restoration process, and *connections* (emphasis mine) must be made between habitat restoration and other social and economic goals. Here we have insightful observations about what conditions will be necessary for restoration to take hold of public imagination in a manner that ensures its long-term success. At this point I maintain that citing these conditions— the axiological context for restoration—is far easier than actually identifying the means of developing methods for public involvement and social connections. What interests me in this chapter are the barriers to the development of those very methods, as well as how innovative program outreach might begin to erode the power of the barriers to block effective restoration.

The principles do attempt to take account of potential sources for outreach. Principle 5: *Restoration plans should be developed at the estuary and watershed levels to set a broad vision, articulate clear goals, and integrate an ecosystem perspective* (RAE-ERF 1999). More precisely, restoration plans must take account not only of scientific information but equally of factors such as cultural and aesthetic values and community and economic interests. These latter values and interests include consideration of public preferences, which can at times work precisely against the gist of scientific reasoning when not carefully integrated with a broader view of the individual's and community's best interests. The delicate question at the heart of this principle is thus: are we to measure public preference as it stands now, that is, under-informed about and unengaged with restoration? Or is the cultivation of public support through information, involvement, and outreach to precede the solicitation of public preference?

For the integrity of both scientific predictions about habitat recovery and allegiance to public preference and human needs, restoration success must be judged against a generous backdrop of time. Principle 2: *Estuaries can be restored only by using a long-term stewardship approach and developing the constituencies, policies and funding needed to support this* (RAE-ERF 1999). Although human requirements must be considered as well as those of other species, over time a balance must be struck between human and ecological needs. Furthermore, restoration advocates must take the long view in attending not only to the ecological dimensions of habitat recovery but also to the development of constituencies that view habitat recovery as a public priority. This does not happen overnight, nor do the cumulative effects of human actions and natural processes reveal themselves immediately. (Figure 6.4) The result of small, seemingly harmless decisions about wetland conversion, small lot development, water use and redirection, sewer extensions and septic system use are apparent in the environmental problems faced by communities today (Ernst, Lee and Desbonnet 1996). Tom Ardito of the Narragansett Bay Estuary Program adds, "It took several centuries of development to produce the coastal environment we know today, and it will require a major collective effort to reestablish even a fraction of the biological value that's been lost on Narragansett Bay (Ardito 1999b)." Echoing the earlier discussion of traditional engineering, David Orr reminds us, "even if humans were able to learn more rapidly, the application of fast knowledge generates complicated problems much faster than we can identify and respond to them. We simply cannot foresee all the ways complex natural systems will react to human-initiated changes, at their present scale, scope, and velocity" (Orr 2002).

FIGURE 6.4 Shore alterations, August 2003 fishkill, Greenwich Bay, Rhode Island. Photo by Thomas Ardito, *Narragansett Bay Journal*, www.nbep.org.

And why can we not see them, at least in the short term? One of the best hypotheses comes from cultural geographer Yi-Fu Tuan. "The lifestyle of a people is the sum of their economic, social and ultra-mundane activities ... Economic and social forces contribute overwhelmingly to the making of lifestyles, but unlike idealistic impulses they lack self-awareness." (Tuan 1990) Recalling Principle 4, which cautions restoration advocates about the need for public support concerning lifestyle changes among other things, Tuan suggests that lifestyles lack self-consciousness: it is only when we reflect on commonplace activities that their original intentional structures reemerge, he claims (Tuan 1977). Changing public perceptions about lifestyles that contribute to habitat degradation thus requires getting individuals to make a transition from unconscious lifestyle habits to conscious ideal choices. It requires reflection, both personal and public. Yet, how does this happen? Tuan, again, argues that meaningful attachment, the feel of a place, "is registered in one's muscles and bones .... Knowing a place clearly takes time" (Tuan 1977). This form of knowing, what philosophers refer to as knowledge by acquaintance (Russell 1997), is served by the last of the four axiological contexts I examine.

Principle 14: *Public access to restoration sites should be encouraged wherever appropriate, but designed to minimize impacts on the ecological functioning of the site.* Of particular note is the part of this principle that encourages hiking, wildlife observation, and school visits as ways to enhance the value people place on restoration sites. In Section IV I shall discuss the efforts of Save the Bay in Rhode Island to redesign their school-based restoration program in order to get children out onto the sites rather than merely studying eelgrass plants in their own classrooms. This accords with the data gathered by child psychologist M.H. Matthews, who notes not only that children tend to rate local water elements highly in their list of valued places (Matthews 1992) but also that giving children responsibility for some planning in real world settings can "set in motion a critical expectation on the part of children that they can exert control over their environment" (Matthews 1992).

Awareness on the part of the public may be, over time, an awareness of places and values instilled during the school years and instilled in a manner that gets children out of the classroom and onto restoration sites. This promotes not only direct knowledge of place but also the opportunity, through teacher-led discussion, for organized reflection and the development of long-term ideals. If restoration is to be more than an abstract policy concern driven by the latest findings of ecological and marine science, then its advocates must try to discern both the gradual development of lifestyle patterns at work in society and efficacious techniques for raising public awareness of their collective impact on estuarine health. Most of all, the public must be persuaded to care about the status of its estuaries and coastal environments such that new ideals in accord with the best scientific information become, over time, well-ingrained but sustainable lifestyle habits.

EPISTEMOLOGICAL UNCERTAINTY

Of course, it is simple to allude to "the best scientific information" casually without pausing to examine the history of scientific discovery itself, which reads much like a book in which each new chapter absorbs and then overthrows the one before it. Against the horizon of time and the specter of uncertainty, how can restoration advocates hope to persuade the public to care about estuaries when the scientific basis for understanding and restoring those systems is still very much under development? And when the development of good science itself takes time, impartiality, and some degree of intellectual freedom from short-term profit and productivity? Three principles, closely related, grapple with this question of ambiguity over time.

Principle 7: *Project goals should be clearly stated, site specific, measurable, and long-term—in many cases greater than twenty years* (RAE-ERF 1999). Stressing once more the necessity for a longer stretch of time over which to gauge the success of restoration efforts, this principle explicitly acknowledges the need for time in terms of uncertainty. Given the uncertainty involved in restoration, goals should be expressed in terms of a scientifically developed range of acceptable outcomes based on appropriate reference habitats. Project timeframes need to allow sufficient scope for natural processes to be reestablished and function restored. Interim goals should include making sure public interest is sustained along the way. But when the public cannot see or perceive change in a short period of time, what hope do restorationists have of sustaining its interest?

This principle dovetails with another that lays a similar stress on uncertainty. Principle 12: *Adaptive management should be employed at as many restored sites as possible, so they can continue to move toward desired endpoints and self-sustainability* (RAE-ERF 1999). Because estuaries are dynamic systems, restoration involves a great deal of uncertainty, and thus restoration activities must be evaluated and changed through adaptive management if they do not appear to be moving toward desired conditions. Here emerges the triad of temporal ambiguity in science. The setting of future goals relies on uncertain data about existing habitats from the past and requires flexibility in the deployment of present interventions. How are we to remain vigilant in cultivating present flexibility? And how do we encourage the patience necessary to examine the past with accuracy and care?

Principle 10: *Scientifically based monitoring is essential to the improvement of restoration techniques and over-all estuarine restoration* (RAE-ERF 1999). The weight within this principle is carried by insistence on a *well thought out, scientifically based monitoring* (emphasis mine) system. This may appear at first glance to be unproblematic, but in truth, any project necessitating deep and extended thought will also require time and, similarly, any project based on the best scientific information will require money and a tolerance for that pure inquiry upon which applied projects come to be based. But consider: We reside in a culture obsessed with speed and productive results, characterized by what philosopher of place Edward Casey calls *dromocentrism,* that is, an obsession with speeded up time. *Dromos* connotes running, race, racecourse; it is the essence of our era, "as if the acceleration discovered by Galileo to be inherent in falling bodies has come to pervade the earth (Casey 1998)." Furthermore, if restoration requires patient monitoring, then it also necessitates a revision of both methods and goals when original projections for success turn out to be mistaken; it requires a suppression of scientific ego. When we refuse to do this—question our goals and techniques in light of emergent information—we fall into what David Orr calls "fundamentalism": where well-educated people fail to ask hard questions about why we do what we do, how we do it, or how these things affect our long-term prospects (Orr 2002).

Indeed, it is Orr who articulates the intersection of the epistemological, axiological, and normative dimensions of restoration most succinctly and eloquently.

How do we quickly capture the imagination of the general public for the slow things that accrue to the health of the entire land mechanism?

It is far easier to describe the general content of such ideas than how they might become powerful in a consumer culture. In one way or another, the ideas we need would extend our sense of time to the far horizon, broaden our sense of kinship to include all life forms, and encourage an ethic of restraint. Not one of these can be hurried into existence. This is not first and foremost a research challenge as much as it is a kind of growing up. It is perhaps more like a remembering of what Erwin Chargaff once called "old and solid knowledge" that has existed in those times when foresight and compassion were cultivated .... But how is this to be made vivid for an entire culture suffering from attention deficit disorder (Orr 2002)?

Referring back to the Section I discussion of the Strategy for implementing the Estuary Restoration Act of 2000, we already see a disjuncture between the goal of 1 million acres restored by 2010, and the timeline of 20 years or more for assessing success put forward in the RAE-ERF principles. Could it be that the scientifically mindful principles take a more realistic view of time with respect to long-term restoration success, whereas the results-oriented Strategy couches its projects in terms of measurable outcomes and relatively quick returns for public investment?

If the latter is true, it must bear some relation to public expectations of prudent policy, which in turn reflects collective values and priorities concerning the function of representative government and the use of public monies. Once again, questions of epistemological uncertainty and normative directives devolve onto axiological contexts for public values and expectations. And these point back to that clouded arena of lifestyle habit and conscious ideal. Can estuarine restoration efforts—as a set of both principled ideals and practical programs for change—compete with the ingrained patterns of water and land use to which our society has become accustomed?

These are overwhelming questions, and sometimes in the face of such daunting and seemingly insurmountable barriers it is a relief to turn away from the more far-reaching philosophical issues beneath restoration advocacy and instead retreat to the concrete particular. The case of eelgrass restoration in Rhode Island offers fine terrain for reflection on the normative, axiological, and epistemological challenges that confront restorationists, for ignorance about eelgrass in the majority of the population stands in stark contrast to the extent and quality of scientific and policy expertise spread throughout the state. This and other paradoxes make the case a rich and intriguing one, one that ultimately grounds our exploration of philosophical speculation in the realm of civic discourse.

EELGRASS RESTORATION IN NARRAGANSETT BAY: PHILOSOPHICAL CHALLENGES WITHIN PRAGMATIC PROBLEMS

Consider the true strength and importance of eelgrass, not only in Narragansett Bay but in other aquatic systems of equal importance. Rachel Carlson, the great naturalist and documenter of unseen effects borne of anthropogenic causes, endows seagrass with both a lineage and an aesthetic presence as she muses on its place in the Florida Keys.

Dark patches like the shadows of clouds are scattered over the inshore shallows of the reef flats. Each is a dense growth of sea grass pushing up flat blades through the sand, forming a drowned island of shelter and security for many animals .... All belong to the highest group of plants—the seed plants—and so are different from algae or seaweeds. The algae are the earth's oldest plants, and they have always belonged to the sea or the fresh waters. But the seed plants originated on land only within the past 60 million years or so, and those now living in the sea are descended from ancestors who returned to it from the land—how or why it is hard to say. Now they live where the salt sea covers them and rises above them. They open their flowers under the water; their pollen is water-borne; their seeds mature and are carried away by the tide. Thrusting down their roots into the sand and the shifting coral debris, the sea grasses achieve a firmer attachment than the rootless algae do; where they grow thickly they help to secure the offshore sands against the currents, as on land where the dune grasses hold the dry sands against the winds (Carlson 1955).

Carlson's vision of a seagrass bed as a "drowned island of shelter and security for animals," as well as her emphasis on the "rootedness" of sea grass as opposed to algae, provides an almost pictorial account of how seagrass might be better understood by the population at large under the right conditions of communication, perhaps through underwater videography or satellite transmission of live dives involving eelgrass transplantation. After all, public imagination is not yet awakened to the place of eelgrass within marine communities. Many people think it is a nuisance for swimming or boats. Even scientifically literate observers of the sea do not always appreciate the role played by the unglamorous, submerged seagrasses. The award-winning Australian science writer Rosaleen Love, for example, provides a confessional account of how she came to be aware of seagrasses through the process of investigating fish and corals.

> Seagrasses are flowering plants that have evolved from the land to live in the ocean ... the roots and rhizomes of the seagrasses trap ... sediments and stabilize them. Seagrasses are like land grasses, absorbing nutrients from the mud; in summer or autumn they flower, and broadcast their pollen to the sea; they produce oxygen; they shelter juvenile fish and prawns, and fleas, lice, crabs, worms, in a grass cape of tiny marvels ....

> When I started researching reefscape, I began with thoughts of corals and fish. Only later did I come to appreciate the reef from the point of view of seagrass communities. For seagrasses in shallow coastal waters, the Great Barrier Reef is the Great Wall of China, holding back the forces of the world outside, while maintaining calm for the inhabitants within (Love 2000).

Love's eelgrass beds have a great reef to shelter them. The eelgrasses of Narragansett Bay have no such protector, which exacerbates the increasingly inhospitable environment in which eelgrass restoration struggles to take place in Rhode Island.

Five distinct yet interrelated problems of a concrete nature face restorationists who wish to see eelgrass flourish in the Narragansett Bay: public ignorance of and indifference to the effects its actions have on existing eelgrass populations and bay health; the degradation of water quality; increasing temperatures in the bay; a lack of specific historical data on Narragansett Bay; and state policy that is both less strict and less long-term than that enacted at the federal level (details related to each of these five pragmatic challenges, as well as a brief ecological portrait of eelgrass, can be found in an appendix to this chapter). Each of these problems, including the particular manifestations of them so far discussed, can be viewed more broadly under the rubric of classifications I have developed for grasping the range of epistemological, axiological, and normative challenges facing restorationists.

EPISTEMOLOGICAL CHALLENGES TO EELGRASS RESTORATION

Substantial *epistemological* challenges face restorationists of eelgrass in Narragansett Bay. Habitats from the past cannot be replicated due to both a lack of historical data and changes among benthic fauna within the bay, posing uncertainties about the efficaciousness of direct eelgrass transplantation in the current environment. Certain trends, such as temperature increases, are not entirely under human control and thus remain obscurely tied to human behaviors. Furthermore, it is difficult to decouple causal lines of nutrient loading when we do know them to be entirely human in origin— such as pollution—as they come from atmospheric, point, and nonpoint land sources. Overall, the real problem for scientists and managers is that we do not know how much nitrogen coastal marine ecosystems can assimilate before there are serious impacts to habitat and fishery resources (Ernst, Lee et al. 1996).

Faced with the systemic impacts of water quality issues, not everyone agrees that restoration constitutes a defensible expenditure of public monies to alleviate these. Some practitioners argue that restoration efforts should focus on water quality improvement exclusively, whereas others think that restoration can accelerate the pace of change when it takes place within the horizon of

FIGURE 6.5 Diver transplanting eelgrass shoots. Photo by Sue Tuxbury, Save the Bay.

simultaneous water quality improvements (Rhode Island Coastal Management Resource Council 2003). Although this is also a normative issue, involving questions of right action, it depends more fundamentally on epistemological questions about how and under what conditions eelgrass can flourish, given the current condition of the bay. Even when organizations can agree that restoration of eelgrass is a worthy project for the expenditure of public monies, decisions about the best methods of transplantation remain in dispute. Which is more effective? The cheaper, community-oriented but labor intensive TERF method (Transplanting Eelgrass Remotely with Frames), where shoots are temporarily tied by biodegradable papers onto weighted frames, lowered by trained community volunteers on boats, and then hand planted in appropriate settings underwater by qualified divers? (Figure 6.5) Or more costly technologies like that being developed by Steve Granger and Mike Traber at the University of Rhode Island, where a boat-pulled sled deposits eelgrass seeds encased in gelatin under the sediment, using a pump to deposit the seeds and a metal flange to cover up the furrows created by the pump (Rhode Island Coastal Management Resources Council 2003)?

Finally, how can "success" in transplantation be measured, when the current rate of site survival for eelgrass restoration hovers between 10–20% and when, again, there is little historical context by which to judge the overall survival of the plants? As Daniel Alongi notes, absolute criteria do not exist. Good preparation, a careful choice of a resilient species, and adequate anchoring and planting method techniques are crucial to achieve stability.[3] And the restoration of eelgrass is but the first step in marine ecosystem recovery; the mammals and other fauna must also return if restoration is to be considered successful (Alongi 1998).

Axiological Challenges to Eelgrass Restoration

These and other epistemological challenges would be daunting enough on their own, given the infancy of restoration science and the exacerbation of uncertainties in larger arenas such as climate change, but when placed against the back drop of the parallel *axiological* challenges facing restorationists, the situation with respect to restoring eelgrass in Narragansett Bay can begin

---

[3] Save the Bay has modified its restoration strategy in response to site success, concentrating its efforts on those sites where previous transplantation has achieved stability. Thus, while fewer sites are being targeted for restoration, more density of transplantation is taking place at those sites where successful rooting and colonization has been established. Many thanks to Bay Keeper John Torgan of Save the Bay for his insight into this issue.

to feel insurmountable. As stressed before now, public ignorance of and indifference to the environmental issues before us fuels the problems occurring in Narragansett Bay. David Orr calls this ignorance or indifference "ecological denial" and has speculated that it is manifest "when large and messy questions about the partisan politics of environmental issues are ignored (Orr 2002)." Included among such reasons are a gap between the powerful rich and everyone else, where scapegoats become the misplaced focus of public ire rather than the true, often corporate forces behind environmental pollution; a defense against anxiety in the face of a truly intimidating problem that could necessitate uncomfortable lifestyle changes; an under-informed populace to whom denial seems a plausible position when faced with hard questions about everyday personal practices and beliefs; and finally, an elite strain in environmentalism that ignores class issues while emphasizing a romantic view of pristine nature.

Because the great majority of people living around Narragansett Bay does not think about eelgrass given its ignorance of eelgrass's existence, function, and importance, they can hardly be expected to respond to isolated, if high-minded, pleas for its preservation and restoration, or to pressure their elected representatives to address the restoration issue legislatively or fiscally.[4] In this sense environmentalists, and often proponents of restoration, sometimes inhabit their own entrenched worlds and remain out of touch with how the general populace views environmental matters. Yi-Fu Tuan believes that all individuals lose sight of the consequences of their actions when those actions become fully entrenched as habit. "Subjectively, even complicated repetitive movements turn into habit; their original intentional structure—envisaging ends and the means to achieve them—is lost. It is only when we reflect on our commonplace activities that their original intentional structures reemerge (Tuan 1977)." Thus, habituation of both viewpoint and consequent action blocks dialogue on eelgrass restoration from both sides of the fence, whether in favor of or indifferent to its merits.

Beset as we are by information overload and what Orr calls "cultural ADD" (i.e., attention deficit disorder), we may be unable, collectively, to distinguish socially important priorities from trivial ones because we are exhausted by consumption and saturated with entertainment (Orr 2002). This is a significant obstacle for those promoting the restoration of eelgrass, for the problem is not so simple as "getting the facts out" to a sympathetic but as yet uninformed public. Rather, public ignorance and indifference are anchored by well-established patterns of daily life in a consumer society—patterns that are implicitly valued by that society even as they contribute directly to the pollution infecting the bay. David Feldman believes that "our most severe water resource problems are caused by a reliance on narrow and often inappropriate acquisitive values that are harmful to nature and to the satisfaction of a wide range of human needs, including biological exigency and living in harmony with nature and in community with other people (Feldman 1991)." Only when a catastrophe of great magnitude occurs—such as the catastrophic fish kill that took place in Rhode Island's Greenwich Bay within Narragansett Bay—can the public be expected to have environmental priorities on its radar screen in an acute manner.

Events such the major anoxia of August 20, 2003, come as a shock to the public: coastal zone resources were once considered to be unlimited and thus there was little, if any, environmental concern for their use (Vernberg and Vernberg 2001; Orr, 2002). Today's resource use patterns might be interpreted as a legacy of such an attitude. In order to promote restoration of eelgrass and voluntary changes in water quality management tied to eelgrass survival, restoration advocates must work at the level of attitudinal change so that cultural assumptions, habits, and values do not clash so forcefully with environmental prudence. Consider but a few of the culprits: cars and factories that

---

[4] Orr warns against presuming that public involvement with environmental matters will occur naturally once the facts are understood, especially as a propensity for unquestioningly following the latest trends—fast knowledge—occludes the public's capacity for reflection prior to action. "Professionalized knowledge is increasingly isolated from the needs of real people and, to that extent, dangerous to our larger prospects. It makes no sense to rail about participation in political and social affairs of the community and nation while allowing the purveyors of fast knowledge to determine the actual conditions in which we live without so much as a whimper (Orr 2002, 41).

produce polluting emissions; lower taxes through allowing the continuance of leeching, cheaper septic systems; golf courses alongside the coast; lawns fertilized in great concentration in suburban coastal communities; stresses on wastewater treatment facilities due to residential and commercial development; and regulations that allow personal docks, dredged inlets to salt ponds and unrestricted motor boat access to inlets. If eelgrass restoration is to capture the public imagination, and consequently gain the support of key legislators who are beholden to their constituents' preferences for a clean bay, it must be linked to things society cares more about than those activities and institutions that contribute to pollution of the bay.

NORMATIVE CHALLENGES TO EELGRASS RESTORATION

This leads to the challenge posed by *normative* issues within eelgrass restoration: What should be done in the face of epistemological uncertainty and axiological impediments to eelgrass restoration? What steps should be taken in relation to restorative activities of benefit to the bay? As discussed in Sections I and II, the Estuary Habitat Restoration Council (established within the Estuary Restoration Act of 2000) and the Principles of Estuarine Habitat Restoration (generated jointly by Restore America's Estuaries and the Estuary Research Foundation in 1999) provide mandates for general as well as specific actions with respect to estuarine restoration. Despite these federal and scientific commitments to restoration, however, supporters of restoration around Narragansett Bay must contend with the less stringent regulations overseen by the Coastal Management Resources Program and a lack of support from the state legislature in securing dedicated funds for widespread eelgrass restoration. Save the Bay has thus far been unsuccessful in its bid to bring state standards into line with federal ones, but continues to work toward harmony among policies and initiatives at all levels. Beneath their efforts is an implicit commitment to the rightness of such harmony and thus to the rightness of the more ambitious restoration program articulated beyond the state level.[5]

As explored in Sections I and II, federal and national guidelines for restoration and estuarine management place heavy emphasis on the cultivation of stewardship as a mechanism by which a greater portion of the community might come to identify themselves as stakeholders in environmental decision making. In this sense stewardship sometimes appears to be a panacea for environmental ills: if more folks get involved, they'll see the virtue of preserving and restoring degraded estuaries. On the one hand such a view seems axiologically and epistemologically naïve, as it appears to presume the efficacy of specific processes, like stakeholder meetings and community gatherings, in promoting a sense of stewardship. On the other hand there can be little doubt that, without more widespread effective investment in what happens to our estuaries, eelgrass restoration efforts will fail to counteract the negative impacts of escalating development (the population is moving to the coastal zone) and water quality degradation that comes with such development. Thus, another normative aspect of the challenge facing restorationists in Narragansett Bay emerges through questions about increasing stewardship among citizens living around the bay.

Some projects have in fact made inroads in the direction of cultivating deeper engagement with restoration on the part of the public around Narragansett Bay. On Friday, December 11, 1998, nearly 100 people attended an open-invitation workshop (called a "charrette" after design-oriented planning sessions) sponsored by the Narragansett Bay Estuary Program (of the department of Environmental Resources) to discuss coastal habitat restoration in the bay. The results of this meeting included increased coordination among restoration and advocacy organizations and the establishment of a "Rhode Island Coastal Habitat Restoration Team," which brought together leading constituent groups of both local and national affiliation in an effort to bring restoration issues to the forefront of

---

[5] For a cautionary tale exploring unforeseen outcomes when state and federal regulations clash, see Foster, Cheryl and Greenwood, Linda Nightingale. 1996. When State and Federal Obligations Conflict. *In Ethical Dilemmas in Public Administration*, edited by Lynn Pasquerella, Alfred Killilea and Michael Vocino, pp. 87–122. Westport, Connecticut: Praeger Publishers. At this writing some hope for more consistency between federal and Rhode Island standards exists in the emergent work of the Coordination Team for Rhode Island's Bays, Rivers, and Watersheds.

public and political opinion. Among their achievements has been the creation of an impressive "Habitat Restoration Portal," a well-designed Website (www.csc.noaa.gov/lcr/rhodeisland/html) devoted to every aspect of estuarine restoration including data and information dissemination for the public and for state and federal agencies; project development; education; and assistance with state planning for restoration.[6] Objectives include the incorporation of the previously discussed RAE/ERF Restoration Principles into habitat restoration activities in Rhode Island. Thus, one very accessible and firmly comprehensive resource for public information has been created and maintained. The question then becomes, what would make members of the public or, more suitably, key legislators, consult this resource in the first place? As Lester Milbrath reminds us in his practical handbook *Learning to Think Environmentally While There Is Still Time*, "we can't depend on technology alone to resolve the environmental problems that arise from the way we conduct our affairs (Milbrath 1996)."

Indeed not. In order for the populace to care deeply about the future of the bay, a sense of firm attachment to the unique place that is Narragansett Bay needs to exist. How does such attachment evolve? Again, Yi-Fu Tuan:

> The feel of a place is registered in one's muscles and bones. A sailor has a recognizable style of walking because his posture is adapted to the plunging deck of a boat in high sea. Likewise, though less visibly, a peasant who lives in a mountain village may develop a different set of muscles and perhaps a slightly different manner of walking from a plainsman who has never climbed. Knowing a place, in the above senses, clearly takes time. It is a subconscious kind of knowing. In time we become familiar with a place, which means that more and more of it can be taken for granted…
>
> Attachment, whether to a person or a locality, is seldom acquired in passing (Tuan 1977).

It may well be that the residents around Narragansett Bay do feel attached to the bay but in a manner that has become habituated and stale; only when an emergency occurs does the public sit up and take stock of what the bay, as well as its sustained health, means to them individually and collectively. Save the Bay has devoted substantial human and financial resources to exploring the factors beyond technology that are required in solving some of the environmental problems in Narragansett Bay. Activities of the organization break down along three lines: advocacy, education, and restoration. One of the activities sponsored by Save the Bay, "Eelgrasses in Classes," involved eelgrass seeding and nurturing undertaken by schoolchildren around the state of Rhode Island. The trajectory of this project and its demise, as well as Save the Bay's resilience in reallocating its energies in more productive directions once the fate of the project became clear, are rich in their implications for the normative and axiological backdrop against which the epistemological challenges explored thus far are placed. What ought to be done with respect to eelgrass restoration is, in part, a function of what has not worked in the past and also of what happens by chance beyond good plans and policies. The following sections discuss the Save the Bay school project (IV) and the more tangled relationships subsisting among science, policy, and value as these inform and inspire action related to eelgrass restoration (V).

SEAGRASSES IN CLASSES: THE CHALLENGE OF AFFECTIVE PUBLIC ENGAGEMENT

As Tuan remarks, affective attachment to place tends to occur at the level of personal involvement over an extended period of time. This has been corroborated by social scientists whose work explores the benefits of direct community involvement with environmental movements. In a paper on the psychological benefits of volunteering in stewardship programs, Robert Grese and his colleagues

---

[6] The site is funded by the National Oceanic and Atmospheric Administration's Coastal Services Center and is supported through the coordination of the Rhode Island Coastal Resources Management Council, the Rhode Island Department of Environmental Management Narragansett Bay Estuary Program, and Save the Bay.

found that a factor analysis of participant attitudes about volunteering led to four general categories of value, with "protecting natural places from disappearing" as that category receiving the highest rating overall (Grese et al. 2000). Similar results accrue from studies focusing specifically on coastal matters. Public education and involvement, especially voluntarism, were found to be crucial factors in the success or failure of Coastal America's efforts to promote positive environmental change among watershed communities (Klesch, et al. 1996). Caroline Davis of the U.K.'s Centre for Coastal Zone Management has similarly observed that estuary management plans drawn up by experts "will be inefficacious without local knowledge and expertise," especially as this pertains to potential conflicts in plan implementation (Davis 1996). Indeed it has been argued that reaching consensus on water quality issues in rural communities where management agencies have trouble reaching small farmers, for example, depends strongly on the commitment of local citizens who undertake the instruction of their peers, thus gaining trust and insuring an ongoing geographical (and community) commitment to water quality improvement (Garitone 1996).

Rhode Island's Save the Bay has long been involved with the cultivation of stewardship through voluntarism but has not been complacent about the manner by which such stewardship takes root. In essence, the organization directs its energies in three distinct yet interrelated directions: advocacy, restoration, and education. The interrelationship becomes clear when one considers the overlap among these directions. Advocacy efforts involve working through civic bodies to promote legislative and policy changes as they concern the Narragansett Bay, but often the lobbying and communications tactics necessary for persuasion involve a form of education regarding the scientific facts of water quality degradation. And while over 100 volunteers annually participate in eelgrass transplantation under the restoration banner, Save the Bay reaches more than 30,000 schoolchildren per year through its "Explore the Bay" educational programs and other public initiatives such as those aboard its research vessel, the *M/V Aletta Morris*.

According to Narragansett Baykeeper John Torgan, Save the Bay continues to refine its efforts as various tactics for change prove successful or not. In terms of advocacy, again, the organization devotes less energy to cultivating broad public attachment to issues up for legislative consideration than it does to working with lawmakers who can influence the fate of policies that will have a tangible influence on environmental matters, matters such as dock construction in sensitive areas or the allocation of dedicated funds for restoration. Yet, Save the Bay does not disregard the need for a longer term strategy wherein the constituents represented by lawmakers are also cultivated, albeit at a slower pace and with an eye towards val ue inculcation. After all, research reports, policy analyses, and open meetings do not necessarily result in legislative changes: politics and the merits of change do not always or perhaps even often coincide (Torgan, personal communication, 2003). In order for scientific merit and politics to find common ground beneath the achievement of environmental goals, lawmakers who represent the citizenry must have an incentive to promote particular legislative agendas. One such incentive can arise from public demand or support for particular modes of change.

To this end, Save the Bay's Education programs target all areas of the populace but are especially linked to school-aged children, which makes sense in light of two things: one, the fact that it takes at least 30 years for changes in nutrient cycling to show up in water quality assessments, so emphasizing issues now may pay off later; and two, research that demonstrates the efficacy of developing affective attachment to the environment in childhood. Young children are particularly susceptible to socially communicated values and are likely to revisit and retain these beyond the experiences that first introduced them (Matthews 1992). Furthermore, in studies asking children to rate environmental features held in high esteem, "most children rated local water elements … very highly (Matthews 1992)." In fact, environmental transactions that emphasize a relationship to environmental features enable children to perceive the environment more acutely, a process instrumental in place-learning and attachment (Matthews 1992). In light of these and other considerations, it makes sense that Save the Bay would invest resources in educating children about eelgrass restoration. This is precisely what occurred during a period between 1998–2002 when the organization sponsored a specific children's program, a program conceived not under the rubric of its education wing but rather under that of restoration.

The restoration division of Save the Bay targets three areas for scientific research: salt marshes, fish runs, and eelgrass. Under the supervision of former Save the Bay Restoration Ecologist Wendy Norden, the organization developed a classroom initiative called "Seagrasses in Classes," whereby selected groups in targeted "agricultural schools" in Rhode Island (i.e., schools with traditional ties to land-based livelihoods and commerce) would raise eelgrass shoots from seed in tanks provided for that purpose. On paper this seemed like an original idea for fostering affective attachment to eelgrass, which might otherwise be ignored or noted only for its bothersome tendency to tickle one's feet when swimming. Children would plant the seeds and then nurture them as they matured into shoots, at which point the shoots themselves would be tethered to frames and transplanted, along with others from different schools, to a selected transplant site in the bay. The physical cultivation of shoots would be complemented by lessons about eelgrass and its role in marine ecosystems such as those in the Narragansett Bay. By taking a hand in growing the grass itself, children would presumably develop a more active and, hopefully, long-term interest in the health of the bay, as indicated by eelgrass restoration. At the very least these children would come to know what eelgrass is, unlike the majority of their parents and peers.

Despite the theoretical beauty of the plan, "Seagrasses in Classes" failed to achieve its primary mission, which was to ignite the children's interest in eelgrass through direct participation in cultivating the shoots. While the emphasis on direct participation was not misplaced, the method by which it was carried out proved to be a disaster for young children. Eelgrass develops slowly from seedling to shoot, and the young restorationists soon wearied of waiting for it to sprout and then grow. Lessons attached to the process thus far outpaced the rate of plant emergence, causing disconnection between the process of plant observation and the place of that plant in the greater scheme of things. And children at this age are naturally temperamental: catching and holding their interest turned out to be far more challenging that the designers of the program had imagined.

Save the Bay was attempting, through "Seagrasses in Classes," to combat the more conventional approach to teaching ecology, where children are dropped into the deep end of complexity through abstract discussion of large, multilayered systems, such as Narragansett Bay in general. Ecologists knew that this conventional approach to galvanizing the scientific imagination does not often work for young children. As J.L. Harper remarks in his study *The Heuristic Value of Ecological Restoration*, system study may be a good way to introduce children to ecology, but "the next stage in teaching ecology *as a science* needs to be the study of the very simplest ecological systems that can be assembled (a population of one species of grass in a tray of soil, for example, or a simple species of alga in a culture solution). This is rarely done (Harper 1987)." Of course, studying a single species of grass—eelgrass—in a tank was *exactly* what Save the Bay was attempting to do through its commitment to "Seagrasses in Classes." So why didn't it work? Why did the children get bored? Why did instructors find it difficult to draw them into ongoing discussion of marine ecosystems? Perhaps one reason was this: the tiny seedling before them grew at imperceptible rates, making it impossible for children to imagine their way into the role played by these seedlings in the marine habitats they heard about more abstractly. The children had no tactile or environmental tether—no sensible reference point—for a connection between the grass in the tank and the wider submerged community of which it would be a part. One might even say that the grasses in tanks did not contribute to a story that the children could grasp and value.

M.H. Matthews has addressed this issue, as well, in surveying studies that support the direct involvement of children in tending to specific environments. "At the age of 8 or 9 years, children could be given responsibility for planning real-world settings …. Inclusion of this type of training in the school curriculum would set in motion a critical expectation on the part of children that they can exert control over their environment (Matthews 1992)." This is similar to what Norden and others at Save the Bay concluded when they decided to terminate the "Eelgrasses in Classes" program. They opted to design a revised initiative that would address the need for children to experience environmental control through direct exposure to ongoing ecological restoration procedures. In the newer version, groups of schoolchildren would help to select transplantation sites and would identify one as "their group's" site. Following this the children would observe actual transplant activities like tying shoots to frames and the lowering of frames onto their selected sites, and would even at times be supervised in carrying out techniques involved in transplantation itself. Finally, experienced children and instructors would mentor newer groups as they, too, joined the process, resulting not only in the peer education shown to be effective in the promotion of environmental change (Garitone 1996) but also in a widening circle of engaged, informed young citizens who would, in the future, form part of public opinion about matters pertaining to the Bay.

Once more, the theory looked promising on paper. But something else was occurring in a parallel restoration universe at Save the Bay. Over a three-year period, restoration ecologists were refining their strategy for the most successful approaches to eelgrass restoration in Narragansett Bay. In year one, 10,000 plants were transferred to a total of 30 sites around the Bay: only three of the sites showed promising results with respect to rooting, growth, and colonization following the initial transplants. Thus, in year two 20,000 plants were transplanted but only to the three sites that had proven amenable to restoration—almost certainly due to a combination of current water quality, wave oscillation and temperature. In year three, now approaching, 30,000–40,000 plants will be transplanted onto the habitable sites with the goal of establishing *not* an increased number of transplant sites in total but rather the *overall acreage* of successful eelgrass colonization.

A restoration education program for children that made site selection integral to the process of affective attachment thus flew in the face of the scientific strategy within which Save the Bay was conducting its broader eelgrass restoration program. This conflict was occurring *within* the restoration component of Save the Bay, between hard science programs (epistemologically driven) and school outreach efforts (axiologically driven). Both sides of the conflict had organizational "policies" in place that supported their missions. How was Save the Bay to interpret and solve this restoration conflict? One way to do so would be to live within a half-truth of student participation, where the alleged transplant sites selected by children would not truly be functional restoration sites but rather models for actual transplantations elsewhere. This, however, undermines the concept of environmental control that seems to be central to the development of genuine and lasting place-values inculcated in childhood. Maintaining a modeling program for student-selected sites might promote community spirit and a sense of group involvement but it would ultimately make a mockery of the very real engagement needed to foster a long-term sense of connection to and control over human interactions with the bay.

Therefore, another way to interpret the clash between the school restoration program (axiology) and the scientific restoration efforts (epistemology) would be to reconceptualize the relationship between the broader educational mission of Save the Bay and the scientific practice of restoration, forming as these do separate aspects of the organization's triadic mission (advocacy, education, and restoration). In other words, Save the Bay could realign its normative stance with respect to how its programs *should* operate, given scientific currency and axiological values. After careful deliberation Save the Bay chose this latter option over the one that would sustain the revised legacy of "Eelgrasses in Classes." This has resulted in an exciting, if subtle, series of changes with respect to the organization's three-part mission. Restoration is now exclusively focused on its mature scientific activities and research, with increased funds coming not from state appropriations but from federal grants and subsidies. Advocacy and Education, in response to the cancellation of the restoration outreach program focused on schools, have beefed up the eelgrass restoration components of their respective programs already in place. Voluntarism continues to form part of Restoration projects concerned with eelgrass transplantation but is highly focused on the labor intensive, practical work of transplanting eelgrass remotely with frames. By contrast, additional resources are allotted for the eelgrass restoration aspects of existing children's programs within Explore the Bay, while restoration has taken on a more marked, and indeed unexpected, role in efforts to advocate for policy change regarding the bay.

The August 2003 fish kill resulted in greater public awareness of potential hazards to sustainable activity in the bay. Images of dead menhaden floating atop the water dominated the front pages of the local press and news stories on television outlets. These came on the heels of numerous Narragansett Bay beach closings throughout the entire summer, some due to shark sightings in the increasingly warm waters but many more connected to inferior water quality (i.e., untreated fecal matter) following sewage runoff after a series of heavy rains. Save the Bay drew on these events in its fall round of newsletters and promotional mailings. It is too soon to tell whether or not the Governor's Coordination Team for Rhode Island's Bays, Rivers, and Watersheds will exercise strong influence on policies and programs that will change the overall pattern of human activity in relation to the bay. Certainly, the catalyst behind the formation of that Team—the horrifying summer of dead fish and feces-infected beaches—galvanized public awareness more dramatically than tempered and patient appeals to science, reason, and prudence had done. When public opinion is galvanized, public officials sit up and take notice. The overall challenge currently facing Save the Bay, as well as restorationists in general, is then this: how to capitalize on recent disasters in promoting the cause of eelgrass restoration in Narragansett Bay? What ought to be done in the wake of public outcry over suffocated inlets and beaches tainted with sewage?

FACT, VALUE, AND ACTION: AFFECTIVE ENGAGEMENT AND CULTURAL NORMS

In approaching this final section of a lengthy and detailed analysis of eelgrass restoration, I have come to conclude that the philosophical dimensions of eelgrass restoration inform and complement the more pragmatic, concrete challenges facing restorationists. Others have also detected a triadic relationship among science and policy, and policy and value—what I call epistemological, axiological, and normative aspects of deliberation. Consider in this spirit the following observation from David Feldman.

> Our understanding of nature and ourselves has simply not kept pace with our knowledge of science and technology. We know how to manipulate some things in this world, but we have had an extraordinarily difficult time deciding why we should manipulate them and understanding why they often resist manipulation (Bennett, 1987: 45–83). The political realm has often deferred to "experts" the major value decisions pertaining to the exploitation of the environment, because it has not been possible to reach political consensus [Feldman 1991].

Or reflect upon this remark, borrowed from Yi-Fu Tuan.

> In the modern world the linking of discrete phenomena by feeling remains strong. Scientists will associate autumn and the setting sun with melancholy, spring with hope, in their offguard moments.

> A symbol is a repository of meanings. Meanings arise out of the more profound experiences that have accumulated through time. Profound experiences often have a sacred, other-worldly character even though they may be rooted in human biology. Insofar as symbols depend on unique events they must differ from individual to individual and from culture to culture. Insofar as they originate in experiences shared by the bulk of mankind they have a worldwide character (Tuan 1977).

Referring to Feldman we can extrapolate thus: What values and priorities drive our personal behaviors, communal expectations, and public commitments? I have attempted to consider some of these through contemplating the axiological challenges facing restorationists, where accepted norms, expressed in cultural practices, contribute directly to water quality degradation in Narragansett Bay. Referring to Tuan, we can go further in this way: To what extent has the bay, or have some features of the bay, entered the symbolic consciousness of those who live around and near it? I have also tried to suggest that direct, affective experiences of bay life contribute to the formation of values that might drive personal and public commitment more energetically if advocacy groups and policy makers could harness those values concretely. Often, people do not grasp the extent of their attachment to a particular environment until disaster strikes and threatens to compromise the relationship. This is what happened in the wake of the August 2003 fish kill in Narragansett Bay. Although most individuals still do not know what eelgrass is, they certainly have developed a heightened sense of what can happen when eelgrass or oxygenation disappear from a place they value. Visual images and media coverage have been instrumental in nurturing this sense of potential loss.

Taking these things together—the split between fact and value, and the need for symbols in the creation of meaningful experience—consider once more those prophetic comments by Rhode Island Governor Donald Carcieri after participating as a volunteer in an eelgrass transplantation exercise during June 2003, two months before the fish kill that ignited public horror over pollution in the bay. "Narragansett Bay is in my blood, and I feel strongly about it," Carcieri said. "I believe we have to do everything we can to preserve and protect the bay." Accompanying the comments was a photograph of the governor helping to lower an eelgrass frame onto a transplantation site, along with additional comments indicating that the Governor recalled "scalloping in the waters of Rhode Island when he was a child—but it's an industry that no longer exists today, due in part to the loss of eelgrass in Narragansett Bay (Emlock 2003)."

Was the governor's voluntarism in support of eelgrass restoration a publicity feature, making the front page of the state paper? Well, sure. Did it draw attention to an environmentally-oriented agenda (something another Republican from Rhode Island, former Senator Lincoln Chafee, did in the wake of his father, the late senator John Chafee)? Of course. But most crucially, did the governor understand that his participation in eelgrass restoration, and his memories of the Bay from childhood, might function as a symbol of civic commitment to workable environmental policies for the bay? Did more folks learn about eelgrass as a result of seeing the governor out there lowering it from a boat, than they would have in numerous articles and reports published in earnest by informed environmentalists?

Media attention given over to the governor's day of voluntarism and his affective attachment to the cause behind it indicates that something as unknown or unvalued as eelgrass can be brought successfully into the public eye under the right axiological conditions. The emphasis on axiology cannot be too strong here: Eelgrass is not conventionally picturesque, nor is it an aspect of that marine environment likely to capture the imagination of the general public without a good deal of encouragement. Yet, attitudes can and do change. "Before being aware of their large-scale environmental and economic benefits, society in general thought that wetlands were at best useless and at worst a dreadful, smelly, disease-ridden hellhole that should be "reclaimed" by draining and filling to create usable uplands. In recent years, laws have curtailed this destructive practice and interest in restoration and creation of wetlands has grown (Vernberg and Vernberg 2001)." Even the nomenclature shifts with value: we hear less of swamps today than we do of wetlands, and the connotative shift is obvious.

Shifts in perception, value, and nomenclature come about through complex and often subtle means, but philosophically it is useful to consider them within a framework put forward by the philosopher Martin Heidegger. Two aspects of that framework have some relevance here. One, Heidegger maintains that the kind of being possessed by humans is characterized by a temporal structure, where what one has been and what one might be are mediated by a present immersion in *discourse*. Discourse is not merely verbal but coveys a much broader concept that means something like this. A person is not merely part of a community of other persons with similar metaphysical structures but rather is actually constituted in part by his or her capacity for interaction with others. What one is can be said to be, in part, a function of the others with whom one engages in discourse. Discourse can be inauthentic or authentic. Inauthentic discourse is characterized primarily by "idle chatter," where there is no real attention to either the matter at hand or reflection on the significance of that matter for the future. Authentic discourse, by contrast, means that one engages in solicitous listening as well as speaking and where one recognizes the importance of the present moment for shaping the path of the future.

Earlier in this chapter I cited Tuan, who makes a distinction between unconscious, ultramundane lifestyle habits (Tuan 1977) and conscious, reflective values where the intentional structures beneath our values and choices come into view (Tuan 1977). His distinction has some kinship with Heidegger's account of discourse, which is either inattentive and unreflective, or attentive and reflective. Around Narragansett Bay, ultramundane habits involving practices that pollute the ecosystem can be thrown into relief by events or images that shake up our sense of complacency and lead us to question how and why certain things happen. The governor volunteering for an eelgrass restoration exercise can thus prompt those examining the related images and stories to *wonder* about the nature and cause of his activity. This would not be so effective if the governor were not the governor, but his symbolic presence as the elected head of state in Rhode Island has a powerful impact on how the public eye gets directed towards specific issues and concerns.

An even more powerful lure for a shift in attention—and in discourse—occurred when images of the massive Narragansett Bay fish kill entered public consciousness. This leads to the second aspect of Heidegger that has relevance here: being made aware of one's present possibilities and the relation of these to the future. For Heidegger, an inauthentic mode of being entails an unreflective path from past to future, where what has been given to one already in a world is unquestioningly assumed in formations of the future. In the present, where change can be effected, idle chatter and unreflective discourse mask the possibility of forming a future not causally chained to a given past. As Eliza Steelwater has put it, "Our everyday being has the characteristic of unselfconscious involvement in a life composed of tasks (Steelwater 1997)." Whereas Heidegger focuses on the structure of being in general, it is possible too to see how, collectively, a society becomes entrenched in its own habits and presumptions about lifestyle: our attachment to cheap sewage treatment, lawn fertilizer, and recreational motorboats occludes awareness of how these contribute directly to the destruction of the marine environment.

Yet, when a "shock" in the present moment occurs—a shock like the near-apocalyptic fish kill—it has the power to impel reflection, or again, wonder, as to how things got to be the way they are. Heidegger links shocking moments such as this to an awareness, in the individual being, of one's own death and the possibility of its occurring at any time. Although it is not necessary to unwrap Heidegger's metaphysics here, for our purposes we can see a more reflective and solicitous attitude toward the natural environment emerging at moments when the possibility of non-possibility—death, of ourselves or more abstractly the environment of which we are a part—throws itself into our path.

Linking this to the more concrete challenges faced by restorationists on Narragansett Bay, we can ask how advocacy groups like Save the Bay might harness unexpected opportunities for discourse in raising public consciousness about water quality and eelgrass transplantation. One could argue that sustaining community participation in eelgrass transplantation exercises by the TERF method, despite the emergence of more efficient technologies using sleds and gel-encased seedlings, or even the placement of artificial grass beds for spawning and nursery purposes, promotes a level of individuated, embodied awareness of the marine environment that cannot be achieved at a distance. Despite the fact that only about 100 volunteers per year actually engage in these exercises, the program using volunteers remains important as a vehicle for establishing a nucleus of committed and concerned citizens willing to act on behalf of the bay. On one level engaging in eelgrass restoration provides that kind of "knowing in one's bones" that comes with direct bodily experience in an environment. Save the Bay sponsors myriad educational programs that get people out onto the water and into direct contact with the marine environment, and maintains its commitment to involving schoolchildren, tomorrow's decisions makers, in activities that foster immediate awareness of marine experience (and, hopefully, nostalgia for that in the future, a la Governor Carcieri!). At another level, however, engagement need not always be direct for everyone. Much can be done in utilizing the mechanisms of participatory democracy once a small group of citizens takes it into their heads to work for policy change. Save the Bay began in just such a fashion and the organization has not lost sight of political advocacy as a central component of its mission.

Baykeeper John Torgan notes that the organization feels it is less important for a majority of the general public to know what eelgrass is or why it is important than for elected representatives to demonstrate a commitment to restoration and water quality. Nevertheless, representatives do not act with decisiveness unless there is a perceived will among their constituents. An ignorant or indifferent populace will not give an impression of such will, so we are still left with the problem of how to create or inspire such will into being among at least some members of the general population—members who do not currently consider restoration issues to be a high civic priority—in order to have a critical mass of persons committed to a cleaner bay. Again I would argue: There are occasional shifts in public awareness, windows of opportunity for advocacy groups and politicians to sustain public interest beyond the moment of disaster that prompted such awareness in the first place. Advocacy groups, informed scientists and policy makers, and committed citizens ought to be vigilant in taking advantage of these opportunities, in keeping the window between past action and future possibility open for change.

Although we would not choose to replicate the stench of sulfur dioxide that permeated the area surrounding Cedar Point, Rhode Island, after the August 2003 fish kill, images reminding us of millions of dead fish on the bay can function as symbols of one possible future for a society whose habits continue to load nutrients into the marine ecosystem. Scientists can cite facts and figures in support of that possible future but, without harnessing the affective concern of at least some as-yet uninformed members of the general population, scientists' efforts will soon fade from public memory and cease to exercise an influence on what the citizenry demands from its elected officials. In her presentation within "Online, Byline, or Redefine—How to Communicate with new Coastal Audiences," Charlene McClellan urges advocates to research how management information systems tools can be optimized in getting educational messages out to relevant audiences. Besides knowing one's own tools and the particulars of one's intended audience, "it is incumbent upon communications professionals to assess how audiences best receive information, and to match the message to be transmitted with what audiences will respond to and actually use (McLellan 1996)."

In other words, advocates of restoration must do far more than keep on top of the epistemological challenges within restoration science. They must do far more than band together to issue normative principles of restoration or Congressional mandates about public participation. No, on top of these necessary tasks advocates must also adopt a less naïve view about the nature and character of public awareness, a view that takes into account a sophisticated understanding of human values and how those get called into question. More practically, they must work to grasp the intricacies of axiological communication such that people pay attention to messages of epistemological complexity. Sometimes the way to communicate about complexity is not itself complex: One image of millions of dead fish, or of frustrated beachgoers, or of effluent floating near one's dock, might be more powerful in gaining public attention than scores of newsletters aimed at the converted.

In Rhode Island, perhaps, some new converts have been made. The fish kill disaster of August 2003 has appeared at this writing to be exerting some influence on policy considerations at the state level. Most encouraging is an emergent cooperation among the executive and legislative branches, nonprofit advocacy groups, institutions of higher education and agencies that manage environmental resources. Central to the success of any efforts to make restoration a public priority, however, will be an enlightened and determined effort to address those cultural values that perpetuate the problems causing pollution. It's not that people must come to care about their relationship to the Narragansett Bay, for many already do, implicitly and nonpolitically. Instead, people must be made aware that they do indeed care about that relationship, enough to consider what will happen if the possibility of maintaining it erodes as pollution exacerbates existing problems.

Symbols and images that portray the rich variety of relationships human beings have with their local marine ecosystem can do a great deal to encourage reflective discourse about environmental and civic priorities. Knowing how to use such symbols and images, as well as when to unleash them, is a difficult matter—one whose expertise requires the intervention of humanists and communications experts alongside the inputs of scientists and policy makers. And a difficult matter it is, to cut across professional and disciplinary discourses, but one that will be crucial to the future of estuarine restoration if our earnest scientific and policy intentions ever hope to be realized.

ACKNOWLEDGMENTS

This chapter owes a great deal to the patient assistance and expertise of several individuals. Peter Lord, environment editor at the *Providence Journal*, first set me on the road to thinking about the relationship between complex science and public communication: I thank him here not merely for his assistance in surveying the August 2003 fish kill but more so for the lucidity and elegance of his continuing contributions to environmental writing. My former philosophy student Mike Traber, until recently a marine research assistant at the University of Rhode Island Graduate School of Oceanography, has been generous and entertaining in his accounts of eelgrass restoration as they happen from the level of benthic fauna on upwards. His colleague and boss Steve Granger has provided a complementary take on current challenges in restoration by discussing with me the long view of nutrient cycling and how many years it will take us to address the problems we have created. John Torgan at Save the Bay helped me to grasp the clash of science and politics in a way that might otherwise have remained abstract and academic for me, while Wendy Norden at the same organization allowed me to understand just how effective one passionate, informed individual can be under the right circumstances.

Many thanks, too, to Tom Ardito at the Narragansett Bay Estuary Program for his insightful newsletters aimed at the general community and for his exquisite images of the August 2003 fish kill. Chris Deacutis and Andrew Lipsky were generous in helping me to procure eelgrass images and indispensable in their assessments of the state of Narragansett Bay. My continuing appreciation goes to Chip Young, director of the Metcalf Institute at the University of Rhode Island, for reminding me of the "So what?" factor in writing for the public. And a daunting troika of thinkers—Scott Nixon, Virginia Lee, and Art Gold at the University of Rhode Island—first educated me about nutrient loading and managed to do so in terms that were clear and accessible. Pete August, director of the Rhode Island Coastal Institute, provided invaluable assistance and mentoring in the daunting arena of integrated coastal science, as did Candace Oviatt, who in my acquaintance is an interdisciplinary thinker without peer. To these, and to my most patient editor Robert France, I extend my warmest thanks. Given the host of expertise on which I have relied, the almost certain errors and obscurities contained herein are thus attributable entirely to me.

APPENDIX: WHAT IS EELGRASS AND WHY IS IT IMPORTANT? *ZOSTERA MARINA L.,* OR EELGRASS

Because less than one-third of the population around the Narragansett Bay actually knows what eelgrass is, it is useful to consider the ecological provenance of this particular plant. An underwater marine flowering seagrass, eelgrass grows in clumps, with green blades measuring ¼ in. wide and up to 3 feet long. Its roots (rhizomes) anchor the plant to sandy, silty, or gravelly substrates in both brackish and salt water. Uprooted grasses, often mistaken for seaweeds on the shore, turn papery and thin, and take on a brown or black color. An important source of food for many plants and animals, eelgrass also provides critical shelter for shellfish and finfish, and functions as a filter of pollutants and a location for nutrient cycling, which guards against shoreline erosion by dampening wave energy and the effects of storms. As Rhode Island's primary seagrass, eelgrass is among 50 different kinds of seagrasses and one of those that lives completely under water at depths from 3 to 20 feet. Its blades capture sunlight to produce oxygen for other life in the estuary and, once decayed, serve as valuable sources of organic matter for microorganisms at the base of the food chain.

Although eelgrass can grow either in small separate beds or large meadows, it requires clear water for photosynthesis and growth, so in the current climate of Narragansett Bay, where pollution occludes sunlight at lower depths, eelgrass cannot thrive at depths lower than 10 feet at low tide (Save the Bay 2003; Rhode Island Coastal Resources Management Council et al. 2003). In addition, eelgrass and other seagrasses require more light for growth than algae or seaweeds, meaning that in conditions of limited clarity and light, eelgrass must compete with more efficient photosynthesizers for the resource available. This is why algae blooms and phytoplankton will often choke off eelgrass plants in areas of high turbidity: Eelgrass can process light less efficiently and thus grows more slowly than algae and phytoplankton, allowing these light-occluding species to flourish at its expense (Alongi 1998). It is estimated that more than 10,000 acres of eelgrass once covered the bottom of Narragansett Bay (Doherty 1997). Today, there are fewer than 100 identifiable acres in total, due to a combination of a mass wasting disease that affected most Atlantic Coast populations in the 1930s and, following a partial recovery from disease up until the 1960s, more recent human impacts on the environment (Rhode Island Coastal Resources Management Council 2003). As an example of how changes in eelgrass populations influence human social and economic life, we can cite the once-flourishing bay scallop industry that used to thrive in the upper portion of Narragansett Bay. The industry has vanished as phytoplankton blooms and macroalgae, caused primarily by increasing population density and development-driven nutrient loading, deprive eelgrass of the light needed to grow and colonize in that part of the bay. Bay scallops utilize eelgrass as primary habitat. With the loss of eelgrass comes the loss of both shellfish populations and the commercial activities that depend on them.

WHAT ARE THE PRAGMATIC, CONCRETE PROBLEMS FACING RESTORATIONISTS WHO WORK IN NARRAGANSETT BAY?

Changes in population growth and the increase in human practices that often accompany such changes have a direct impact on the health of eelgrass populations. It is nevertheless unlikely that local zoning boards or property developers take into account the effects these changes will have on the Bay, especially with respect to deceptively humble and entirely submerged species like eelgrass. Yet the need for such consideration is increasingly urgent. Today about 50% of the U.S. population lives in coastal areas, which account for less than 10% of the landmass area in the 48 contiguous states. Demographic indicators predict that, by the end of the first third of the 21st century, about 70% of the U.S. population will live in the coastal zone (Vernberg and Vernberg 2001). As Yi-Fu Taun notes, "Economic and technological factors explain the accelerating volume of the movement to the sea, but not why people should have found it attractive in the first place" (Tuan 1990). As the coastal population with its attendant practices exerts pressure on the marine ecosystem, so, too, might the appeal of the coastal zone diminish in proportion to that increase: The coastline does not offer aesthetic or recreational appeal when thousands of dead shellfish float atop its inshore waters or when algae blooms prevent swimming and healthy fishing. So, unreflective lifestyle habits perpetuate the threat to eelgrass and continue to undermine both bay health and, eventually, human well-being and prosperity. This, in turn, means that existing eelgrass beds will disappear at a rate greater than that by which we might restore them. Thus, the most significant problem facing eelgrass restorationists is public ignorance of or indifference to the effects its actions have on existing eelgrass populations and bay health.

This problem manifests itself in a set of practices that scientists know to be harmful to eelgrass and thus the bay through a second problem, the *degradation of water quality*. Among the most obvious of these practices is the continued use of decaying septic systems that directly abut the coastline. Such systems constitute widespread, nonpoint (indirect) sources of nutrient loading to the bay, which in turn encourages the growth of algae and phytoplankton blooms that prevent sunlight from reaching eelgrass beds. In addition, when increased nitrogen is taken up by eelgrass shoots, these shoots can withstand significantly lower tensile forces and hydrodynamic drag (Kopp 1999) and also manifest frequent signs of chronic wasting disease (Kopp 1999). In other words: nutrient loading not only effects existing eelgrass beds but also lessens the capacity of new shoots to remain rooted or healthy in tidal waters—an obvious implication for restoration efforts involving the transplanting of shoots.

Effluent overflow from wastewater treatment systems during and after rainstorms constitutes a different (point source) but still threatening source of nutrient loading in addition to that of septic intrusion. Even so, according to Steve Granger, a marine research associate and restoration specialist at the University of Rhode Island, if we replaced all existing septic systems today and upgraded denitrification processes in sewage treatment as well, it would still take about 30 years to see a significant reduction of nutrients in the bay (Granger 2003). In addition, implementing solutions such as these is expensive and often unpopular with local populations. Alongside septic and sewage issues, other destructive practices favored by human beings can also be counted among those factors exacerbating water pollution in the bay. Fertilizers used on lawns and golf courses provide an additional source of nutrients, which enter the bay through groundwater contamination and runoff during storms. This dimension of the problem around Narragansett Bay is heightened by the parallel occurrences of the eelgrass breeding cycle in the late spring and summer, and the increase in human population spurred by the arrival of tourists from out of state beginning at the same time—tourists contributing to those activities that exacerbate pollution in the first place. Finally, atmospheric deposition—the depositing of pollutants into the bay from the air—presents another sort of problem for eelgrass restorationists, not least because such deposition cannot easily be decoupled from land-based nonpoint sources of pollution, or even from nonanthropogenic causes of change in the bay. As such, public dependence on practices and lifestyle habits contributing directly to the mitigation of water quality presents an additional problem to that of more general ignorance for eelgrass restorationists.

A third kind of problem involves *increasing water temperature* in the bay. The slime molds that have compromised eelgrass growth and colonization in the past, as well as under certain conditions now, have been tied to increases in both temperature and salinity. With reference to temperature increases, and their potential links to global warming more generally, scientists cannot be entirely certain of why specific areas of the Atlantic Coast have been affected while others have not. Since the 1920s the average daily temperatures of the coastal waters between New Hampshire and Virginia have been increasing. Areas south of Virginia do not show this trend, while those north of New Hampshire show less consistent changes and certainly not a clear trend (Granger 2003). Scientists do not yet have a firm grasp on the relationship between, say, atmospheric pollution and coastal warming, nor can such warming be entirely attributed to anthropogenic causes with certainty. The problem of increasing water temperature—which is lethal to eelgrass—thus poses a third and more complicated problem for restorationists, as the reasons for the warming cannot be fully isolated and understood at present.

A fourth problem relates to but is conceptually distinct from that of increasing temperature: a *lack of specific historical data* on Narragansett Bay makes it difficult for scientists to untangle knots involving anthropogenic vs. natural change in climate, atmospheric deposition vs. nonpoint sources of runoff from the land, and animal populations in eelgrass beds today vs. 100 years ago. For example, did previous eelgrass populations flourish when the bay temperature was cooler? If so, restorationists may need to breed shoots that are less susceptible to decimation at today's higher temperatures. Another unknown: a vicious circle exists between benthic fauna and eelgrass growth. Eelgrass was once more abundant than it is presently. Its decimation coincides with an increase in some benthic populations, like crabs. The increase in crabs, in turn, leads to increased consumption of remaining eelgrass shoots, making it difficult to determine the full range of causal sources for continued eelgrass disappearance (Traber 2003). Adding to this, scientists possess only estimates of eelgrass bed acreage before the mass wasting decimation of the 1930s. Some of these are based on oral history while others speculate using data-driven modeling instruments, though scientists continue to develop a mechanism for more precise ecological derivations of previous populations (Nixon 2002; Doherty 1997). Furthermore, data on sediment quality and composition from the past is virtually nonexistent. In essence, the science of eelgrass restoration must struggle against a number of significant variables about which we have little to no concrete, reliable information. The data develops as the restoration efforts do, slowing down our capacity to make useful predictions about the effect of restoration on the function of current and future eelgrass beds in the bay.

Finally, in Rhode Island, a less intractable but more immediate problem emerges at the intersection of federal and state regulations regarding permissible activities in the Narragansett Bay: State policy is less strict than that enacted at the federal level and the state legislature has resisted making a stronger commitment to eelgrass restoration. Despite consistent efforts by the advocacy organization Save the Bay to implement a "no eelgrass lost" policy at the state level, Coastal Management Resources Program regulations addressing issues like dredging and filling the bay, dock construction, and residential development permit activities that are destructive of eelgrass. Although eelgrass beds are afforded extra protection under the Federal Clean Water Act, state policy undermines the capacity of agencies to implement such protection because local regulations presently allow activities known both to destroy existing eelgrass beds and to mitigate new eelgrass growth and colonization (Save the Bay, 4). In addition, state advocacy groups are still struggling to persuade state representatives to approve dedicated funds for eelgrass restoration in the bay (the state senate, by contrast, has been amenable to such a move). Although there are currently state funds allocated to restoration in Narragansett Bay, funds derived from settlements with industry after oil spills, these are tied to efforts relating only to oil-spill mitigation. Dedicating part of those funds to eelgrass restoration would provide a level of state-sponsored regulation that would insure the long-term priority of improving the health of the bay. It would also bring Rhode Island into line with certain national standards that would then allow the state to access federal funds, which currently remain out of reach due to the disparity between state and federal policy.

LITERATURE CITED

Alongi, D. 1998. *Coastal Ecosystem Processes.* Boca Raton, FL: CRC Press.

Andersen, T. 2002. *This Fine Piece of Water.* New Haven, CT: Yale University Press.

Ardito, T. 1999. The Narragansett Bay Estuary Program: Coordinating Coastal Habitat Restoration in Rhode Island. *Water Connections* 16(2): 14–15.

Carson, R. 1955. *The Edge of the Sea.* Boston, MA: The New American Library.

Casey, E. 1998. *The Fate of Place.* Berkeley, CA: University of California Press.

Davis, C. 1996. The Process is as Important as the Product: An Examination of the Role of Public Participation in Estuary Management Plan Development in the U.K. In *Seeking Balance; Conflict, Resolution and Partership: Conference Proceedings of the 15th International Conference of The Coastal Society,* pp. 280–285. Alexandria: The Coastal Society.

Doherty, A. 1997. Historical Distribution of Eelgrass in Narragansett Bay, Rhode Island. Presentation at The State of Our Estuaries: 14th Biennial Estuarine Research Foundation International Conference, October 12–16, 1997, Providence, RI.

Early, J. How much more can Greenwich Bay take? *The Providence Journal.* Monday, August 25, 2003, p. B–04.

Emlock, E. 2003. Governor Joins Effort to Transplant Eelgrass. www.projo.com. June 30, 2003.

Ernst, L. M. and Lee, V., and Desbonnet, A. 1996. The Cumulative Impacts of Management Decisions on Nitrogen Loading to the Rhode Island Salt Pond Region. In *Seeking Balance; Conflict, Resolution and Partership: Conference Proceedings of the 15th International Conference of The Coastal Society,* pp. 491–500. Alaexandria: The Coastal Society.

Feldman, D. L. 1991. *Water Resources Management: In Search of an Environmental Ethic.* Baltimore, MD: Johns Hopkins University Press.

Foster, C. and Greenwood, L. N. 1996. When State and Federal Obligations Conflict. *In Ethical Dilemmas in Public Administration,* ed. Lynn Pasquerella, Alfred Killilea and Michael Vocino, pp. 87–122. Westport, Connecticut: Praeger Publishers.

Garitone, J. E. S. 1996. Achieving Consensus on Water Quality Issues in Rural Communities: A Peer Education Model. In *Seeking Balance; Conflict, Resolution and Parternship: Conference Proceedings of the 15th International Conference of The Coastal Society,* pp. 376–380. Alexandria: The Coastal Society.

Grese, R. E., Kaplan, R., Ryan, R. and Buxton, J. 2000. Psychological Benefits of Volunteering in Steward-ship Programs. In *Restoring Nature: Perspectives from the Social Sciences and Humanities*, ed. Paul H. Gobster and R. Bruce Hull, pp. 265–280. Washington D.C.: Island Press.

Granger, S. 2003. Conversation with the author, June 6, 2003.

Harper, J.L. 1987. The Heuristic Value of Ecological Restoration. In *Restoration Ecology: A Synthetic Approach to Ecological Research*, edited by William R. Jordan III, Michael E. Gilpin, and John D. Aber, pp. 35–45. Cambridge: Cambridge University Press.

Heidegger, M. 1977. The Question Concerning Technology. In *Basic Writings*, ed. David Farrell Krell, pp. 283–318. New York: Harper and Row.

Heidegger, M. 1985. *Being and Time* (trans. John Macquarrie and Edward Robinson). Oxford: Basil Blackwell.

Jenkins, V. S. 1994. *The Lawn: A History of an American Obsession*. Washington D.C.: Smithsonian Press.

Klesch, W. and Edwards, N. 1996. The Coastal America Partnership: Lessons Learned. In *Seeking Balance; Conflict, Resolution and Parternship: Conference Proceedings of the 15th International Conference of The Coastal Society*, pp. 225–229. Alexandria, VA: The Coastal Society.

Kopp, B. S. 1999. *Effects of Nitrate Fertilization and Shading on Physiological and Biomechanical Properties of Eelgrass (Zostera Marina L)*. Doctoral dissertation in Oceanography, University of Rhode Island.

Lee, K. 1999. *The Natural and the Artefactual: The Implications of Deep Science and Deep Technology for Environmental Philosophy*. Lanham, MD: Lexington Books.

Love, R. 2000. *Reefscape.* Australia: Allen and Unwin.

Matthews, M.H. 1992. *Making Sense of Place: Children's Understanding of Large Scale Environments*. Savage: Barnes and Noble Books.

McLellan, C. 1996. Online, Byline, or Redefine: How to Communicate with Coastal Audiences. In *Seeking Balance; Conflict, Resolution and Parternship: Conference Proceedings of the 15th International Conference of The Coastal Society*, pp. 14–15. Alexandria: The Coastal Society.

Milbrath, L. 1996. *Learning to Think Environmentally While There Is Still Time*. New York: State University of New York Press.

Narragansett Bay Estuary Program. 2003. *Final Report: Coastal Habitat Restoration Charrette*. Online material available at www.nbep.org/pubs/index.html.

Nixon, S. 2002. Prehistoric nutrient inputs and productivity in Narragansett Bay. *Estuaries* 20(2): 253–261.

Orr, D. 2002. *The Nature of Design*. Oxford: Oxford University Press.

Ortiz, E. 2003a. Carcieri appoints panel to preserve Narragansett Bay. *The Providence Journal*, Wednesday, October 22, 2003, p. B–03.

Ortiz, L. 2002a. Draft Request for Comments on the Draft Estuary Habitat Restoration Strategy Prepared by the Estuary Habitat Restoration Council. *The Federal Register*, Vol. 67, No. 86, Friday, May 3, 2002, Notices/44215–44221.

Ortiz, L. 2002b. Estuary Habitat Restoration Strategy Prepared by the Estuary Habitat Restoration Council. *The Federal Register*, Vol. 67, No. 232, Tuesday, December 3, 2002, Notices/71942–71949.

Polichetti, B. 2003a. Greenwich Bay hit by huge fish kill. *The Providence Journal*, Thursday, August 21, 2003, p. A–01.

Polichetti, B. 2003b. "Perfect recipe" for fish kill. *The Providence Journal*, Friday, August 22, 2003, p. A–01.

Public Law 106–457—November 7, 2000. *Estuary Restoration Act of 2000*.

Restore America's Estuaries and Estuary Research Foundation. 1999. *Principles of Estuarine Restoration: Working Together to Restore America's Estuaries—Report on the RAE-ERF Partnership, Year One*, pp. 6–11.

Rhode Island Coastal Resources Management Council, Narragansett Bay Estuary Program and Save The Bay. 2003. *Rhode Island Habitat Restoration Portal*. Online material available at www.csc.noaa.gov /lcr/rhodeisland/index.htm.

Rhode Island general Law Title 46 (Waters and Navigation) Chapter 31: The Rhode Island Bays, Rivers and Watersheds Coordination Team.

Russell, B. 1997. *The Problems of Philosophy*. New York: Oxford University Press.

Save the Bay. 2003. *Eelgrass: A Critical Narragansett Bay Habitat*. Online material at www.savebay.org /bayissues/eelgrass.htm.

Shea, N. Pollution, rain created a toxic bathtub in Greenwich Bay. *The Providence Sunday Journal*, Sunday, August 24, 2003, p. A–01.

Steelwater, E. 1997. Mead and Heidegger: Exploring the Ethics and Theory of Space, Place and the Environ-ment. In *Space, Place and Environmental Ethics*, ed. Andrew Light and Jonathan M. Smith, pp. 189–208. Lanham: Rowan and Littlefield.

Torgan, J. 2003. Conversation with author, December 2, 2003.

Traber, M. 2003. Conversation with author, May 13, 2003.

Tuan, Y. 1990. *Topophilia: A Study of Environmental Perceptions, Attitudes and Values.* New York: Columbia University Press.

Tuan, Y. 1977. *Space and Place: The Perspective of Experience.* Minneapolis, MN: University of Minnesota Press.

Vernberg, F. J. and Vernberg, W.B. 2001. *The Coastal Zone: Past, Present and Future.* Columbia: University of South Carolina Press.

7 Bottom-Up Community-Based Coral Reef and Fisheries Restoration in Indonesia, Panama, and Palau

Tom Goreau and Wolf Hilbertz

CONTENTS

ABSTRACT

The Global Coral Reef Alliance (GCRA) works with community-based management efforts to restore severely damaged coral reefs and fisheries using its Biorock® technology. In Bali, Indonesia, the village of Pemuteran used traditional village law to organize patrols to stop reef blast and cyanide fishing in their waters, establish marine protected areas for ecotourism and fish sanctuaries, and built more than 40 Biorock coral reef nurseries with a total length of over 500 meters. Their bottom-up efforts have won many national and international awards for community-based coastal zone management and underwater ecotourism.

The visibly obvious buildup of fish populations has made fishing villages across Indonesia realize that they can grow corals and harvest fish and shellfish, farming reef ecosystems instead of hunting fish and destroying corals. Our workshops are training fishermen, dive shops, university students, and government fisheries agencies in the new methods of reef restoration. In Ukupseni, Panama, the Kuna Indian community has established solar powered Biorock coral nurseries, lobster habitats, and breakwaters to increase habitat for lobster (the economic basis of the region), and to protect their low lying islands from eroding. The Kunas are preparing their own coastal zone management plan to establish lobster hatcheries and create habitats to increase shelter and food for lobsters at all stages of their life cycle. In Palau, GCRA works with the Hatohobei State Government in the most isolated atolls. These islands suffered catastrophic mortality of corals and fish after record high temperatures in 1998 and are undergoing severe erosion. Projects are being planned to grow solar-powered Biorock breakwaters to save the islands and their culture, coral and fisheries habitat, and revive ancient techniques for fish habitat that have not been practiced for generations. The tools for large-scale community-based restoration of coastal marine habitats and fisheries now exist, but their implementation is being blocked by the lack of government policies backing community-based restoration and lack of funding for them from international agencies.

INTRODUCTION

THE DILEMMAS OF COASTAL ZONE MANAGEMENT: BY WHOM AND FOR WHAT?

Coastal resources are in severe or catastrophic decline almost everywhere. Discussions with the oldest fishermen invariably reveal a former wealth of living resources, which have declined so severely that young fishermen find it hard to believe their elders and assume that they are senile or making it up. Scientific studies of coastal resources almost everywhere postdate the worst of the decline, and so shed little light except for documenting the disappearance of the last vestiges. Modern coastal zone management has largely been a top-down imposition by outside agencies (whether from the national capital or foreign "experts"), rather than a genuine outgrowth of community needs, and tends to override or ignore local long-term management concerns (Walley, 2004).

Furthermore coastal zone management is generally based on ideologies that value nature not for its own sake but only insofar as it is exploited to yield immediate returns. The dogma that nature exists only for humans to exploit it is common to "modern," "universal" ideologies, including globalism, capitalism, communism, Christianity, Islam, and the "monetary value is the only measure of worth" theorizing of economists. These "missionary" ideologies often believe they possess the only universal values, and feel compelled to force all to follow their own view of the world. In contrast, a multitude of ancient traditional cultures view humans as an integral part of nature, with a reciprocal responsibility for humans to maintain nature's overall balance and responsibly nurture it to the most healthy possible state (Ereira, 1990). These "native" or "aboriginal" traditions are largely viewed by outsiders as "backward," and are being actively exterminated, forcibly displaced, marginalized, or at best undermined, by the "universal" forces of modernist ideologies that value resources only insofar as they can be converted into money.

Any economist can "prove" that the "optimal" use of any natural resource is to "realize its monetary value" (i.e., extract it and sell it), invest the money, and live off the interests and profits, and that this strategy inevitably maximizes monetary return and hence all possible "utility" and "satisfaction." Because the purpose of such "management" is simply more efficient exploitation, it reinforces top-down relationships of external power and economic control over indigenous communities as well as nature. For example, until one generation ago Australian aboriginal people were legally classified under the Wildlife Act, with all the legal protection from hunters it entailed!

These ideological frameworks primarily benefit outside groups at the expense of both local ecosystems and local human communities, and almost never really empowers locals to manage their own resources for their own long term benefit, much less that of the entire ecosystem, despite often-made claims that this will result. Top down projects often act to destroy local community-based management practices and potential (Colchester, 2003; Chapin, 2004; Walley, 2004). Management based on contempt for local cultures and ecosystems alike is hardly likely to improve the lot of either.

Conservation Versus Restoration

The major current tool of conventional biodiversity and fisheries management is conservation based on the Marine Protected Area (MPA), which is widely claimed to increase both (Gubbay, 1995; National Research Council, 2001). By sealing off an area and barring its traditional users, it is claimed that naturally "resilient" ecosystems will automatically bounce back by themselves. Nevertheless, even though many MPA proponents say that this will happen anywhere, it is obvious that such policies can only work where ecosystem health, water quality, and carrying capacity are excellent. Yet every coral reef ecosystem is now degraded, and every coral reef MPA is also undergoing drastic decreases in live coral cover (and fish populations) due to global scale stresses (global warming, new diseases, land-based sources of pollution, etc.) that are beyond the control of any MPA. Because MPAs cannot control the real root causes of coral death, or of decreasing habitat quality and declining carrying capacity for fishes and shellfish, they cannot protect coral reefs and fisheries in the long run (Goreau and Hilbertz, 2005). There are some remarkable cases of short-term MPA success on a small scale at locations where environmental conditions are excellent, such as Apo Island in the Philippines, but it is clear that the same successes will not take place in the vast majority of reefs, which are already badly degraded.

The most lavishly funded MPAs in the world are in the richest countries, yet long-term monitoring programs of the Great Barrier Reef Marine Park Authority and the Florida Keys National Marine Sanctuary show that live coral cover is now down to only about 20% (AIMS, 2004) and 7% (CREMP, 2003), respectively, and steadily falling. Yet these two MPAs are the focus of publicity claiming that these are "well managed," "resilient ecosystems," and that poor countries can achieve the same results by following their example! If the rich can't protect their reefs now, and will be even less able to do so as global warming, pollution, and new diseases intensify, it is clear that imposing such policies on those without vast financial resources is even less likely to work. Yet this failed strategy remains the primary management objective of governments and international funding agencies. Because so little reef now remains in pristine condition, and none can be protected from globally-intensifying stresses, conservation efforts focus on finding the last good patches and, in effect, trying to build walls around them, ignoring the vast degraded areas around them on which local communities must rely. No funding is allocated to effective restoration, only towards methods like gluing and cementing broken corals, but these will die whenever the water gets too hot, dirty, polluted, or when diseases break out and pests infestations occur.

We take the antithetical approach that active, large-scale, community-based habitat restoration is the *sine qua non* of effective biodiversity and fisheries management in a world that is changing so fast that the conventional top-down conservation strategies can no longer work. We take the view that the proper management of natural resources is simply to maximize its quantity and quality, not to isolate the last examples in preserves, and that the best results for ecosystem health and maintenance of natural services will result from local optimization, writ large. Local communities are the best placed to value and protect their habitat, but they only can if they are allowed to hold responsibility for managing their own surroundings, obtain the knowledge and tools to do so effectively in the context of their own traditions, and convince their neighbors of the benefits by example. Unfortunately, local communities are losing or have lost control of their resources almost everywhere, often to allow them to be exploited by outsiders. Thus, it is essential to re-empower communities to control the resources providing their livelihood for long-term management to be possible.

In this paper, we describe collaborative coral reef and fisheries restoration projects with traditional communities in Indonesia, Panama, and Palau who have maintained control of their coastal zone from national and outside forces despite increasing economic pressures on their resources.

TOP DOWN VERSUS BOTTOM-UP MANAGEMENT

The Global Coral Reef Alliance (GCRA) works closely with a wide range of local partners around the world, including community groups, local governments, divers, environmental organizations, and hotels. Their common feature is that they have recognized that their local reefs are vanishing, and realized that if they wait until they can get funding from top-down programs it will be too late. They further understand that conservation alone has become inadequate, and that immediate efforts to grow corals and restore fisheries habitat are urgently needed to make a meaningful difference, and must be started whether or not outside funding can be found (Goreau and Hilbertz, 2005).

GCRA works closely with local partners to assess current and historical health of sites, and trains them to design, construct, install, monitor, maintain, and repair restoration projects. Because to date no government or large funding agency has supported meaningful coral reef restoration efforts, all projects have been supported purely by small donations, mostly in-kind donations of materials, food, and lodging by local groups. This local support has increased local participation and control of projects, but of course the very small amounts of funding available have greatly limited the size and number of projects. We have been unable to respond to urgent requests for help from many communities around the world due to lack of funds. This situation will remain until governments and funding agencies change their policies and choose to support bottom-up community-based ecosystem restoration programs instead of the top-down efforts that have consumed so much money with so little results.

Similar efforts at community-based management of whole watersheds and coastal zones and reef and fisheries restoration were first developed in Jamaica (Goreau et al., 1997). However, these failed because outside funding agencies, which came in only after local community management plans had been developed, paid for, and controlled the management implementation agendas and did not see community-based management of entire ecosystems as their goal. Instead they focused on making tourism "parks" to repay foreign exchange debts. Instead of supporting detailed pre-existing community-based management plans, developed over years of meetings in every community in the watershed, they funded MPAs in the tourism areas as separate units administratively isolated from the human populations in the up-stream watersheds that affected them. As a result, the community's role in management was marginalized. The reefs could not be protected from the land-based sources of pollution that were killing them, and proceeded to deteriorate even more rapidly after money was spent to "manage" them (Lapointe and Thacker, 2002). After 10 years, the failure of these "conservation" efforts to protect corals and fish are obvious, and once the foreign funds and consultants that dictated their mismanagement were exhausted and left, local communities are again requesting help from GCRA to restore their crippled reefs and fisheries. Ironically, only after bottom-up community-based efforts to integrally manage whole watershed and coastal zones had been effectively prevented by top-down funding impositions, these pioneering efforts in Jamaica became the direct inspiration for top-down "ridge to reef," "hill top to ocean," and "white water to blue water" programs by large international funding agencies.

METHODS

We use GCRA's Biorock® Ecosystem Restoration Technology (Hilbertz and Goreau, 1996). Biorock uses safe, low-voltage, direct current, which can be provided by solar panels, windmills, and tidal current turbines to grow solid limestone structures of any size and shape in the sea. Biorock is the only coral reef restoration method that increases coral growth rates (typically 3–5 times), coral healing from breakage (more than 20 times), coral survival from lethal conditions of high temperature, sediments, and nutrients (16–50 times increase in survival over adjacent reefs after the 1998 bleaching in the Maldives), coral settlement (by 2–3 orders of magnitude), and coral reproduction, while greatly increasing fish and shellfish populations, especially juveniles (Goreau & Hilbertz, 2005). Using these methods, coral reef organisms can be kept alive where they would die, and coral reefs restored in record time where they cannot recover naturally. The Maldives, one of the countries most threatened by global sea level rise and global warming (Gayoom, 1998) has been a major focus in the development of Biorock technology to save coral reefs from global warming (Goreau et al., 2000). Biorock reefs in front of severely eroding beaches in the Maldives produced an ecotourism attraction full of corals and fish in a barren area in front of the beach, and the reduction of wave energy caused the beach to grow by 50 feet (15 meters) in a few years (Goreau et al., 2004). These projects were awarded the Sperry Award (top prize for Innovators and Pioneers from the Society for Ecological Restoration), and the Maldives Environment Award. Although energy and materials were needed for the Biorock reef breakwater, the energy consumption was less than the beach lights, and total costs per meter of protected shoreline were a few percent of the concrete breakwaters that surround the nearby capital island, Male. Those massive concrete walls have increased scouring of the shoreline while producing no ecological benefits.

PEMUTERAN, BALI, INDONESIA

HISTORY AND SOCIAL CONTEXT OF FISHERIES

Indonesia has the largest area of coral reef and the highest marine biodiversity of any country in the world (UNEP, 2001). The village of Pemuteran lies in one of the few areas of Bali that are too dry to grow rice, the staple food, and its people were forced to rely on the sea, feared as the home of the evil spirits in Balinese culture. The shallow offshore banks nearby have the largest area of coral reef in Bali that are free of ferocious tidal currents, giving it the richest reef fisheries on the island. Most fish catch was used for subsistence or traded for rice, and the population was among the poorest in Bali, sharing mud-floored huts with their animals. Because Pemuteran is in the most remote corner of Bali from the tourist entry points in the South, it was the last coastal area to develop tourism. The large, lush, and current-free reefs made it an exceptional location for diving, and a small diving industry developed followed by small hotels. These deliberately avoided the large-scale mass tourism of the South, giving it a tranquil setting and attracting Indonesian and foreign residents of Bali and Java eager to escape the congestion of the south as well as tourists straying off the beaten track to find more peaceful surroundings. The development of tourism created jobs other than subsistence fishing for the first time, creating new alternatives for educated local young people. The result has been a great increase in village standards of living, without loss of tradition.

Traditional village law, Adat, remains very strong in Bali and many outlying parts of Indonesia. It has been weakened in areas of the country where Muslim or Christian influences have displaced traditional culture and law, but is institutionalized in Bali because the ancient village laws have been incorporated into the island's Hindu-Buddhist culture, which focuses on maintaining harmony between natural forces. The founders of the Pemuteran tourism industry were careful to work closely with the village council, led by the Kepala Adat, the interpreter of village traditional law, and the Kepala Desa, the elected village leader. In order to protect the fish habitat in the fringing reef in front of the fishing beach and the snorkeling reefs for hotels, the village declared the area in front of the beach to be protected from all forms of fishing. A ban was also placed on use of bombs and cyanide for fishing in the entire reef area offshore from the village. It is important to note that this was done using village law alone: no permission was sought from national authorities, nor was any needed.

IMPACTS OF DESTRUCTIVE HARVESTING

During the Indonesian economic crisis of 1998, when millions of displaced workers and farmers took to fishing as a means of survival, maintenance of the ban on offshore reefs was no longer enforced and bomb and cyanide fishing went out of control. One would hear 5 or 10 bomb blasts a day in Pemuteran Bay, and since diving was no longer safe, the local diving business collapsed. The bombers destroyed almost all of the shallow reefs on the offshore banks, leaving only deep waters and the area in front of the beach untouched. Most of this destruction was blamed on fishermen from the islands of Java and Madura, and on nearby villages in Bali inhabited by fishermen from Madura who emigrated because the fisheries on their own island had collapsed.

As the vibrant reef was reduced to dead rubble, the fisheries collapsed due to lack of habitat. Shocked by rapidly declining food and tourism income, the village decided to once again enforce their ban on destructive fishing methods on their offshore reefs, but the damage was already done. However, this time the hotels and dive shops arranged with the village to donate 5% of their profits to the village in order to organize boats, engines, and personnel for the village Pecalang Laut (Sea Guardians) to monitor fishing activity and enforce village laws. If fishermen are found to persist in using banned methods, their boat and all their gear are seized and they are thrown in jail. Although this is done by village law, Indonesian legal authorities recognize the arrests and prosecute the cases. As a result of this ban the devastated areas offshore are steadily recovering.

COMMUNITY RESPONSE

GCRA began working with dive operators in Pemuteran to start Biorock coral reef restoration projects in 1998. From the first small start, these have exploded in size and popularity, now stretching over about half a kilometer of Biorock reef in 40 separate structures of different shapes. Studies by Putra Nyoman Dwija have shown that corals grow 4 times faster and heal from damage more than 20 times faster on Biorock than nearby controls (Dwija, 2002). Naturally broken coral fragments on nearby reefs that would roll around and die in the mud and rubble have been transplanted onto the Biorock reefs.

Not only have hard corals and other attached marine organisms like soft corals, sponges, and clams grown at phenomenal rates, they have attracted huge numbers of reef fish into an area that had been fairly barren. The types of fish attracted to structures depend on the size and shape of spaces created, and the structures can easily be built to have much more space for habitat than a natural reef, as it has many more layers. As a result fish populations have bounced back in the bay, and the reefs have attracted tourists from all over the world. The village has won Indonesia's most prestigious environmental award, the Kalpataru/Adiputra Prize, the KONAS Award for best community-based coastal zone management project in Indonesia, the SKAL Award for best Underwater Ecotourism Project in the World, the Association of South East Asian Nations Tourism Agencies Award for Excellence, and the Pacific Asia Travel Association Gold Award.

Three international training workshops in coral reef restoration have been held in Pemuteran in which students learned the theory and hands-on practice of design, construction, installation, monitoring, maintenance, and repair (Goreau and Hilbertz, 2004, 2005) (Figures 7.1–7.4). The first Pemuteran Workshop attracted around a dozen participants, the second attracted two dozen, and third had around 60. Only one Indonesian attended the first workshop, but the Indonesians made up the majority of the rest, with the remainder coming from all over the world. Biorock students have started their own projects in Java, Sulawesi, Lombok, Flores, and other parts of Indonesia, as well as in many other parts of the world. The fourth workshop was held in November, 2006, on the island of Gili Trawangan, Lombok, where around 70 participants built some 20 new structures.

Because of the dramatic buildup of fish populations in the Biorock reefs, including large swarms of juveniles, fishermen all around Bali have requested similar projects as a means of restoring their coastal fisheries and reef habitat. Besides fishermen, divers, hotels, and tourists, the projects have been visited by top officials of the Bali Island government and succeeding Indonesian government ministers of Environment, of Marine Affairs and Fisheries, and of Tourism, as well as by the executive director of the United Nations Environment Program. However, until very recently the project has received no funds from any government or large funding agency. It has been entirely supported

FIGURE 7.1 Second Pemuteran Reef Restoration Workshop participants and the newly built Big Manta Reef, Bali.

by small private donations, many of them local in-kind donations of materials, and food and lodging for participants, along with donations by impressed tourists. The project's success has attracted press and television coverage around the world.

The first funding to the village came at the very end of 2005, when the Bali Recovery Fund, a group funded by the Australian Aid Agency to aid economic recovery of areas economically depressed by the collapse of tourism following terrorist bombings of tourist targets in South Bali, supported the new Reef Gardeners of Pemuteran Program. This program, whose motto is "Protecting and Preserving the Reefs Now … for the Future," has trained 10 young people from the village to be full-time paid reef restorers, focusing on the offshore bank reefs. The Reef Gardeners, who are young fishermen without the education and language skills needed to get a job in the tourism sector, have been trained as divers, are learning English, and taking courses in business management. A barge was built to house a mobile power supply used to start Biorock reef restoration projects to speed up recovery of the outer reef. They have sunken several old boats on the edge of one of the banks, and are turning them into Biorock reefs. They have constructed an underwater snorkeling trail, with sculptures in the design of a traditional Balinese temple garden. As the funding is a short

FIGURE 7.2 Getting ready to float the Little Manta Reef to its site, Bali.

FIGURE 7.3 Attaching naturally broken coral fragments to Big Manta Reef, Bali.

term startup, and will not be continued, the village has introduced a $2 fee for divers and snorkelers using the area, and local dive shops and hotels encourage guests to participate. It is hoped that these fees will pay for maintenance of the project and the salaries of the Reef Gardeners. To our knowledge these are the first professional paid reef restorers anywhere in the world.

Recently, for the first time, the Government of Indonesia has officially decided to seek large-scale funds from the German Government Debt Relief for Nature Swap to implement community-based Biorock coral reef and fisheries restoration projects in four different regions of Indonesia, and to train fishermen and students in the new techniques. If this funding materializes, Indonesian fishermen will begin to switch from being hunters of the last wild fish to coral reef farmers who increase their productivity by actively restoring reef habitat and improving its habitat quality and carrying capacity. It is hoped that Indonesia's lead in developing policies in support of active restoration will result in a transformation of Indonesia's fisheries and will be followed by other governments and funding agencies.

FIGURE 7.4 Biorock reefs have prolific coral growth and dense fish populations: Ibu Karang Reef, Bali.

KUNA YALA, PANAMA

HISTORY AND SOCIAL CONTEXT OF FISHERIES

The Kuna Indians of Panama are unique among native peoples of the Americas in having never lost their land, culture, and laws (Howe, 1998). There are no roads to their land, Kuna Yala, and they do not allow outsiders to own land, settle, or deforest the jungles for the cattle ranching that has destroyed forests and soil fertility throughout the Americas (Ventrocilla, Herrera, & Nunez, 1995). The Kuna live on an archipelago of 365 islands in the Caribbean, 50 of them inhabited, but also own the entire watershed of northeastern Panama from the coral reefs right up to the top of the mountain ridges separating the Atlantic from the Pacific. Despite their poverty, Kuna culture is vibrant and confident; they have a unique appreciation for knowledge of all kinds, and produce an exceptional number of university students. Their leaders are elected based on intelligence and knowledge through an elaborate participatory democracy. Books written about their love of freedom and equality in the 1600s by Lionel Wafer and William Dampier, English pirates who were stranded there following failed raids on the Spanish gold fleets passing through Panama to Spain from Peru, were the direct inspiration for French theorists of *"Liberte, Egalite, et Fraternite,"* not the rigid class structure of Athenian "democracy."

Due to their isolation and traditional way of life, the Kunas have few products they can sell besides coconuts (which are of practically no economic value) and *molas*, the unique embroidery art of Kuna women using a reverse appliqué technique with multiple layers of cut cloth of different colors. Because they have the largest area of coral reef in Panama and because their coastal mangroves are intact, they had exceptionally high populations of spiny lobster and several species of edible crabs. Their entire economy is based on the export of lobsters and crabs, and the Kuna shellfisheries make up 70% of Panama's marine exports by value. However only a very small fraction of this goes to the Kunas, most being made by the retailers and middlemen who fly them from Kuna Yala to Panama City and then on to Miami. Because lobster and crab are the only available source of money for almost all Kunas, virtually the entire male population dives for lobster. SCUBA diving is strictly banned by Kuna law, even by tourists, because it would lead to too rapid depletion of lobsters. Kuna fishermen specialize in free diving, and are so good at this that they have migrated all over Panama, working in diving fisheries.

IMPACTS OF OVERHARVESTING AND EUTROPHICATION

Although the lobster and crab fisheries are the driving economic force in Kuna Yala, they are in an advanced state of decline. Fishermen admit that the populations are vanishing, but have no other way to earn money. The chief of one village told us "I am very worried for the future of the children because when I was young we would jump in the water right in front of our house in the village, pick up 10 or 15 lobster, and have them for dinner. Now we must paddle our canoes very far out at sea, dive very deep, and the lobsters are few and small." The head of the lobster fishermen in this village said that "Every day there are more fishermen and less lobster. If we don't restore the lobster populations our way of life will be finished."

Besides over-fishing, the entire reef habitat is undergoing serious eutrophication. Inhabited islands are very densely populated, and are ringed with outhouses over the water. The corals and fish are gone around inhabited islands, whose shores are covered with massive growths of algae species that are indicative of severe pollution and garbage. In contrast, uninhabited islands have dense coral populations, but even these are being steadily smothered by algae where affected by water flows from populated areas (Goreau et al., 1997). The health of the reef, and its ability to maintain large populations of fish, lobster, and crabs is steadily declining. All the islands are only slightly above sea level, and suffer strong erosion on their northward sides during high waves for 4 months early each year. Traditionally, the Kunas mined live corals to build up the eroded sides of their islands, to expand their area, and even to build new islands. When there were few people and many corals, the reefs were not noticeably depleted, but this is no longer the case.

COMMUNITY RESPONSE

GCRA has worked with the people of Ukupseni, Kuna Yala, since 1994. Due to lack of electricity, our projects there largely use solar panels to power a variety of coral nurseries, lobster nurseries, and breakwaters. The panels used were part of the world's largest solar power plant, built in Michigan in the 1970s as part of President Carter's Solar Energy Research Initiative, which was sold to a large oil company by the Reagan administration and dismantled for scrap. GCRA bought the last crate of the old panels for a fraction of their value and recycled them. Despite their age, their output still exceeds the original specifications. We have made breakwaters that are slowing down the waves next to one of the village eco-resorts, and a breakwater for the village hospital so that patient's lives can be saved by bringing them directly to the hospital by canoe, which had been impossible during the rough weather season. Coral nurseries are growing around 15 species of hard corals, plus soft corals and sponges, on a 20 ft (7 m) diameter dome over a muddy area where no corals were growing before (Figures 7.5–7.7), and the reef has attracted a large resident barracuda (Goreau et al., 2005).

GCRA has also brought educational books and videos, and given lectures to all the schoolchildren on the importance of protecting and restoring coral reefs. GCRA's Children's Program donated masks, fins, and snorkels to the school so that the students, who swim like fish but are too poor to own snorkel gear, can go on field trips to learn the differences between healthy and sick reefs, and participate in reef restoration projects (Figure 7.8) (Goreau, Goreau, & Solis, 2003).

The small private donations used for these projects have dried up completely, so we are unable to maintain or expand them, and no support for reef restoration has yet been provided from Panamanian funding sources. If funding can be found, programs in Kuna Yala will increasingly focus on creating habitat to increase populations of lobsters. Biorock structures can be built in any shape and size, and we noticed in Jamaica, Mexico, and Panama that structures built to provide spaces of the right size and shape are densely packed by lobsters. Lobsters suffer severe mortality due to lack of hiding places, especially the juveniles which live in sea grass beds, so their populations can be greatly increased by building shelters of the right sizes in mangrove, seagrass, and coral reef habitat used by lobsters at different stages of their life cycle. This technique has been widely developed in Cuba, where it has greatly increased production. Cuban "casitas," made from marine plywood, ferro-cement, or roofing material, provides shelter alone. Use of Biorock shelters is likely to be even

FIGURE 7.5 Assembling solar panels for lobster nurseries, Panama.

FIGURE 7.6 Transporting the Akabiski Galu Reef, Panama.

FIGURE 7.7 Akabiski Galu Reef, Panama.

FIGURE 7.8 Kuna children getting ready to snorkel on Akabiski Galu, Panama.

more effective as it provides a substrate for growth of barnacles, clams, and other food for lobsters. Future work will build solar powered lobster shelters of suitable sizes in different habitat in order to increase lobster populations and allow sustainable harvesting at higher production levels.

Successful efforts have been obtained in Kuna Yala despite the remoteness and poverty of the inhabitants, and very little funding. Now we are working to set up fisheries, mariculture, and reef restoration projects in other parts of Panama, such as restoring damaged reefs in Marine Protected Areas, building habitat for the release of juvenile snapper released from hatcheries, and oyster culture. These projects will be done with fishermen's cooperatives, mariculture operations, research hatcheries, the Autoridad Maritima de Panama, the Autoridad de Medio Ambiente, and in programs planned for the newly formed Universidad Maritima Internacional de Panama, if funding can be found. During 2005 there were record high sea surface temperatures in Panama, and severe coral mortality is likely to have taken place, making the need for such projects more urgent than ever.

HATOHOBEI, PALAU

HISTORY AND SOCIAL CONTEXT OF FISHERIES

The Southwest Islands of Palau consist of six very low islands stretching from Palau almost to Indonesia, the Philippines, and New Guinea. All are small islands completely surrounded by fringing reef with no lagoon or passages, except for one, which consists of a large atoll with a single small sand cay. They are inhabited by people speaking a different language from Palauan, called Tobian or Sonsorolese, which is related to the languages of Yap (Friends of Tobi Island, 2005). In his book on traditional fishing cultures in Palau, *Words of the Lagoon*, Robert Johannes described them as the master fisher folk of the Pacific, having a wide variety of handmade hooks designed for specific fish species that were unknown elsewhere (Johannes, 1981). Johannes was never able to reach the Southwest Islands, and based his comments on the notes of P. W. Black, an anthropologist who had lived there. The Southwest Islanders lived in isolation until forced to labor in coconut plantations on Tobi by German colonial authorities, and later as mine and construction workers on other islands by Japanese colonial authorities during the Second World War. Following independence of Palau almost the entire population migrated to the capital, Koror, where they have established their own village. This was done because there were limited jobs, educational opportunities, and medical services on their islands, but small communities of subsistence fishermen and their families remain on the southwest islands. Migration between the southwest islands and Koror is an active, dynamic, vibrant, and ongoing population mobility resulting in cultural, political, economic, and social exchanges in both directions. Southwest islanders' residences in Koror village are not considered permanent homes but habitual ones due to the long period of wait between trips—due to the islands' distance from the main island of Koror. Although the Southwest Islands are surrounded by some of the richest tuna fishing grounds in the Pacific, the islanders do not have the boats to exploit it, and access to these resources is leased by the Palau Government to foreign fleets.

The coral reefs of the single atoll, Hotsarihie (meaning Reef of the Giant Clam, but usually called Helen Reef by outsiders), have the greatest diversity of corals, fish, and mollusks recorded on any Pacific oceanic island (that is, excluding Indonesia, the Philippines, and New Guinea) (Maragos, 1993). The island has been used by the people of Hatohobei Island (also known as Tobi) from ancient times, but was not permanently inhabited due to lack of drinkable groundwater on the small sand spit that is emergent at high tide. The lack of permanent inhabitants made it a mecca for poachers from Indonesia, the Philippines, and even Taiwan, who plundered the richest giant clam and valuable trochus shell populations in the Pacific until little remained. The unique marine diversity of the atoll, and its huge bird and turtle nesting populations, make it potentially valuable for diving and nature ecotourism, if it can be protected.

FIGURE 7.9 Helen Reef is a sandbar full of turtle nests and birds overhead, Palau.

IMPACTS OF GLOBAL CLIMATE CHANGE

In 1998 the waters around Palau and the southwest Islands were extremely hot, and the vast majority of the corals died (Goreau et al., 2000). Following the death of most of the corals the Tobi fishermen noticed a dramatic decline in fish populations. The single small sand spit on Helen Reef is extremely vulnerable to global sea level rise (Figure 7.9). It is only about 10–20 centimeters above the normal high tide mark, and one can clearly see that sand waves have repeatedly passed right over it in storms. The island is unstable, and is moving southeast at about 15 m a year. The remains of a concrete floored structure built on the island by Japanese troops now sits in shallow water about 200 m from the shore. Coconut trees planted on the eastern shore of the island collapse into the sea on the opposite side of the island before they can bear, because the island has completely shifted its position eastwards in this time (Figure 7.10). Now that most of the surrounding coral is dead, the rate of erosion is likely to increase as global warming and sea level rise accelerate, placing control of all the natural resources of Helen Reef at risk, as well as the surrounding tuna fisheries.

FIGURE 7.10 Trees collapse into the sea on one side of Helen Reef, Palau.

COMMUNITY RESPONSE

The Hatohobei people feel that their top environmental priorities are to restore their coral reefs and fisheries and to stabilize the island on Helen Reef so that it can be developed as a base to protect the atoll's resources and for ecotourism. The Hatohobei State Government succeeded in getting a grant from a very large U.S. private foundation to station rangers on Helen Reef. However, instead of being allowed to control the funding to use to solve their own pressing problems, they found that the control of the money was entirely handed over to a U.S. organization. This is using it for their own consultant's salaries, travel from the U.S., and to tag turtles, instead of restoring habitat and protecting their islands from erosion, as the Tobi people wish. The attitude of the agency funding this project appears to be that local people don't know what their own problems are, and should not be able to control any funding, i.e., that indigenous people are not to be trusted to act on their own behalf.

The Hatohobei State Governor Sabino Sackarias learned about GCRA restoration and shore protection technology by chance in 2000 and immediately invited a collaborative project. Several large funding agencies with community-based management programs of the top-down kind all refused to help. It took 4 years to find a small amount of funding from a private donor to start these efforts. Funds were obtained for solar panels and materials to build a breakwater to stabilize Helen Reef Island, the top priority expressed by the governor. Due to the extreme remoteness of Helen Reef, there are only a few trips a year to it by the single supply ship that serves all the Southwest Islands. So there was no chance to assess the site beforehand, and all work had to be done during a single scheduled supply trip. However, the ship broke down, and had to be sent for repair to Manila, which took far longer than expected. By the time the ship was fixed, the good weather season was over, but as the funding had a time limit, there was no choice but to go ahead. Although Helen Reef is almost on the equator, and well outside of the hurricane belt, a typhoon formed in the area, unusually early in the season and far to the south of the area normally affected, delaying departure.

When the island was reached, extremely strong wind, waves, and rain indicated an exceptionally early start to the monsoon season. Despite very hard working conditions and inadequate materials and tools, the team of Hatohobei and GCRA volunteers succeeded in building a 450-ft-long breakwater along the most erosion prone coast, and powering it by 32 solar panels (Figures 7.11–7.12). At this point a Super-Typhoon developed, and the team was unable to leave the island until after exhausting all supplies except for one bag of rice. Due to inadequate funding, there was no money for a generator or a welding machine, so the entire structure had to be wired together by hand. Although the structure worked as planned (Goreau et al., 2004), the fact that it was finally built only at the start of the rough season, not at the start of the calm season as originally planned, meant that it was not

FIGURE 7.11 Jetting solar panel into the sand with a SCUBA tank on Helen Reef, Palau.

FIGURE 7.12 Building the breakwater in fierce winds and rain, Palau.

strong enough to withstand damage from huge tree trunks from Indonesia, the Philippines, or New Guinea that were smashed over the reef by monsoon and typhoon waves. As a result the structure was damaged before it was strong enough to withstand the logs. The solar panels were rescued, to be reused when we can find more funding to rebuild it using stronger materials and tools. There is no conceptual difficulty to saving the island, only a funding issue. The donor that supported the initial effort is not interested in follow-through, and so far no new funding has been found, despite efforts by the former governor, Sackarias, and the new governor, Crispin Emilio.

If more funding can be found we plan to rebuild the breakwater, and grow coral, fish, and giant clam habitat to restore the reef and fisheries. The Tobi people also want to reinvigorate an ancient artificial reef method they have not used now for several generations, which was unknown to Johannes. In the old days they would construct reefs from coral rubble with a lot of hiding spaces for fish, dismantle it after several years, catch the fish in the new habitat they had created, and rebuild the reef nearby. The intention is to compare these revived methods with the new ones. If suitable funding can be obtained to use tidal energy turbines, we plan to build a large island on the barren reef flat right next to the tidal pass into Helen Reef, which has extremely strong currents. The tools now exist for the Tobi people to save their islands, restore their reefs and fisheries, and create a new, much larger island to live on using purely renewable energy. The will and skills are there if the international funding community responds to their appeal to save their unique islands and culture from disappearing beneath the waves.

SUMMARY: CURRENT PROGRESS AND FUTURE STEPS

New tools now exist to allow fishing communities to restore their coral reef habitat and fisheries and protect their coastlines from erosion habitat even under conditions where it cannot recover naturally, and to greatly increase carrying capacity of fish and shellfish using structures specifically designed for certain species. Most Biorock restoration projects have been done with minimal funding, and in remote places that cause extreme logistic difficulties, but they show that even very isolated and poor fishing communities can easily learn and apply new skills. These will allow them to make the transition from being hunters wiping out the last big game to reef farmers who grow reefs to greatly increase populations of desired fish and shellfish species. This will finally bring the Neolithic revolution to the oceans, 10,000 year after big game hunters on land were forced to become farmers to survive. For this transformation to happen worldwide is not a matter of lack of knowledge or suitable methods, it is simply a result of the fact that governments do not invest in reef fishermen in the way that they invest in subsistence farmers to allow them to become cash crop farmers. Fishermen in tropical countries are usually the poorest element of society, and unless coral reef countries choose to invest in them to allow them to apply new skills to restore their environment and sustenance, the deterioration will simply continue apace, no matter how much is spent on MPAs.

Unfortunately all funding being spent by governments and international funding agencies now goes towards implementing methods that have failed to protect reefs on a large scale, and will do even worse in the future as global warming and pollution increase. This strategy is based on "expert" opinion that if nature is left to take its course, "resilient" reefs will grow right back no matter what we do. This advice suits the interests of the rich countries while preventing developing countries from saving their vanishing reefs, fisheries, beaches, tourism, and shorelines. Indonesia has made a pioneering move in supporting community-based restoration, and it is to be hoped that other governments and international funding agencies follow its lead on a large scale before it is too late.

ACKNOWLEDGMENTS

We thank the people of Pemuteran, Ukupseni, and Hatohobei for their constant support and determination to improve their environmental quality for future generations. We thank the photographers who took the images accompanying this paper: Frank Gutzeit (Figures 7.1–7.3), James Cervino (Figure 7.4), Wolf Hilbertz (Figures 7.5–7.6), Gabriel Despaigne (Figure 7.7), Marina Goreau (Figure 7.8), Caspar Henderson (Figures 7.9–7.12). Special thanks go to those who played key roles in implementing the projects described and planning for the future of their communities: Randall Narayana Dodge, Agung Prana, Komang Astika, Chris Brown, Cody Shwaiko, Putra Nyoman Dwija, Gabriel Despaigne, Roque Solis, Lucio Arosomena, Ultmino Avila, Marina Goreau, Paliwitur Sapibe, Sabino Sackarias, Huan Hosei, Crispin Emilio, Caspar Henderson, A. Azeez A. Hakeem, and far too many other members of these communities to list here. Funding for the Pemuteran projects were raised by local hotels and dive shops, especially Taman Sari Resort, Pondok Sari Resort, Yos Dive Shop, Archipelago Dive Shop, Reef Seen Aquatics, and Werner Lau Dive shop. The Reef Gardeners Program was funded by a grant from Ausaid for the reconstruction of economic opportunities in Bali following terrorist bombings in the main tourist areas. The Panama projects were supported by a Pew Fellows Award to Tom Goreau and donations from Leslie Jones and from Roland Pesch. The Hotsarihie Project was funded by an anonymous donation from a private foundation.

N.B.: Color versions of all of these black and white figures can be seen at the Global Coral Reef Alliance Website: www.globalcoral.org.

LITERATURE CITED

AIMS (Australian Institute of Marine Sciences), 2004, results of long term monitoring of the Great Barrier Reef, http://www.aims.gov.au/monmap/monmap.asp?yearcode=&asector=&fcoast=off.

Chapin, M., 2004, A challenge to conservationists, *World Watch*, November–December, 17–31.

Colchester, M., 2003, Salvaging Nature: Indigenous Peoples, Protected Areas, and Biodiversity Conservation, 135 p., World Rainforest Movement, Montevideo.

CREMP (Florida Coral Reef Evaluation and Monitoring Project), 2003, Executive Summary, http://www .floridamarine.org/features/view_article.asp?id=23627.

Dwija, P. N., 2002, Studi Pertumbuhan Karang dan Pelekatan Spat Karang di Pantai Pemuteran, Kecamatan Gerokgak, Buleleng, Bali, Thesis, 45 p., Udayana University, Denpasar, Bali, Indonesia.

Ereira, A., 1990, *The Elder Brothers: A Lost South American People and Their Wisdom*, Vintage, New York.

Friends of Tobi Island, 2005, Tobian Language, http://cas.gmu.edu/~tobi/.

Gayoom, M. A., 1998, The Maldives: A Nation in Peril, 128 p., Ministry of Planning, Human Resources, and Environment, Male.

Goreau, M., T. J. Goreau, & R. Solis, 2003, Children's Environmental Education Program: Kuna Yala, Panama, Phase II, http://globalcoral.org/Children's%20Environmental%20Education%20Program,%20Kuna% 20Yala,%20Panama,%20Phase%20II.htm.

Goreau, T.J., L. Daley, S. Ciappara, J. Brown, S. Bourke, & K. Thacker, 1997, Community-based whole-watershed and coastal zone management in Jamaica, Proc. 8th International Coral Reef Symposium 2: 2093–2096.

Goreau, T. J., G. Despaigne, W. Hilbertz, L. Arosemena, U. Avila, & R. Solis, 2005, Coral Reef Restoration and Shore Protection Projects in Ukupseni, Kuna Yala, Panama: 2005 Progress Report, http://global coral.org/Coral%20Reef%20Restoration%20and%20Shore%20Protection%20Projects%20in%20Uk upseni,%20Kuna%20Yala,%20Panama.htm.

Goreau, T. J., & W. Hilbertz, 2004, Second Biorock Coral Reef Restoration Workshop Report, http://global coral.org/Biorock%20Coral%20Reef%20Restoration%20Workshop%20Report.htm.

Goreau, T. J., & W. Hilbertz, 2005, Third Biorock Coral Reef Restoration Workshop Report, http://www .globalcoral.org/Report%20of%20the%20Third%20Pemuteran%20Biorock%20Coral%20Reef%20Res toration%20Workshop.htm.

Goreau, T.J., & W. Hilbertz, 2005, Marine Ecosystem Restoration: Costs and Benefits for Coral Reefs, World Resource Review, 17: 375–409, http://global24.fatcow.com/WRR%20Goreau%20&%20Hilbertz%202005.pdf.

Goreau, T. J., W. Hilbertz, & A. Azeez A. Hakeem, 2000, Increased coral and fish survival on mineral accretion reef structures in the Maldives after the 1998 Bleaching Event, Abstracts 9th International Coral Reef Symposium, 263.

Goreau, T. J., W. Hilbertz, & A. Azeez A. Hakeem, 2004, Maldives Shorelines: Growing a Beach, http:// globalcoral.org/MALDIVES%20SHORELINES.%20GROWING%20A%20BEACH.htm.

Goreau, T. J., W. Hilbertz, S. Sackarias, & H. Hosei, 2004, Hotsarihie (Helen Reef) Project Report, http:// globalcoral.org/Hotsarihie%20(Helen%20Reef)%20Project%20Report.htm.

Goreau, T. J., T. McClanahan, R. Hayes, & A. Strong, 2000, Conservation of coral reefs after the 1998 global bleaching event, *Conservation Biology*, 14: 5–15.

Goreau, T. J., A. Tribaldos, A. Gonzalez-Diaz, L. Arosomena, & M. Goreau, 1997, Water quality in Panamanian Caribbean coral reefs, Proceedings of the Association of Marine Laboratories of the Caribbean.

Gubbay, S (Ed.), 1995, *Marine Protected Areas: Principles and Techniques for Management*, 229 p., Chapman & Hall, London.

Hilbertz, W. H., & T. J. Goreau, 1996, Method of enhancing the growth of aquatic organisms, and structures created thereby, United States Patent Number 5,543,034, U. S. Patent Office (14 pp.).

Howe, J., 1998, *A People Who Would Not Kneel: Panama, the United States, and the San Blas Kuna*, 390 p, Smithsonian, Washington D.C.

Johannes, R. E. 1981, *Words of the Lagoon*, 245 p., University of California Press, Berkeley, CA.

Lapointe, B., & K. Thacker, 2002, Community-based water quality and coral reef monitoring in the Negril Marine Park, Jamaica: Land-based nutrient inputs and their ecological consequences, pp. 939–963 in J. W. Porter & K. G. Porter (Eds.), *The Everglades, Florida Bay, and Coral Reefs of the Florida Keys: An Ecosystem Sourcebook*, CRC Press, Boca Raton, FL.

Maragos, J., (ed.), 1993, Natural and Cultural Resources Survey of the Southwest Palau Islands of Palau, 62 p., Bureau of Resources and Development, Republic of Palau.

National Research Council, 2001, Marine Protected Areas: Tools for Sustaining Ocean Ecosystems, 320 p., National Academy Press, Washington D.C.

United Nations Environment Programme, 2001, *World Atlas of Coral Reefs*, 424 p., University of California Press, Berkeley, CA.

Ventrocilla, J., H. Herrera, & V. Nunez, 1995, *Plants and Animals in the Life of the Kuna*, 150 p., University of Texas Press, Austin.

Walley, C. J., 2004, *Rough Waters: Nature and Development in an East African Marine Park*, 308 p., Princeton University Press.

8 Coastal Ecosystem Restoration through Green Infrastructure

A Decade of Success in Reviving Shellfish Beds with a Stormwater Wetland in Massachusetts

Mark Rasmussen and Stephanie Hurley

CONTENTS

INTRODUCTION

In the world of watershed restoration, many landscape projects arise with fanfare, garnering local community and volunteer support, scientific inquiry, notoriety among participating agencies and funding sources, and discussion among design and engineering professionals. Yet, these same projects inevitably outlast the attention spans of these audiences, and seldom receive review beyond the requisite final grant reports. In June of 1997, a "Final Report on Monitoring, Community Involvement and Education" was submitted by the Buzzards Bay National Estuary Program to the Massachusetts Environmental Trust, summarizing the successes of the Spragues Cove stormwater remediation constructed wetland project in Marion, Massachusetts (Buzzards Bay National Estuary Program (BBNEP) 1997). At this time, the stormwater wetland was 2 years old and its construction had already led to the reopening of shellfish beds that had been closed for harvest due to "nonpoint source" pollution. Nearly a decade later, the Spragues Cove wetland remains fully functional as both an ecological solution to water quality problems and as a valued landscape for the local community. As is the case with many landscape remediation and restoration projects, the project's longevity exceeded the timelines for grant reporting requirements. Thus, the 1997 "Final" Report is not the last word on the Spragues Cove wetland.

In this chapter, we examine a decade-old coastal constructed wetland project that was originally designed to lower fecal coliform bacteria counts and rehabilitate shellfish beds in Buzzards Bay, Massachusetts, but which has proven to do much more, functioning in turn as a valuable habitat reserve, a community-tended landscape, and a catalyst for subsequent projects around the estuarine watershed. Our intention is to review one of New England's oldest "green infrastructure" and environmental restoration projects, considering issues such as the longevity of grading and planting designs, maintenance strategies, environmental and economic benefits, and long-term community support that can be achieved through landscape alternatives to traditional stormwater engineering projects.

BACKGROUND

BUZZARDS BAY AND ITS RESOURCES

Buzzards Bay is located in southeastern Massachusetts and bordered on the south by the Elizabeth Island chain and to the east by Cape Cod (Figure 8.1). The Bay contains approximately 228 square miles of surface water with a watershed nearly twice that size. Seventeen municipalities are located either totally or partially within the bay's watershed, ten of which front directly on the bay. The Buzzards Bay area's population in 2000 was estimated at 373,690 people (U.S. Census data) and rapid growth is occurring in the region at three times the state average (SRPEDD 2002). Population growth has been correlated with increased suburbanization of the terrestrial landscape surrounding

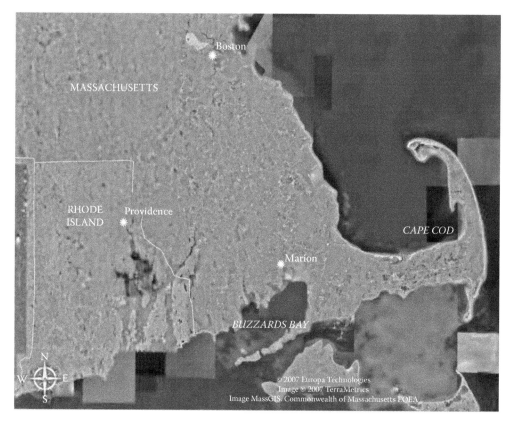

FIGURE 8.1 Map of Town of Marion, Buzzards Bay, Massachusetts. Background photo courtesy of GoogleEarth.

the Bay. Forty-two percent of the Buzzards Bay area residents live within a half mile of the coastline (Costa 2006). Undoubtedly, careful land management is a critical component of ecosystem health within and around the bay. (Haupert and Rasmussen 2003; Coalition Buzzard's Bay 2004.)

Buzzards Bay's aquatic environment provides estuarine habitat that affords numerous economic, aesthetic, recreational, and wildlife benefits for the coastal communities located within the watershed. These benefits include a viable commercial shellfishery as well as a tourism industry that is dependent upon the Bay's scenic vistas and relatively good water quality. Notably, with the exception of localized contamination issues in the vicinity of New Bedford Harbor, Buzzards Bay has comparatively fewer baywide water quality problems than other more urbanized coastal watersheds on the eastern seaboard. Nevertheless, land development and a growing population have degraded the bay's natural resources, particularly in its 30 small coves and harbors where impacts to the estuary have been attributed to poorly planned development. Specifically, the bay ecosystem is threatened by bacterial contamination from improper sewage disposal and stormwater runoff, as well as excessive nutrient loading generated by wastewater disposal and suburban sprawl development. All of these threats are directly related to the region's population growth and land use.

In recognition of these fragility of Buzzards Bay—and its intrinsic ecological, aesthetic, cultural, and economic values—the bay was designated an Estuary of National Significance by the United States Congress in 1985. The federally funded Buzzards Bay National Estuary Program (BBNEP) was established as a result of this 1985 Congressional designation and remains at the forefront of regional restoration planning. The BBNEP has focused its restoration efforts upon the projects and initiatives set forth in the 1991 Comprehensive Conservation and Management Plan (CCMP), which outlines a vision for the future of Buzzards Bay. The CCMP, currently being updated, is based on the scientific and technical information gathered by the BBNEP and analyses of the present regulatory programs designed to protect the Bay.

Shellfish Beds in Buzzards Bay

From the beginning of the CCMP development process, the cleanup of economically and culturally-important shellfish beds have been a top priority for bay restoration. Species of significance to commercial and recreational harvests in Buzzards Bay include the quahog, or hard-shelled clam (*Mercenaria mercenaria*), soft-shell clam (*Mya arenaria*), bay scallop (*Aequipectens irradians*), and oyster (*Crassostrea virginica*); quahogs dominate the commercial catch.

The closure of shellfish beds to harvest marked one of the first visible signs of an expanding pollution problem in Buzzards Bay. The first closures came in 1904 following a typhoid fever outbreak in New Bedford. Between 1900 and 1903, 565 cases of typhoid were found in the city among individuals consuming shellfish. Of these cases, 93 people died. A 3-year study of the epidemic by the state Department of Health followed in which it was determined that sewage entering the bay had caused the typhoid outbreaks. The following year, the waters around the city were closed to shellfishing indefinitely. Even after the closure, between 1904–1910, 503 cases of typhoid fever with 73 deaths were reported in New Bedford; many of these deaths were attributed to shellfishermen continuing to harvest in closed areas (Germano 1992).

Although not as dramatic as the New Bedford closures, over the following decades closures of shellfish beds occurred in all of Buzzards Bay's towns due to pathogen pollution from cesspools, sewers, runoff from urban areas, and animal wastes. Mattapoisett's first closure in 1926 was followed by closures in Wareham in 1936; Gosnold, Westport, and Bourne in the 1970s; and Marion and Falmouth in the 1980s. The expansion of closures beyond zones adjacent to the denser urban areas and points of sewage discharge tracks with the dissemination of development away from cities and into more rural stretches of the coastline.

By the end of the 1980s, 15,000 acres of shellfish beds were closed due to pollution around the bay. This number peaked in September 1990 when 16,583 acres, approximately 73% of all the bay's once harvestable shellfish beds were closed. A little more than a decade later, in response to efforts to eliminate sewage and stormwater discharges to the bay, shellfish beds closures had been reduced by 43% with 9,392 acres closed (Coalition for Buzzards Bay (CBB) 2003). This action represents a very important, as yet unfinished, bay-management success story.

Although cleanup of direct sewage discharges and failing home septic systems drove the majority of early bed reopening, the majority of contaminated shellfish beds today are closed due to stormwater runoff from urban and suburban roads and other impervious surfaces. The 9,392 acres of shellfish beds remaining closed represent 41% of all commercially-harvestable beds and therefore significant economic losses for the region (CBB 2003). Whereas once the battle to reopen shellfish beds targeted sewage, today the focus is on elimination of polluted runoff from the more than 2,600 pipes that discharge to the Bay when it rains (BBNEP 2003, NRDC 1999).

FOCUS ON STORMWATER

In 1991, the 13 municipalities comprising the bulk of the bay watershed signed the Buzzards Bay Action Compact to voice their support for the goals and objectives contained within the Buzzards Bay CCMP, committing the Bay community to the implementation of the recommendations in the plan. Although progress was made on many of the action items listed in the plan—including establishing boat no-discharge areas, updating of sewage treatment plants, rewriting of the state's septic system rules, and the promotion of nitrogen removal septic systems (Costa 2006)—no area received more attention than those efforts aimed at reopening shellfish beds through the remediation of stormwater discharges. Throughout the 1990s, the BBNEP, local municipalities, and advocacy groups such as The Coalition for Buzzards Bay, determined prioritizations of sites for designing stormwater treatment systems throughout the Bay watershed and worked toward their construction.

The largest and most successful of these early projects occurred in the towns of Wareham and Bourne in the eastern part of the watershed that is dominated by well-drained sandy soils. Stormwater treatment infrastructure in these areas typically utilizes leaching catch basins and other forms of direct infiltration to eliminate direct discharges of at least the "first flush" of rainfall (i.e., the first half-inch of rainfall that generally transports 80–90% of contaminants). These simple systems function by separating primary solids out within chambers in the catch basins, and allowing infiltration of the remaining flow through perforations in the basin or through an adjacent leaching field (often constructed beneath the road surface and therefore not requiring additional land area). Construction of these systems has often been timed with regular road maintenance or other utility projects to save on cost and has now become virtually standard practice in these communities.

By focusing on the filtering and infiltration of the first flush, these strategies were successful in reducing the amount of hydrocarbons, metals, and pathogens in the runoff, thus triggering the reopening of shellfish beds for harvest. Notably, these methods were not designed for, nor did they achieve, significant reductions in nutrients (phosphorous and nitrogen), which are a significant concern in nearshore areas throughout Buzzards Bay.

In the western portions of the watershed, from Westport to Marion, stormwater remediation projects have been much slower to develop. One reason for this is that few locations around the bay are suitable for infiltration. Due to the dominance of poorly-drained, glacial till soils within the region, the palette of potential forms of stormwater treatment has been limited to large surface detention ponds and wetlands. Designers of these systems have aimed to capture the "first flush" and allow pollutants to settle out of the water column prior to discharge. Although such projects have the technical potential to remove the sought-after pollutants, their achievement has been compromised by several on-the-ground factors, ranging from spatial to social challenges. These include:

- Lack of available land area, e.g., for surface detention ponds near roads in existing suburban neighborhoods
- Lack of local precedents or models, e.g., to educate officials about "softer" infrastructure techniques

• Citizen concerns about the aesthetics and perceived public safety risks of creating wetlands in their neighborhoods

Recognizing that a large demonstration project was needed to combat these challenges, and that the town of Marion was eager to try something to improve water quality in an area adjacent to their public swimming beach, the BBNEP embarked on the largest and most important stormwater remediation project around the bay at Spragues Cove.

SPRAGUES COVE STORMWATER REMEDIATION PROJECT

Spragues Cove is a three-acre estuarine shellfish bed in Sippican Harbor within Buzzards Bay in the town of Marion, Massachusetts (Figure 8.1). It is adjacent to Silvershell Beach, Marion's primary public swimming beach. Spragues Cove is fed by a small creek that drains a 64-acre suburban residential watershed with an average housing density of one-half acre per home. In the 1950s, the landscape of the cove was converted from a salt marsh to a filled parking area, using materials from dredge spoil from adjacent Sippican Harbor. Throughout the 1990s, fecal coliform bacteria counts in stormwater runoff flowing into the cove from road discharges into the creek resulted in the closure of shellfish beds for harvest, and elicited the concern of Marion's town residents and the (BBNEP 2006).

The nature of the stormwater conveyance system and residential development pattern in the area around Spragues Cove was typical of many Buzzards Bay neighborhoods—poorly-drained soils, a high water table, significant watershed conversion due to the filling of wetlands and loss of natural forest cover, and 50- to 100-hundred-year-old road and drainage networks designed to move rain flows as quickly as possible to the nearest waterway without any regard for pollutant removal. The commonality of these characteristics reinforced the value of investing in the development of an innovative solution at this site as a model demonstration project for other Buzzards Bay towns.

Together, the town of Marion and the BBNEP, who served as the lead coordinating entity for a host of federal and state government agencies, studied the bacterial contamination issue and evaluated several treatment alternatives, including chemical treatments such as chlorination, UV light, ozone, and reverse osmosis; and physical methods such as infiltration, settling, and constructed wetlands. Most of the former treatment options would require a significant capital investment; of the latter options, infiltration was determined impossible due to the site's hydric soils. A constructed wetland with settling areas was determined to be the most viable options to meet economic and ecological needs.

The BBNEP was greatly assisted at this stage of the project by the USDA's Natural Resources Conservation Service who donated staff to develop the engineering designs for the system and worked with the BBNEP and local advocates to present the designs to the town meeting and officials.

Community support was achieved by a combination of outreach and education beginning in the early 1990s. The BBNEP, The Coalition for Buzzards Bay, and interested citizens used a combination of brochures, newspaper articles, and informal breakfast meetings to generate public support. In addition, the BBNEP constructed a 1 in. = 20 ft scale model of the constructed wetland, which appeared at numerous meetings and discussions and helped clarify the project's details. The model served as a visual tool to explain both how the wetland system would work and how it would look when completed. After extensive conversations, the majority of the citizens of Marion believed that concerns had been adequately addressed, and they overwhelming approved the project by vote at the town's 1994 Town Meeting (BBNEP 1995).

The town of Marion decided to allocate nearly half of the approximately four-acre beach parking lot to the wetland project, including a basketball court and "prime" waterfront parking. Aside from the land value, estimated at around $200,000 in 1993 dollars, the total project budget amounted to $64,000, which was comprised of a combination of federal and local funds that included $19,000 from the town of Marion (further details on project costs are discussed later in this chapter).

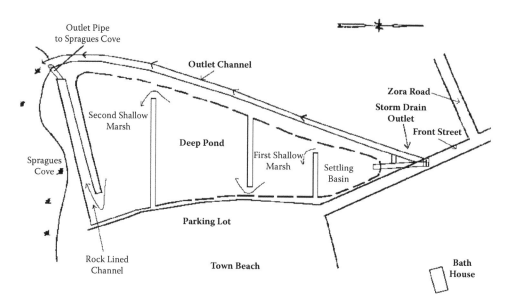

FIGURE 8.2 Diagram of Sprague's Cove Wetland Design. Courtesy of BBNEP 2006, htttp://www.buzzardsbay .org/sprafact.htm.

WETLAND DESIGN: GRADING, VEGETATION, AND THE NEED FOR FLEXIBILITY

Although the land area donated to the project was significant, so is the stormwater volume being discharged to the cove from 64 upland acres of suburban land. The constructed wetland project is designed to have a 14-day retention time for every 1 in. of stormwater, and is thus able to treat the "first flush" of pollutants (EPA 2006). The spatial layout of the project consists of a forebay settling basin—which is 6 ft deep on average and provides the bulk of treatment benefits—and three large cells: two shallow marshes that sandwich a third deep marsh (Figure 8.2).

Water flows through the project in a snaking S-curve, shaped by the linear vegetated dikes that separate the marshes from each other. The deep marsh in the center serves as a micropool (approximately 6 ft deep) for fish refugia as well as significant storage capacity for the project (Figure 8.3). The linear—in fact, nearly orthogonal—grading of the wetland design was largely a product of spatial constraints and calculated treatment volume, rather than an aesthetic decision (Taber 2006). Ultimately, the treated water is discharged through a small culvert beneath the last dike into Sprague's Cove.

One planting design challenge was to select freshwater vegetation that is tolerant to salt spray as well as occasional inundation with brackish tidal flows moving back up the culvert into the constructed wetland project. Project plantings included aquatic plants that have filtration potential, such as soft-stemmed bulrush and narrow-leafed cattail in the shallow areas, as well as lilies and other floating plants along the shore (Figure 8.4). Notably the cattails and bulrushes, though initially scattered evenly throughout the marsh have "self-selected" local niches, and now the individual species dominate in one area or another. The dike side-slopes were seeded with a "generic conservation seed mix" (Taber 2006). Project planting was largely achieved by the volunteer efforts of town residents.

Notably, human habitat is scarce in this project. Due to initial safety considerations with regard to the deepwater areas, as well as the close proximity of neighboring beachfront properties, people were largely designed out of the wetland. Shrubs, including native roses and beach plum, were part of the planting design along the edges of the wetlands to deter people from entering the wetland. A split-rail fence also signifies that people are meant to be excluded, although an interpretive sign outside the fence provides some educational value (Figures 8.5 and 8.6). Only one of the dikes—that which is closest to the shoreline—is accessible without clambering over the fence, and shrubs block the "path" along the top of the dike. This inaccessibility, though understandable due to space constraints

FIGURE 8.3 Photograph of Deep Pond/Marsh section of Sprague's Cove Constructed Wetland (April 30, 2006). *Photo by S. Hurley.*

and perceived safety concerns, is one of the few aspects of the Sprague's Cove project that could be considered a weakness. Numerous constructed wetlands projects the world over have successfully integrated habitat restoration and environmental problem-solving with compatible human recreational uses (see, e.g., Bays 2002). Allowing foot traffic within constructed wetland projects creates opportunities for *in situ* environmental education that signage alone cannot achieve.

The Sprague's Cove project illustrates the need to be flexible in designing constructed wetlands, allowing vegetation assemblages and maintenance regimes to shift over time and in accordance with unpredicted site conditions. For example, soon after construction there were contamination problems created by ducks and geese landing on the newly-created openwater "ponds." The Department of Public Works had initially been mowing the dikes to keep the grass down within the wetland,

FIGURE 8.4 Photograph of aquatic vegetation and shrub-covered berms (April 30, 2006). *Photo by S. Hurley.*

FIGURE 8.5 Photograph of split-rail fence at project edge (April 30, 2006). *Photo by S. Hurley.*

but this led to wider areas of open water for waterfowl, so the maintenance response was to "let it go natural" (Taber 2006). Another complication was a late 1990s drought that killed significant numbers of the Sprague's Cove project plants. Fortunately, townspeople rallied to replant them and the vegetation—a mix of planted species and volunteers—is now thriving. Project maintenance has also had to combat the growth of *Phragmites* spp.; the BBNEP initially tried to control the invasive reed using paintbrush 'Roundup' applications, but has since discontinued efforts, accepting the presence of the invasive plant with the knowledge that it does offer equivalent water-quality benefits. Costa (2006) notes that the *Phragmites* problem was more of a concern early after the wetland construction because then, in some areas the invasive plant was out-competed by shrub species that established on the berms. Further, the geese (considered more of a challenge than the *Phragmites*) have been discouraged by the fear of predators who have found habitat among the shrub vegetation (Costa 2006). The discussion of these interrelationships between flora and fauna, both volunteer and

FIGURE 8.6 Photograph of interpretive sign in parking lot (April 30, 2006). *Photo by S. Hurley.*

managed, underscores the need to allow functional dynamics and interactions among organisms in created ecosystems (e.g., constructed wetlands) to develop over time.

PROJECT BENEFITS

Environmental Benefits

Completed in June 1995, the project has, all in all, been considered a success and remains a hallmark of stormwater remediation initiatives in the Buzzards Bay area. As a direct result of the project, fecal coliform bacteria counts in Sprague's Cove are now low enough to allow commercial and recreational shellfishing. Bacteria counts within a year of construction we recorded as low as < 10 organisms per 100 ml where previously they had been known to be in excess of 20,000 organisms (BBNEP 1997). As an added benefit of the wetland project, the town no longer has the frequent pipe maintenance problem associated with the old stormwater outfall.

Whereas before stormwater flowed as a channelized stream, prone to flooding near its outlet, runoff is now slowed and filtered via a designed combination of grading, soils, and vegetation, leaving contaminants behind as it discharges into the Bay. The forebay, alone, significantly removes sediments from the water column. Even in storm surges the flows of runoff are halted upon arrival at the forebay, causing particles—and the contaminants adsorbed to them—to settle to the bottom of the sedimentation basin. Soon after construction, more sediment than expected entered the forebay. This has since been explained; there was a lot of sediment because the former stormwater discharge pipe was at a 0.5% grade, so that when the pipe was redirected into the stream the change in slope effectively allowed the pipes to clean themselves out over the first three or four rain events (Taber 2006). Even after a decade of receiving storm flows, the forebay still successfully sequesters suspended solids and heavy metals, leaving the micropools of grading and wetland vegetation downstream to filter out remaining bacteria, hydrocarbons, and metals.

Interestingly, while it was a recommended maintenance procedure at the time of project construction, dredging of the forebay has never occurred; it is possible that if the forebay were to be dredged, it would adversely affect the plant community now assembled there (Taber 2006). We speculate that the hydrophytic vegetation that has been able to establish within the forebay is quite tolerant of rapidly fluctuating water levels as well as adverse soil conditions.

Although the project was aimed at improving habitat for the shellfish in Sprague's Cove, the constructed wetland's intrinsic habitat value should not be overlooked. Compared to the previously existing parking lot, the Sprague's Cove wetland is a haven for wildlife. Beyond the reopening of the adjacent shellfish beds, other fish repopulated the new wetland system on their own, including mummichogs in the wetland's deep pool. Another benefit of the decision to provide a deep pool for the maintenance of small fish population has been the effective natural control of mosquitoes in the wetland. The wetland plants offer food for migratory birds, and numerous amphibians live in the marshes.

In addition, the Sprague's Cove project has become one piece in a larger puzzle wherein a coastal landscape has begun to function more according to natural hydrology and less according to engineering. The stormwater wetland is demonstrative of the notion of restoring watershed functions in a suburban setting, where such functions may tend to be hidden within infrastructure. Further, by creating a landscape that enables water to seep slowly through soil and plant roots, ameliorating conditions of anthropogenic pollution, and offering economic restoration to at least a small part of the local shellfish industry, the designers have also constructed a forum for environmental education in Marion's back yard.

Project Funding & Economic Benefits

The pollution of Sprague's Cove from stormwater runoff presented an ongoing economic loss to the Town of Marion, most clearly in the form of reduced commercial shellfish harvests. Beyond direct

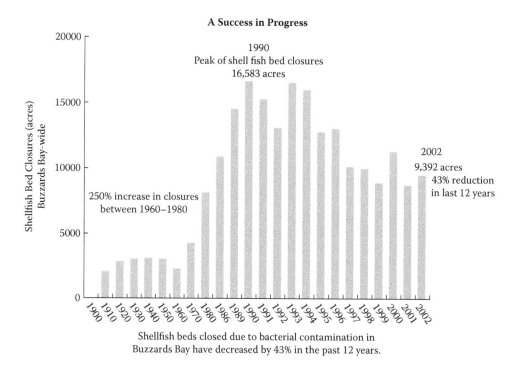

FIGURE 8.7 Decline in shellfish bed closures. Courtesy: Coalition for Buzzards Bay 2003, http://www
.savebuzzardsbay.org/pdf/state_of_the_bay.pdf.

economic calculations is the fact that shellfishing is also an important cultural value on the Buzzards
Bay coast, and bed closures due to pollution therefore diminish overall quality of life.

Moreover, there existed the potential that adjacent Silvershell Beach, the town's primary and
very popular public beach, could be closed to swimming if pathogens were detected in regular water
sampling. In a seaside community that relies on summer tourism and vacation homes for a large part
of its economic activity, town officials seek to avoid the stigma of beach closures.

In response to these potential economic losses, the Town of Marion invested more than $219,000
in the construction of the Sprague's Cove Stormwater Wetland. Approximately $200,000 of this invest-
ment came in the form of the dedication of roughly 2 acres of the existing beach parking lot to the project, waterfront land which was valued at $100,000/acre in 1993 dollars. The remaining $19,000 was
appropriated for the construction by the town's voters at its annual Town Meeting (BBNEP 1995).

Additional funding for the project came from a combination of local, federal, and private funds.
Marion received $25,000 in federal funding through the EPA Section 319 grant program. Another
$10,000 came from the U.S. Fish and Wildlife Service's Wetlands Restoration Grant Program,
and $10,000 from a local, private family foundation (BBNEP 1995). The total project budget was
$264,000 with actual construction costs at $64,000.

These cash figures do not include hundreds of hours of labor on the part of the town's Public
Works Department, federal and state agency staff, and donated citizen volunteer hours invested in
the project. In addition, plant donations continued and additional postconstruction funds for community involvement, water quality monitoring, and plant maintenance were provided by the Massachusetts Environmental Trust.

The benefits of this combined investment in the restoration of Sprague's Cove were visible
immediately to town officials involved in the project and became very clear to the broader public
within a year. As an added restoration bonus, sand and gravel excavated from the parking lot (originally dredge material from the harbor used to fill the saltmarsh in the 1950s) was put to local use

to replenish sand on adjacent Silvershell Beach and to prevent coastal erosion on nearby Planting Island Causeway, providing a significant cost savings to the town.

Most exciting, however, was the decision by the Massachusetts Division of Marine Fisheries to reclassify the Sprague's Cove beds as open for harvest only a year following construction of the stormwater wetland. Water quality sampling done by the division as well as the Buzzards Bay National Estuary Program clearly documented dramatic improvements in the quality of the storm-water being discharged into the cove from the new wetland system.

Finally, it is worth mentioning the economic benefit (avoided costs) of never having to deal with any impacts that would have been associated with a water-quality-related beach closure at Silvershell Beach, a draw for town residents and tourists. In addition, other indirect benefits that have not been calculated may include real estate enhancement value for the adjacent waterview homes that had formerly overlooked a neglected parking lot but now overlook the attractive constructed wetland.

COMMUNITY BENEFITS

Beginning in the early 1990s, members of the Marion community who were interested in the health of Buzzards Bay were able to follow the project from its inception to its construction and thus were able to witness the changes in the coastal restoration over time. From early town meetings led by the BBNEP and The Coalition for Buzzards Bay, through the volunteer days of project planting and "emergency" replanting after the late 1990s drought, many members of the town citizenry—including activists from a town-based "Save Our Seas" group—have come to learn more about their home landscape (and waterscape), and have established ties with each other, as members of the same community.

Beyond the aesthetic values of replacing a wide swath of parking with vegetation and wildlife, the recreational values of improved water quality and views have also been bolstered by the project (Figures 8.8 and 8.9). Where the naturalistic project aesthetic may not have been enough, a few

FIGURE 8.8 Aerial view of Sprague's Cove parking lot, before the wetland construction. Background photo courtesy BBNEP 2006, http://www.buzzardsbay.org/sprafact.htm.

FIGURE 8.9 Aerial view of Sprague's Cove after construction. Background photo courtesy: BBNEP 2006, http://www.buzzardsbay.org/sprafact.htm.

members of the community have sought to add flourish to the edges of the wetland project: daffodil bulbs, an indicator that the project is cared for (Figure 8.10). Thus, despite the minimal physical access into the project, the views—and perhaps the knowledge that the swimming beach and shellfish beds are cleaner—seem to have satisfied local groups, who consider the project a valuable

FIGURE 8.10 Daffodils planted along inside of fence at Sprague's Cove (April 30, 2006). Photo by S. Hurley.

FIGURE 8.11 Aerial photo of Sprague's Cove project (arrow points to constructed wetland) and Silvershell Beach in Marion, Massachusetts. Courtesy: Coalition for Buzzards Bay 2004, http://www.savebuzzardsbay .org/bay-info/BBGRP/PDFs/15.pdf.

use of public land. Figure 8.11 shows the visible integration of the Sprague's Cove wetland with its surrounding landscape.

DISCUSSION: REPLICATION OF SPRAGUE'S COVE'S DESIGN

Two years after construction, Ray Pickles, the town of Marion's executive secretary, had this to say about the effort:

> The Sprague's Cove drainage remediation project presented the designers and engineers with the Buzzards Bay Project with some unique challenges .... These challenges were met head-on with careful study and planning, which has produced a very effective drainage remediation project. The results are better than anyone expected. In addition to being an efficient treatment of urban runoff, the Town is left with an extremely pleasing site filled with wetland plants and flowers, the return of native wildlife, and a pleased citizenry. The introduction of fish in the deep lagoon and improved habitat has reduced the summer mosquito population and has not resulted in a single neighborhood complaint. This is a unique cooperative project in that it combined the resources of the NRCS, Massachusetts DEP, MCZM, U.S. Fish and Wildlife Service, and the Town of Marion.
>
> **—Ray Pickles, Executive Secretary, Town of Marion (1997 Biennial EPA Review of the Buzzards Bay National Estuary Program)**

Though the project is over a decade old, his words still hold true today. As Pickles' suggests, the combination of good design, community acceptance and volunteerism, and the cooperation of multiple federal and local institutions and agencies (in both funding and construction efforts) have enabled the success at Sprague's Cove. Without any of these elements, the project could not have been achieved.

Evidence of the successful application of a landscape based "green infrastructure" project can perhaps be found in its imitation elsewhere in the watershed. Because the Sprague's Cove wetland has proven to meet water quality restoration goals as well, and garnered widespread community support, the town of Marion has embarked upon another stormwater remediation project. In the summer of 2005, the town completed a new $203,000 "rain garden" stormwater retention system at Island Wharf Park. Water quality problems in the area around Sippican Harbor near Island Wharf mirrored those plaguing Sprague's Cove ten years earlier. Bacterial contamination from suburban stormwater runoff discharged into the harbor from two pipes has been linked to shellfish bed closures and threatened an adjacent public beach.

With a smaller contributing watershed of only 7.5 acres (BBNEP 2005) less land area was needed to address the problem; however, sites were scarce in the center of Marion's densely developed downtown village. Island Wharf Park, a small waterfront park that hosts the town's performance band shell, Harbormaster's Office, and marina parking, provided the only available site to construct a stormwater project. Marion chose to excavate a nearly 1-acre area of the park's lawn to construct a rain garden bioretention basin to treat the first 1-in. of stormwater runoff from the village. From the surface, the rain garden appears to be a large depressed landscaped island (Figure 8.12). Although, unlike Sprague's Cove, the rain garden is not designed to retain open water, it provides another local example of applying "green infrastructure" techniques for solving stormwater problems.

The choice to construct a vegetated, soft-infrastructure solution to improve water quality and the adjacent shellfish beds is again significant because like Spragues Cove, the project effectively reduces the useable area of public park and recreation space. Where public space is limited, the public's willingness to utilize open space for a project aimed at water quality restoration is noteworthy. As the town's Public Works Superintendent Robert Zora commented: "The public's positive response to Sprague's Cove made it possible for us to consider a similar alternative at Island Wharf. We are committed to cleaning up pollution problems in the Harbor and these projects are helping us reach that goal" (Zora 2006).

Unfortunately, with the exception of the Island Wharf project discussed above, Buzzards Bay communities have been slow to embrace the use of surface retention and detention basins or constructed wetlands like Sprague's Cove for improving stormwater hydrology and water quality. Stormwater remediation projects overall have slowed in the bay watershed in recent years as compared to the large number of projects undertaken in the mid-1990s. Further, where projects have gone forward, they tend to lean toward hard solution (catch basin and piping) approaches that can take advantage of highly permeable soils for infiltration predominantly in the eastern portions of the watershed.

We believe the reasons behind this trend are a combination of the lack of availability of suitable land proximate to problem stormwater discharges, and the lack of community awareness of and

FIGURE 8.12 Rain garden at Island Wharf Park. Courtesy: BBNEP 2005, "Island Wharf Biofilter" flyer.

acceptance of options such as naturalistic wetland and rain garden treatment systems. However, in towns like Marion—where the public is familiar with these landscape approaches—new projects are proceeding; this points to the need for additional models and demonstration projects throughout the watershed.

We hope that by highlighting the ecological, economic, and community benefits of projects like those in Marion, Massachusetts, we can contribute to fostering awareness about landscape solutions to problems that traditional stormwater engineering has been unable to solve, and that creative "green infrastructure" solutions such as those demonstrated at Sprague's Cove may continue to spread throughout Buzzards Bay and beyond.

ACKNOWLEDGMENTS

We offer sincere thanks to the Buzzards Bay National Estuary Program, especially Executive Director Joe Costa and Stormwater Specialist Bernie Taber, for their time and feedback. Without their valuable input on the content of this chapter, and their integral roles in facilitating the Sprague's Cove constructed wetland project's success over the last decade, the telling of this story would not have been possible.

LITERATURE CITED

Bays, James S. 2002. Principles and Applications of Wetland Park Creation. *Handbook of Water Sensitive Planning and Design* (R.L. France, Ed.).

Buzzards Bay National Estuary Program. 2003. *Atlas of Stormwater Discharges in the Buzzards Bay Watershed.*

Buzzards Bay National Estuary Program. 1991. Comprehensive Conservation and Management Plan (CCMP).

Buzzards Bay National Estuary Program. 1997. Final Report on Monitoring, Community Involvement and Education to Massachusetts Environmental Trust, June 1997.

Buzzards Bay National Estuary Program. 2005. Island Wharf Stormwater Project flyer.

Buzzards Bay National Estuary Program. 1995. Section 319 Final Report, Spragues Cove/Front Street Storm Water Remediation Project, Marion, MA. Nonpoint Source Project Number C9001214-92-0. September 1995.

Buzzards Bay National Estuary Program. 2006. Spragues Cove fact sheet, updated December 15, 2006: http://www.buzzardsbay.org/sprafact.htm, accessed January 13, 2007.

Coalition for Buzzards Bay. 2004. *Research Planning, Inc. Buzzards Bay Geographic Response Plan.* Sippican Harbor, Summary Sheet May 2004, http://www.savebuzzardsbay.org/bay-info/BBGRP/PDFs/15.pdf, accessed January 13, 2007.

Coalition for Buzzards Bay. 2003. *State of the Bay 2003,* http://www.savebuzzardsbay.org/pdf/state_of_the_bay.pdf.

Costa, Joseph E. 2006. BBNEP Executive Director. Personal communications. December 2006.

EPA 1997. Biennial Review of the Buzzards Bay National Estuary Program.

EPA 2006. Wetlands to the Rescue-Spragues Cove Stormwater Remediation Project. http://www.epa.gov/owow/nps/Section319II/MA.html, accessed January 13, 2007.

Germano, Frank. 1992. History of Shellfish Beds Closures in Buzzards Bay: 1900 to 1992.

Haupert, C. and Rasmussen, M. 2003. State of the Bay 2003, The Coalition for Buzzards Bay.

Massachusetts Department of Environmental Protection. 2005. Indicative Project Summaries, Section 319 Nonpoint Source Competitive Grants Program. FFY 2001–2005. Bureau of Resource Protection.

Natural Resources Defense Council. May 1999. Stormwater Strategies: Community responses to runoff pollution. Ch. 6 Strategies in the Northeast, http://www.nrdc.org/water/pollution/storm/chap6.asp, accessed January 13, 2007.

SRPEDD. 2002. Southeastern Regional Planning and Economic Development District Vision 2020, http://www.srpedd.org/background.html, accessed January 13, 2007.

Taber, Bernie. 2006. Buzzards Bay NEP. Personal communications. May 2006; December 2006.

Zora, Robert. 2006. Public Works Superintendent. Personal communication, December 04, 2006.

Intermezzo

Com'era, Dov'era: Battling Water, Time, and Neglect with MOSE and Other Techno-Fix Reparations in Venice[1]

> We must remind the worshipers of the lagoon *com'era dov'era* (as it was and where it was) that it has been continually reorganized throughout the centuries in order to adjust to the requirements of Venice's life and prosperity.
>
> —**A. Rinaldo**, *in* **Musu, I. (Ed.)** *Sustainable Venice: Suggestions for the Future* (2001)

INTRODUCTION

The sustainable future of Venice does not look good. Indeed, Fay and Knightly concluded their groundbreaking 1976 book with the pessimistic statement that "there is no hope of saving her." And almost 30 years later, one of the prefaces to Fletcher and Spencer's (2005)[2] comprehensive tome about Venice's environmental challenges posits that time seems to be running out for this loveliest of cities. Rather than succumbing to apathy, however, Venetians and the international community have implemented a series of interventions designed to preserve this Queen of the Adriatic.

ON-SITE DEFENSES AND REPAIRS

BUILDINGS

If there is one constant in Venice, blatantly obvious to both tourist and resident alike, it is the chronic state of disrepair of much of the city's buildings. Preservation of architectural heritage is really only a recent phenomenon in Venice. Research on the island of Torcello, for example, has shown that the early lagoon dwellers had periodically torn down and built new structures on top of the old foundations in much the same way as an archeological tell. This practice stopped, however, with the fall of the Republic, at which time Venice began to evolve into the fossilized museum of decrepit charm (or shame) that it is today.

Venetian restoration has until recently been synonymous with a preoccupation in preserving the city's building heritage (Sabaeadv 1987). British intellectuals and artistic elite—the self-imagined heirs of Byron, Turner, and Ruskin—decided in the 1870s that it was their particular mission to "save" Venice (Pemble 1995). The fall of the Campanile in the Piazza San Marco in 1902 served as a powerful metaphor for the crumbling of Venice and an overt reminder of the inability of Venetians to effectively deal with the problem. This served to inspire a campaign of restoration, often by international donors, that has continued ever since (Pertot 2002). Emphasizing the city as no longer being the patrimony solely of Venetians but rather part of the cultural heritage of all has meant that huge amounts of funding has poured in from abroad such that there are now over 50 different organizations working on building restoration in addition to UNESCO.

---

[1] An expanded version of this essay will appear in the book *Waterlogged: Environmental Reflections on Venice.*

[2] See *Finale* section for complete reference listing.

Today's Venice is a city of scaffolding, buttresses, and dry-barriers, all part of the constant rescue and renovation efforts necessary in the city's struggle to survive against increasingly high waters and decades of neglect. One of the most obvious band-aid measures taken to address the symptoms of flooding rather than the causes are the ubiquitous presence of *paratia* or temporary metal barriers placed across doorways. More subtle measures include construction of underground cisterns to hold water and thus protect the ground floor of buildings from flooding (Pertot 2004).

PEDESTRIAN WAYS

Today it is, of course, impossible to demolish and rebuild Venice's time- and water-ravaged buildings. However, the raising of pavements, walkways, embankments, and bridges was identified as early as the start of the 18th century as a viable solution to the subsidence problem (Spinelli and Follin, *in* Fletcher and Spencer 2005). Raising sidewalks and the sidewalls of canals are, therefore, believed to go a long way toward "saving" Venice from the sea. Experts consider that this single effort, alone, could in fact solve one half to two thirds of the city's flooding problems and should therefore be a top priority (Nova 2002: Pertot 2004; Fletcher and Da Mosto 2004). Since the mid-1990s, these *insulae* or islets have been created by building up ground levels (going from 100 to 120 cm above sea level) and making them impermeable through changing the underground drainage system. To date about 20% of the 100,000 m² earmarked for raising has been completed. The problem here, of course, is the obvious one that there are limits to how much the ground can be raised without affecting the look of or access to buildings.

One of the most visible of these on-site renovations has occurred in Venice's most busy location. Water invades the Piazza San Marco in three ways: flowing from the lagoon, rising up through the drains, and seeping through the subsoil. Therefore, raising the outer edge, isolating underground conduits to prevent back-flowing, building a new rain runoff collection system, installation of a conveyance system with pumps out to the lagoon, and laying down a bentonite sealant to prevent soil seepage have all been implemented as part of the renovation (Biotto, *in* Fletcher and Spencer 2005).

CANALS

A 1939 urban plan stated that "Venice is above all a city requiring reclamation" (Pertot 2002), and called for reopening buried canals to allow easier communication of districts to the historic center. Despite this, most one-time canals remain in their buried *rio terra* state. The extent of the restoration works are nowhere near as dramatic as has occurred elsewhere for stream "daylighting" projects. For the first time in many decades, canals are finally being dredged and cleaned (sometimes of up to a meter of contaminated sediment) and their sidewalls stabilized. The bordering water frontage of buildings and bridges are also being restored as well as the underground infrastructure (Pertot 2004). And some have suggested the installation of small mobile barriers to temporary close off canal inlets from the rising waters of the lagoon, though no steps have been taken in this direction beyond conceptual plans.

POSTINDUSTRIAL REGENERATION

Attempts to rejuvenate the crumbling architecture and economy of Venice have been underway for years. The hulking bulk of the long-abandoned Arsenale, the one-time naval shipyard of the Republic and the largest structure in the old city center, has long been a favorite subject of design charettes and studios in architecture schools around the world. One of the more intriguing ideas would transform part of the Arsenale into an ecotourism museum operating as a new doorway to the lagoon (Forum per la Laguna 2007). The project goals are to redirect tourist flow from the Piazza San Marco towards the museum and then out into the lagoon to promote sustainable tourism, and to recover the aquatic heritage of the formidable building.

Recently, Venice has begun to create a strategic plan for redeveloping many of the abandoned warehouses and other structures in Porto Marghera and Mestre as the workforce and industrial output of the area continues to decline. Preliminary plans now exist from a suite of real estate developers to reuse these complexes for research, entertainment, housing, and education purposes, to be accompanied by creating a series of large parks and open spaces (MIPM 2006). Another promising proposal is to reclaim the central lagoon island of S. Giorgio in Alga, which had originally been a Benedictine monastery, then a political prison, and finally a powder magazine until it was abandoned in the middle of the last century. The idea advanced by the Forum per la Laguna (2007) is to create a study center on "Environment and the City" based on restoring the ruins as an interpretive center along with administration offices, a restaurant, and guestrooms, and all surrounded by re-landscaped open space.

LAGOON DEFENSES AND REPAIRS

The 1971 special issue of *Architectural Review* was significant in that it was really the first attempt in the international arena to widen the understanding of the complex myriad of problems facing Venice, in particular the need to move beyond Ruskin and his ideological descendants' artistic preoccupation with saving individual buildings in order to consider larger topographical, technical, oceanographic, financial, and political issues (Plant 2002). UNESCO began to look at water quality in Venice's canals and stated that "the preservation of a single work of art in Venice increasingly implies the repair and consolidation of the many elements of its environment just as the man-made city of Venice itself can only be meaningfully restored in the context of the preservation of her much larger, more complicated, and more sensitive lagoon" (UNESCO 1979). A decade and a half later, the high visibility organization, Venice in Peril, realizing that there was little point in rescuing individual buildings when the whole city remained precariously threatened, reinvented itself from an organization focused on restoring monuments to one dealing with the waters of the lagoon. Today, almost all recognize the need to regard Venice and its lagoon as a single entity. Venice is thus very much a "diffuse city" (Caniato, *in* Fletcher and Spencer 2005), with the lagoon neither being the frame nor the backdrop for the city (Fletcher and Da Mosto 2004).

SEA WALLS

Armored seawalls along the outer edge of the barrier islands of Pellestrina and the Lido were constructed in Medieval times and have been continually added to ever since (e.g., another 12 miles being built along the Lido from 1744 to 1782 with a 14-m thick base). In the 18th century, one such seawall constructed at great expense included a plaque with the following inscription: "The Guardian of the Water has set this colossus made of solid marble against the sea so that the sacred estuaries of the city and the Seat of Liberty may be eternally preserved" (Fay and Knightly 1976).

During their occupation in the 19th century, the Austrians spent about $2 million per year on seawall maintenance and added 15,000 tons of marble a year as reinforcement. In contrast, by 1960, the Italian government in Rome (which has generally ignored Venice) was spending on average only $1 per yard per year to maintain Venice's sea defenses (Fay and Knightly 1976). Consequently, today's seawalls are in obvious need of repair. Other protective measures have been implemented: 45 km of beaches have been reconstructed and 11 km of breakwaters reinforced (MITVWACVN 2006a).

For some, the only long-term solution to Venice's flooding problems is to extend the seawalls across the mouths of the channels between the outer islands, thus permanently sealing off the lagoon from the Adriatic (Nova 2002). Construction of such permanent walls or dikes would transform the lagoon into a freshwater lake but would also assure that it would never flood again. Obviously incredibly controversial, its adherents steadfastly argue that only implementation of extreme corrective measures can solve extreme problems. In this regard, the concluding commentary in the 2002 Nova documentary simply stated: "The Venice you know today cannot be preserved as it is today." Others have, however, proposed corrections that, though certainly more extreme in cost, might not be as much so in terms of their environmental consequences compared to the strategy of permanent barriers.

MOSE Barrier

About 400 million m³ of water is exchanged daily between the lagoon and the sea with an average tidal height of about 70 cm. With the threat of sea level rising, most experts agree that Venice's only hope for the future is to seal itself off from the Adriatic when the need arises (Fletcher and Da Mosto 2004). In fact, plans for building a system of outer gates go back to the 17th century (Caniato, *in* Fletcher and Spencer 2005). Today, about the only thing generally agreed upon in terms of building a flood barrier is that the challenge to pull this off successfully is very big indeed. The system of gates must be able to deploy rapidly and with reliability, require little maintenance, not negatively impact navigation and commercial port activities, have no or only short-term limited effects on lagoon water quality and ecology, and when not in use, be unobtrusive and thereby not affect site aesthetics (Fletcher and Da Mosto 2004; Spencer et al., *in* Fletcher and Spencer 2005c).

The first mention of building a system of mobile gates, believed to be the answer to satisfy the aforementioned concerns, arose in the early 1970s, only to be followed by years of plans, programs, political bickerings, and procrastinations (Keahey 2002). The Consorzio Veneza Nuova, a consortium of large Italian engineering and construction companies, was created by a special law in 1984 to undertake construction of the flood barrier. In 1988, a full-sized module of a single gate was hauled along the coastline to gain public support, with the assembled crowd clapping as the gate was raised in place and attached to a large derrick (Nova 2002). Referred to then as the *Modulo Sperimentale Elettromeccanico*, the acronym MOSE is now what the entire project is called, the name being a playful reference to Moses (*Mose* in Italian), the Biblical divider of waters. Unfortunately, the demonstration project became a huge public relations disaster with the confused public erroneously believing that the large red towers used to suspend the test gate would be replicated en mass for the final barrier project which would be comprised of many such gates.

In point of fact, the current design consists of 78 gates to be joined in three major groups at their base and situated across the mouths of the three channel inlets (Lido, Malamocco, Chioggia) between the outer barrier islands (MITVWACVN 2006a,b). Accompanying each bank of gates would be a system of breakwaters and seawalls needed to reduce inflows of water by 15 to 28% at each channel mouth. A key feature of the design is the mobility of the gates, allowing them to swing back and forth with waves, thereby dissipating energy to the sides and foundation supporting structures which, as a result, would not have to be as large, as they would have needed to be otherwise.

With an expected lifespan of a century, 455 million Euros will be spent for the first phase of MOSE with the estimated total cost to be 2.5 billion Euros. Construction is expected to take 8 years and will provide over a thousand direct jobs per year with 4,000 indirect or spin-off jobs, as well as about 150 specialized maintenance jobs per year thereafter (Scotti, *in* Fletcher and Spencer 2005).

Each gate will be about 20 m wide, 2 m thick, 18 to 30 m tall, and will be designed to hold back tides of about 2 m in height (Nova 2002; Eprim *in* Fletcher and Spencer 2005; MITVWACVN 2006a,b). When not in use the gates will lie prone on the bottom of the channels anchored to their rigid steel bases. With warning of an expected *acqua alta*, compressed air will be forced into the hollow gates, displacing the water that had until then filled them. This will cause the structures to quickly rise up to block the incoming tidal surge. After the tide abates, water flows back into gates displacing compressed air and causing them to sink back down out of sight into their beds. Modeling exercises based on flooding events between 1955 and 2002 estimated the average closure duration to be 5 hours over this reference period, or less than 1% of the time (Eprim et al., *in* Fletcher and Spencer 2005). This amounts to 2 to 3 closings per year of which 1 to 2 were false alarms. With a projected sea level rise of 20 cm due to climate warming over the next half a century, the average number of closings per year may rise to 25, of which 10 would be estimated to be false alarms.

Factors influencing the effectiveness of MOSE (Eprim *in* Fletcher and Spencer 2005; Eprim et al., *in* Fletcher and Spencer 2005) include direct rainfall on the lagoon, watershed inflows, wind patterns in the lagoon, and total storm surges, as well as the structural operation of the gates. To investigate the latter, a $15 million physical model of the lagoon was built in a warehouse in Padua in which to test how the gates would perform under a set of varying hydrodynamic conditions. This research did uncover the problem that when waves of a certain pattern hit the barrier, adjourning gates rocked in the opposite direction, thereby creating gaps in which floodwaters could flow. As a consequence, this discovery led to a change in the gate dimensions and angle of repose (Nova 2002).

An international group, a *collegio*, of experts was appointed by the Italian government to review their internally produced environmental impact study. In terms of effectiveness, CELI (1998) concluded that the MOSE proposal was "an effective way of protecting the city against high water," and the best option was to go forward with the project in unison with continuing the ongoing diffuse *insulae* projects. Alternative measures suggested by those opposed to MOSE, such as opening up the fish farms in the lagoon, reducing channel inlet sizes, redesigning breakwaters, or creating/restoring mudflats and sandbanks, were considered to have a very limited effect on the water level in the lagoon. Other alternatives such as raising low-lying areas of the city up to a height of 120 cm (above projected *acqua alta* incursions) would be just as costly as MOSE, take too long to construct, and not be as effective. Perceived operational problems due to out-of-phase resonance and gate wobbling, sudden collapse of the entire barrier system, difficult closure of the last gate against the pressure of inflowing water, or measurable leakage between the gates, were concluded to not be issues of concern.

In terms of environmental effects, CELI (1998) decided that the predicted closure frequency of about 12 per year, for a total of 42 hours, mainly during the winter, would have negligible effects on the lagoon. Only the prediction in relation to severe climate change models—suggesting that gate closures might increase up to 70 times per year by 2050—were believed to produce noticeable effects on the chemistry and ecology of the lagoon. On the other hand, presence of the gates would slow the inevitable process of climate change and their operational flexibility would allow the lagoon ecosystem to be able to gradually adjust to sea level alterations.

In terms of economic effects, CELI (1998) concluded that the gates would provide measurable benefits equal to or exceeding their projected operational costs. However, with further sea level rise, they acknowledged that the effects of frequent gate closure on shipping would need to be reexamined in terms of lost revenue to port activities.

Dithering about implementing MOSE ended in the winter of 2000 when, after a series of severe storms, another notable *acqua alta* occurred, which put 93% of the city under water, the worst flooding to occur since 1966. Empirical models showed that the MOSE barriers would have been closed for 9 hours and restricted waters in the lagoon to just over half a meter above normal, and once the gates were reopened, natural tidal flushing would have quickly and effectively cleaned out all pollution that might have built up during this limited period (Nova 2002). In consequence, and with support of a pro-business national government, construction of MOSE finally began in 2003, over three decades after it was first proposed.

Groundwater Injection

The petroleum industry has considerable experience in the deepwell injection of water to float hydrocarbons upwards, as does the wastewater industry in terms of removing contaminants downwards. This has led some to suggest that the injection of seawater into the brackish aquifer located 600 to 800 m below the lagoon could actually raise the city of Venice upwards by about 25 cm over a decade (Spencer et al. *in* Fletcher and Spencer 2005b). Problems might result, however, in differential elevation responses due to the presence of heterogeneous substratum, which could destabilize buildings. Obviously, more research needs to be invested in this potentially promising yet hitherto undeveloped intervention strategy. Groundwater injection was tried on a small lagoon island in the 1970s but modern, more sophisticated measures are being examined such as a series of 600 to 800 m deep wells in a 5 km radius around Venice to have a uniform rise of about 30 cm relative to mean sea level (Fletcher and Da Mosto 2004). Other engineers remain, however, very skeptical that such a strategy would be effective.

ECOLOGICAL RESTORATIONS

The Venice Lagoon contains a diverse typology of salt marshes (Bonometto *in* Fletcher and Spencer 2005). Restoring the salt marshes are believed to be the key to achieving a future resilience of the entire lagoon ecosystem (Fletcher and Da Mosto 2004). As a result, over the last 15 years, 6 million m³ of sediments from canal maintenance dredging have been utilized within the lagoon for the reconstruction of nearly 1000 ha of salt marshes and tidal mudflats, including creation of over 50 dredge islands of 0.4 to 57 ha in size at a half a meter above sea level (MITWVACVN 2006a; Cecconi, *in* Fletcher and Spencer 2005; Scotti, *in* Fletcher and Spencer 2005; Bonometto, *in* Fletcher and Spencer 2005).

Salt marsh restoration in the Venice Lagoon involves special challenges. First, due to a limited tidal cycle there is a narrow elevation window in which to build marshes with the dredged sediments (Cecconi *in* Fletcher and Spencer 2005). Second, due to reduced sediment inputs as a result of historic river diversions (redirection of the Brenta and Piave rivers back into the lagoon is not feasible due to their present level of contamination) ground compaction is a problem that leads to permanently submerged areas, which are in need of maintenance in terms of active sediment nourishment (Cecconi *in* Fletcher and Spencer 2005). For example, boats are now used to actually spray sediment on the marshes. And finally, as a result of motorized boat traffic, the borders of the marshes require armoring with gabions, wood pilings, or use of bioengineered fences made of wood bundles to entrap sediments for stabilization, all serving as breakwaters to reduce erosive wave action (Cecconi *in* Fletcher and Spencer 2005).

Reconstructed marshes now represent about 15% of total marsh area present in the lagoon. This is an amount that is comparable to that which has been lost due to erosion and relative sea level rise over the same period. Salt marsh reconstruction and creation is therefore is an ongoing project just to stay in balance with the rate of loss.

POLLUTION ABATEMENT AND MITIGATION

During the 1980s, images of gondoliers wearing face masks brought much adverse international publicity and finally forced city officials to admit that there was a pollution problem in Venice (Buckley 2004). During the last decades of the 20th century, an active harvesting program removed thousands of cubic meters of macroalgae whose proliferation was related to the eutrophication of the lagoon (Scotti *in* Fletcher and Spencer 2005). The lagoon-wide improvement in water quality is one of the few success stories concerning Venice's environment.

Pollution problems have decreased due to reduction in nutrient inputs from a phosphate ban within the lagoon (Marcomini et al. *in* Fletcher and Spencer 2005), which incidentally occurred decades after similar bans elsewhere in the world. And construction of thousands of septic tanks throughout the historic center (Keahey 2002) has removed much of the dumping of raw sewage from the city of Venice into the canals. Industrial waste from Porto Maghera which had been directly dumped into the lagoon has recently been stopped. Fish farms have been reopened to improve flushing and consequent water quality (Scotti *in* Fletcher and Spencer 2005). In addition to the pollution control strategy of reducing new discharges, a program of containing old toxic (and leaking) waste dumps has been implemented: by 2002, 5 of the 6 major lagoon dumps had thus been sealed (Scotti *in* Fletcher and Spencer 2005). In addition, 322,000 m³ of polluted sediment from 32 km of industrial canals which had been contaminating the water column due to wind and boat traffic resuspension have been removed through dredging (MITVWACVN 2006a; Scotti *in* Fletcher and Spencer 2005). And to stop the disastrous possibility of a spill, oil tanker traffic has been rerouted from the inner lagoon. There are still serious issues that need to be addressed, however, such as the continual discharge of untreated graywater from sinks and baths into the canals of the old city, and the input of terribly polluted water to the lagoon from the Po River.

Another encouraging development is that Venice is finally waking up to the possibility of using green infrastructure such as the subsurface flow treatment wetlands created to clean the waste from a community of about 50 individuals on the island of Lazzretto Nuovo (Chang 2006). The wetland system, instigated as a pilot project that will hopefully inspire other lagoon communities, arose through a small consortium of grassroots organizations and is composed of individual septic tanks, phytoremediation beds, a rainwater collection network, wastewater collection and reuse systems, and a final polishing pond.

The obvious solution to the possibility of a buildup of pollution when the MOSE gates are closed and tidal flushing is prevented is simply to collect and treat the sewage from as much of the population as possible (Harleman *in* Fletcher and Spencer 2005). To this end, the 2000 master plan for the region called for the construction of a new wastewater treatment plant on the mainland in the city of Mira/Fusina. The constructed wetland will treat the 4000 m^3 of municipal effluent as well as rain runoff through phytoremediation and then reuse this purified water as a substitute for the present industrial withdrawal of Sile River water (which is of a higher quality and therefore needed elsewhere). Additionally, the 100 ha treatment wetland (the largest of its kind not only in Italy but in all of Europe), unlike the one on Lazzretto Nuova where space constraints were an issue, will be a surface flow system which will include open water zones and islands that will attract and support a wide variety of wetland animals (Albano et al. 2007). Also, the treatment wetland is intended to be surrounded by a landscaped park which in addition to providing public education related to wetland ecosystems and natural water treatment systems, will provide an area of passive recreation for the community (J. Bays, CH2MHILL, pers. comm., 2007). The creation of such treatment wetland parks (France 2003a), one of the most exciting and innovative developments in contemporary water management (France 2006), will provide much-needed open space. This is an important project attribute given the scarcity of public land in Venice (Dolcetta *in* Fletcher and Spencer 2005). Although the planning process for development of the treatment wetland park has evolved with the same glacier-like rapidity characteristic of most Venetian decision-making exercises, once it is finally built (construction is to take place from 2007 to 2009), this environmental intervention could very well represent an important new direction in how Venice advances into its next century of uncertain existence. And the fact that the project is based on converting a post-industrial dredge spoil basin facing the Venice lagoon into an ecologically functional wetland with many beneficial uses makes this the most important regenerative landscape design project currently underway in Venice.

Part 4

Communities

9 Detroit [Re]Turns to Nature

Stephen Vogel

CONTENTS

ABSTRACT

This chapter tells the story of the 50-year process of deterioration and depopulation of the City of Detroit. The Detroit condition is an extreme example of the abandonment of postindustrial American cities. In the 40 square miles of vacant land and buildings within the city, nature in the form of flora and fauna has reclaimed its primal position and is erasing the man-made environment. In this environment live hardy people who are re-creating an agrarian and craft-oriented society. Based upon this context and other cultural clues found within Detroit, the chapter proposes an alternative settlement pattern for a 3,000-acre portion of the city that is largely abandoned and is the former watershed of Bloody Run Creek. This regenerative new settlement is called *Adamah*. This chapter tells the story of *Adamah*.

DETROIT [RE]TURNS TO NATURE

Detroit is perhaps the ultimate postindustrial city. It has suffered social and physical deterioration at a scale unique in the world with the possible exception of cities ravaged by war or the depopulation of former socialist cities. The city has lost half its former population of 2 million inhabitants, and one-third of its 140 square mile area is either completely vacant or occupied with deteriorated vacant structures. The most graphic illustration of the abandonment of the city is the disinterment and relocation of hundreds of human remains a year from Detroit cemeteries to the suburbs.[1]

With a persistent inevitability, nature has filled the physical void of disinvestment and disenfranchisement (Figure 9.1). Hardy flora and fauna is erasing the handiwork of humans and creating a new, mystical environment that is neither urban nor suburban nor even rural. This new "frontier" has brought to the city or caused to stay in the city men and women who are carving out a "brave new world" that is on the one hand preindustrial and on the other hand portends a new future.

---

[1] Crumm, David, Dead Join the Flight to the Suburbs: Families Removing Remains from Detroit, *Detroit Free Press*, October 7, 2000, p. 1A.

FIGURE 9.1 Abandoned railroad right-of-way near downtown Detroit. Photo: Will Wittig, University of Detroit, Mercy School of Architecture.

The condition of the city does not easily lend itself to urban planning textbook answers for its future—free market economic systems do not know how to cope with shrinkage as opposed to growth. In the past decade it has been popular to speculate on possible physical solutions to the Detroit condition—invariably these solutions come from academics unfamiliar with the reality of the condition or the people who live with the condition every day. Whether well intended (e.g., the creation of a vast new park system) or cynical (e.g., the creation of urban boot camps for the training of firefighters), these proposals miss the mark and enrage the long-time residents of the city.

It is within the context of speculation that this chapter postulates an alternative paradigm based upon historical, cultural, and environmental conditions that exist within the City of Detroit today. It will show how 3,000 acres of the lower east side of Detroit might embrace a new concept for community. This new community is called *Adamah*—Hebrew for "of the earth."

A TRADITION OF FARMING AND MAKING

The *Ville de Troit* (city on the straits) was founded by the French explorer Antoine Laumet de Lamothe Cadillac in 1701 because of its strategic position between Lake Huron and Lake Erie. By 1752 Detroit was a thriving farming community with the long, narrow "ribbon" farms typical of the French, orientated perpendicular to the Detroit River—the means of transportation, communications, and defense (Figure 9.2). The early French settlers are recognized today by the many street names that reflect the owners of farms from which the streets were carved—e.g., Beaubien, Jos. Campau, Riopelle, and DuBois. Fort Ponchartrain protected the city and was accessible to the farmers from River Road that extended from the fort in two directions along the river. Several creeks drained into the river including Parent's Creek, which River Road crossed by means of a wooden bridge (indicated by the arrow, Figure 9.2).

Among the artifacts of the French period were a series of windmills that lined both sides of the Detroit River, including one at Windmill Point at the mouth of the river. Wind power was used for milling grain and pumping water—the millstones from the Windmill Point windmill were found at the bottom of Lake St. Clair in the late 1970s.

As a result of victories in the French Indian wars, the English took control of Detroit in 1760. In 1763, Pontiac, the great Ottawa chief and an ally of the French, lured the English garrison from

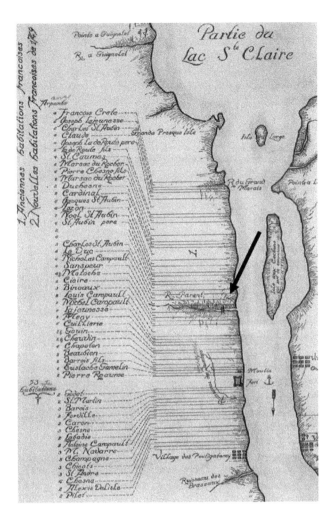

FIGURE 9.2 Detail of a map of the Detroit River from Lake Erie to Lake St. Clair, a 1935 clarification of a 1749 French map. Clements Library, University of Michigan, Ann Arbor, MI. (map division, 6-N-3). Arrow Indicates Parent's Creek and the Wooden Bridge Crossing the Creek.

the fort, along River Road to the bridge crossing Parent's Creek where they were attacked and a total of 23 British soldiers killed and 34 injured.[2] As the soldiers fell off the bridge into the creek, the creek turned red with blood. From this point forward the creek became known as Bloody Run Creek. Regardless of the temporary defeat, the English maintained control over Detroit until 1796 when the Northwest Territory was ceded to the United States.

Detroit grew as a trading port, at first because of the fur trade centered at Mackinac Island, Michigan. After a fire in 1805 destroyed the city, the thriving port town expressed its new-found importance with a city plan by Judge August Woodward, modeled after the plan for Washington, D.C., by Charles L'Enfant. Although the grand plan was only pursued until 1818, Detroit was on the move and envisioned itself becoming a great city. With the opening of the Erie Canal in 1825, the port of New York was less than two weeks away. Traders could transport farm produce, furs, lumber, and other goods from Detroit to New York and to markets in England and continental Europe.

An 1825 map of the city illustrates the growing community and the growing port (Figure 9.3). The map also shows a new Jeffersonian orthogonal grid legally defining the expanding city and

---

[2] Woodford, Arthur M., *This is Detroit: 1701–2001*, Wayne State University Press, 2001, p. 30.

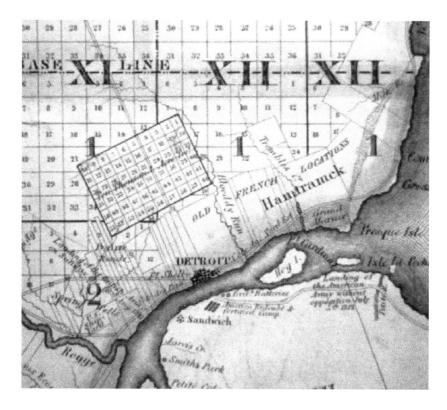

FIGURE 9.3 Detail of 1825 Surveyed Part of the Territory of Michigan by O. Risdon. Clements Library, University of Michigan, Ann Arbor (Map Division 6-0-3).

overlaying the old French Farm and military coordinates of the 18th century. Through these geometric patterns flows the now-named Bloody Run Creek.

During the mid- to late-19th century Detroit evolved from farming and the fur trade to the beginnings of an industrial center, of which it would eventually gain worldwide fame. The manufacturing of ships, bicycles, stoves, railroad cars, and other products foretold the future automobile industry. Detroit became a center of "making" and fabrication, whether by hand or machine, which it remains today.

Small machine shops, cottage industries, and neighborhood industries dominate the Detroit landscape and create a proud heritage and capacity for metal molding and fabrication.[3] It is not surprising that Detroit became the center of the labor movement. Today, the City of Detroit's government is the most unionized of all American cities, including New York.

By the time of the Civil War, Detroit's riverfront is industrialized, including a burgeoning railroad system that will soon replace shipping as the major mode of transportation in America. Bloody Run Creek runs through the industrial waterfront and is essentially an open sewer.

Like other growing American cities attempting to combat cholera and similar epidemic diseases, Detroit began to install a sewer system that carried sanitary and storm runoff from the streets and buildings of the city directly into the river. In 1875 the process of burying Bloody Run Creek into a sewer started, beginning at the river and moving north.[4] By 1915 Bloody Run Creek was no longer visible except for a short stretch through Elmwood Cemetery through which it exited and re-entered the sewer system, and does so today.

At the beginning of the 20th century, the automobile industry was introduced in Detroit by Henry Ford and others. Spurred by this new industry, Detroit grew rapidly, attracting immigrant workers

---

[3] Poremba, David Lee, *Detroit: City of Industry,* Arcadia Publishing, Chicago, 2002.
[4] City of Detroit, Water and Sewerage Department, 19th century sewer maps.

FIGURE 9.4 Henry Ford in 1920 designed worker housing, Dearborn, MI. Photo: Stephen Vogel, University of Detroit, Mercy School of Architecture.

from the south as well as from Europe. To African Americans, Detroit represented a place where they could get a decent wage without the apparent discrimination of the South. The new automobile factories often provided housing for its workers so they could easily walk to work or be a short trolley ride away. Factories became parts of neighborhoods. Using the high wages from these industries, workers purchased their own homes as well as cars. Single-family homes became the typical living style in Detroit (Figure 9.4). As a consequence, Detroit never produced, in any large quantity, row houses, apartment buildings, and high-rise housing typical of its eastern counterparts. With no natural barriers, Detroit expanded rapidly to the north and west in subdivisions of single-family housing.

By the 1920s, Detroit was one of the largest cities in America and growing rapidly. Ninety five percent of all downtown skyscrapers were built in a five-year period between 1923 and 1928.[5] The borders of the city enclosed over 130 square miles and extended from the Detroit River to Eight-Mile Road (Base Line). The stock market crash of 1929 brought city expansion to a halt, and through the 1930s little physical growth occurred. The dreams of a great city, with a downtown expanding from the river to the New Center, 3 miles away, were put aside.

World War II caused one more burst of energy and growth. By converting automobile plants, Detroit provided the majority of the military hardware for the war effort and, consequently, became known as the "Arsenal of Democracy." At the Willow Run automobile factory, Henry Ford exceeded the challenge of the U.S. government by manufacturing B-24 Liberator Bombers at the rate of one every 63 minutes.[6]

The postwar era return of veterans created a huge demand for jobs and housing. A reborn automobile industry provided the jobs. Immigrants continued to flow into the city. By 1952 its population peaked at nearly 2,000,000 people—the majority of whom lived in single-family houses. Detroit became one of the lowest density large urban cities in the world—great population living on a large land area. It became known worldwide as the home of the automobile and an industrial giant.

DISINVESTMENT AND DECENTRALIZATION

Post WWII saw the beginnings of decentralization of urban America and the ultimate disinvestment in the City of Detroit. Disinvestment occurred for many reasons, but one of the most striking was a result of the atom bombs dropped on Japan that brought an end to World War II.

---

[5] 1999 interview with Carl Roehling, President, SmithGroup, architects of the majority of downtown skyscrapers.
[6] Woodford, p. 134.

FIGURE 9.5 Renaissance Center, Detroit, MI. Photo: Stephen Vogel, University of Detroit, Mercy School of Architecture.

The government realized that a single bomb could destroy a city like Detroit and thereby wipe out a significant portion of the military industry. Government incentives caused the automobile industry to decentralize.[7] As the plants moved to the suburbs, then out of state, then out of the country, the workers left behind in Detroit either had to follow the jobs or lose out on the economic boom. Those who could afford to leave left. Freeways were built. The Interstate Highway System, initiated through the Highway Act of 1956 as a means to move military resources rapidly, ironically helped to speed the suburbanization of the region. It is no coincidence that Detroit had the first concrete road (Woodward), the first freeway (Davidson), and the first regional shopping center (Northland).

As this movement away from the city accelerated into the 1960s and 70s, the people left behind were the poor, the disenfranchised, people of color, and others who had been "redlined" as marginal and undesirable by the majority.

A shrinking city in decline, without jobs, caused huge frustration. This frustration reached the boiling point in 1967 when the African-American community rebelled after a police shooting. For days, looting and fires besieged the city patrolled by National Guard troops with tanks. The act of burning buildings led to insurance fires, where vacant buildings that had lost their value were intentionally burned to collect insurance money, a practice that has continued. The drug culture grew; street gangs moved into the neighborhoods. Though the city began shrinking long before the riots, the postriot negative perception of the city helped to accelerate both people leaving the city as well as suburbanites refusing to come back into the city, even for civic and cultural events. Eventually, the population of Detroit became 85% African American. Many white suburbanites refused to cross the "Eight Mile" divide between city and suburb.

The abandonment of the city was not just by individuals and families, but also by businesses and corporations. Corporate guilt created high profile projects in the city while simultaneously making major investments in the suburbs or even out of state. Henry Ford II, whose personal persuasion made the project happen, championed the 1978 development of the Renaissance Center, the largest private investment project in the United States at that time (Figure 9.5). The project, which suffered multiple

---

[7] Sugrue, Thomas, *The Origins of the Urban Crisis*, Princeton University Press, Princeton, NJ, 1996, p. 140.

FIGURE 9.6 Bloody Run Creek area, Detroit, MI. Photo: Chris Pomodoro, University of Detroit, Mercy School of Architecture.

financial crises until the 1998 purchase by General Motors for its world headquarters, was miniscule compared to the economic investments of former Detroit corporations beyond Eight Mile Road.

Today, the shrinking city's population has dipped below 900,000 and even with the economic boom of the 1990s, there is no apparent end in sight. Within one or two decades, a city of 750,000 is easily imagined. For every new housing unit built ten are torn down; 5,000–8,000 demolitions a year create vast wastelands (Figure. 9.6). Spectacular demolitions have occurred, including the 2-million-square-foot former Hudson's department store removed from the core of downtown and the 3-million-square-foot Uniroyal factory on the riverfront. These demolitions are among the largest in the world. Vacant skyscrapers, derelict factories, and boarded up storefronts have become the reality of the abandoned city.

The impact of disinvestments is readily seen in the watershed area of the now buried Bloody Run Creek. A 1949 aerial photograph shows a built-out city, without vacant lots (Figure 9.7); compare with a 1996 survey by University of Detroit, Mercy School of Architecture students indicating the amount of vacant buildings and lots (Figure 9.8).

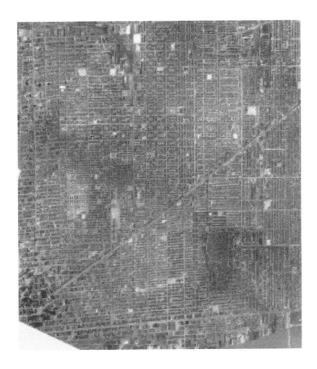

FIGURE 9.7 1949 Aerial photograph, Bloody Run Creek area. Source: DTE Energy, Detroit, MI.

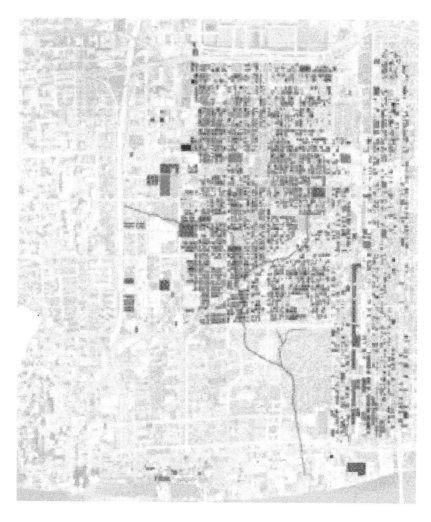

FIGURE 9.8 Vacant property and buildings from 1996 Windshield Survey. Source: University of Detroit, Mercy School of Architecture. Photo: Studio of Sonja Ewing, 1998.

THE RETURN OF NATURE

With the amount of vacant land, abandoned railroad right-of-ways and empty buildings, it is inevitable that nature has begun to take back the city. Superimposed and enhanced satellite images show that between 1972 and 1997, there were dramatic increases in vegetation in the central city whereas dramatic drops in vegetation occur in the suburbs. An increase in vegetation also leads to concurrent increases in fauna. The most famous of these fauna is not the occasional deer that wanders into the city via abandoned rail right-of-ways, but rather the ring-necked pheasant. The number of pheasants are so large, the State Department of Natural Resources has been arresting people in Detroit for game poaching of pheasants for food—an ironic use of public resources given the notorious crime statistics of the city.[8] Approximately 15 years ago, the State of Michigan Department of Natural Resources began trapping the readily available Detroit ring-necked pheasants and releasing them in Northern Michigan to breed with imported Szchuen black pheasants to create a pheasant more suited to survive the harsh winter conditions of the north country.[9] Peregrine falcons hunt pigeons

---

[8] Greenwood, Tom, Walk on the wild side turns up pheasants, peregrines, raccoons, *Detroit News*, September 3, 1997.
[9] 2003 interview with Rod Clute, Wildlife Division, Michigan Department of Natural Resources.

FIGURE 9.9 Farm at Catherine Ferguson Academy, Detroit, MI. Photo: Stephen Vogel, University of Detroit, Mercy School of Architecture.

from their perches in abandoned skyscrapers. Raccoons and opossum join dogs as the most common road kill on city streets.

Given the increase in flora and fauna, it should not be surprising that urban farming and gorilla gardening are widely practiced in Detroit, including in the Bloody Run watershed area.[10] Vacant single-family lots are used for high-density food producing gardens, whereas the consolidation of vacant land is used for farming on a larger scale (Figure 9.9). Alfalfa is a popular crop because there is a market for hay and because alfalfa removes toxins from the contaminated soil and purifies the land in a relatively short period of time.[11] Much of the farming and gardening efforts illegally utilize city-owned land, although the city has generally ignored this violation.

Evidence of the growth and importance of farming is the increasing use of the Michigan State University Extension Service and the popularity of the 4H Club. When quizzed about the popularity of farming, a resident of the Bloody Run Creek area stated, "Our parents were share croppers in the South. Farming is in our blood."[12] The ability to be self-reliant and grow your own food has attracted a hardy class of individuals, both black and white, to the inner city.

FEUDAL KINGDOMS AND SELF-RELIANCE

As the shrinking city loses its tax base, it becomes more and more difficult to maintain even the most basic city services. The inability of Detroit Public Schools to provide a reasonable education is well documented and is countered by the growth of charter and home schooling programs and

---

[10] Wildon, Charles, Asphalt Eden, *Preservation Magazine*, May–June 2002.
[11] Michigan State University—verify source.
[12] Interview with Lee Burns, 1997.

the continued success of faith-based schools. The Detroit Water and Sewerage Department, which provides water and sewer service to the entire Metropolitan Detroit region, is under widespread suburban political attack because of its apparent inability to pay for and maintain its aging utility system. The only form of public transportation in Detroit, other than a downtown people-mover, is the bus system, which struggles to maintain schedules and keep its fleet repaired and service ready. That the bus system does not interface with the suburban express system is one of the most blatant examples of poor suburb/city cooperation. Former Mayor Dennis Archer's largest political crisis was the inability of the city to plow streets after a major snowstorm. The glorious park system of the 1920s, including the unique island park, Belle Isle, now stand in poor repair due to lack of maintenance. Finally, the Detroit street light and traffic signal system is continually failing. The growing anarchy of the streets is most evidenced by the current practice of ignoring stoplights, whether they are working or not.

The lack of functionality of the city has caused local community organizations to be leery and distrustful of the city's ability to maintain services. This has created a "feudal" system of powerful neighborhoods surrounded by "no man's land." As an example, the North Rosedale Park neighborhood of middle-class homes has private police, a private park, a private community center, pays for snowplowing, pays for median and park lawn mowing and maintenance, runs a legitimate theater organization and utilizes private schools in large percentages. This is not atypical for middle class neighborhoods in the city but it is also practiced in poorer neighborhoods, where cooperative citizen band patrols safeguard the street, occasionally with vigilante tactics, or deals with garbage pickup through volunteer "trash and clean-up" days. Some block clubs or neighborhood organizations have developed their own street light system (typically using natural gas).

Efforts by the city to institute a city-wide master plan and visioning effort have been stymied because of the lack of trust in the city's intentions on the one hand and the unwillingness of neighborhood leaders to give up the hard-won power that comes from solving their own problems. Forty years of poor services cannot be overcome quickly, especially with continuously diminishing resources.

It should not be surprising that the city's feudal system attracts hardy and self-reliant people who see Detroit as a "frontier" where there is a peculiar freedom away from the watchful eye of government rules and regulations. This freedom is supported by the relatively low cost of land and buildings in many areas of the city. The building boom of the 1990s, which saw the construction of casinos, sports stadiums, and housing projects, caused land speculation with resultant rising land costs. However, it is still possible to buy vacant 30 ft × 100 ft residential lots at tax auctions for under $500 in many areas of the city; buy unimproved vacant buildings for $1 per square foot; and occupy upper floors of low-occupancy skyscrapers at incredibly low rates.

One of the results of this system has been the creation of what might be called "extreme adaptive reuse." The easy availability of buildings, even excellent historical examples of architectural periods, has caused unusual adaptive reuse at all levels of the spectrum. These examples include the use of former single-family residences as barns; abandoned Catholic churches as gymnasiums; and 1920s movie palaces as parking garages (Figure 9.10). The latter example has received national attention through publications such as Camilo Jose' Vegara's *The New American Ghetto*.[13]

The wide availability of building material has also created a new aesthetic of found materials as urban entrepreneurs utilize these materials to reconstruct their environment and attempt to recapture the embodied energy inherent in the material. In 1995 under the leadership of Associate Professor Tom Dubicanac of Carleton University, University of Detroit Mercy and Carleton University students adaptively reused, designed, and built the community offices of the Southwest Detroit Business District Association using entirely found materials from vacant lots, buildings, and junk yards, especially an airplane junk yard. The "airplane" building is now notorious in the neighborhood. Former Adjunct Instructor Ashley Kyber of the University of Detroit Mercy utilized vinyl car seats to create growing "pods" for a new urban garden (Figure 9.11).

---

[13] Vegara, Camilo Jose', *The New American Ghetto*, Rutgers University Press, New Brunswick, NJ, 1995.

FIGURE 9.10 Michigan Theater Parking Garage, Detroit, MI. Photo: Dan Pitera, University of Detroit, Mercy School of Architecture.

The conditions of the city have created a population that is hardy, creative, practical, risk-taking, self-reliant, and entrepreneurial. They are also stubborn and leery of outside interests, especially government, land speculators, and real estate developers. They are focused on reusing the abundant resources that surround them in the abandoned city.

FIGURE 9.11 Ashley Kyber's "Pods," Bloody Run Creek Area, Detroit, MI. Photo: Ashley Kyber.

THE SEARCH FOR A NEW MIDDLE LANDSCAPE

A review of the demographics of Detroit comes to the inevitable conclusion that the shrinkage of the city will continue, regardless of the political unwillingness to address the issue. What does a city do, given these conditions? To encourage development, it offers vacant land that, with great cost in assembled and environmentally cleaned sites, is given over to new development. Two automobile plants have been built in the city in the last 20 years at the cost of 3500 homes and businesses and hundreds of millions of dollars of site clearance.[14] These ventures have not created the number of jobs promised at their initiation.

The builders, financiers, and developers of Victoria Park, the first new subdivision in Detroit in over 30 years, decided to use a suburban model of development—i.e., curved streets, large lots, garages facing the front. Victoria Park, the new auto factories, strip shopping centers are all examples of the suburbanization of Detroit. Even while the public press decries the effect of suburban sprawl, we are imitating the suburbs in the city. If Detroit becomes indistinguishable from the suburbs, then why have a city?

The suburbanization dilemma asks the question: What are appropriate settlement patterns in a shrinking city like Detroit? Do you take an urban, high-density approach, a suburban low-density approach, or even a lower density rural approach to new development within the city? Given 40 square miles of vacant land, all of these approaches are possible. But which are appropriate?

Each of these approaches has inherent issues. High-density approaches ignore the declining population of the city and the inability of the city to maintain vacant land or even park land. Suburban approaches destroy the identity of the city and repeat the mistakes that are taking place on the fringes of the urban center through consuming excessive raw materials and ignoring embedded history and infrastructure. Rural approaches seem to be in opposition to the concept of an urban center for a 5-million-person metropolitan region. Detroit has always been relatively low density. However, Detroit is not a suburb. We need to search for a new "middle landscape" that is appropriate to this city and is not based upon traditional models of town planning.

ADAMAH: A NEW EQUITY FOR DETROIT

The search for a new middle landscape has led to the creation of a concept for the lower east side of Detroit called *Adamah* (Figure 9.12). The physical impetus for this new landscape is the daylighting of Bloody Run Creek. In the late 1970s and early 1980s, a Detroit firm of architects and landscape architects, Schervish Vogel Merz, was creating a recreational plan for the east riverfront of Detroit and three major riverfront parks referred to as the Linked Riverfront Parks Project. The research for the project included an investigation into the overflow sewers that during periods of intense rain would dump raw sewage into the Detroit River. One of these sewers had a name—Bloody Run Sewer. The original design of the park system, never implemented, included creating a water feature extending from Jefferson Avenue to the river that generally was a representation of Bloody Run Creek, although not following its original course.

Over the years several schemes have proposed the daylighting of Bloody Run Creek. In 1993, Schervish Vogel Merz proposed to the then-new Mayor Dennis Archer that a state park be developed that opens the creek as an incentive for new development. Although well received, the scope of the project—3000 acres—was beyond the city's ability to contemplate. In 1995, Stephen Vogel and the Detroit Collaborative Design Center at the University of Detroit Mercy presented "Unearthing Detroit" that proposed the daylighting of Bloody Run Creek as part of the exhibition "Empowering the City: New Directions in Urban Architecture."[15] The city-sponsored Community Reinvestment

---

[14] *Detroit Free Press*, Detroit: A Special Issue: Poletown, September 8, 1985.

[15] Exhibition entitled "Empowering the City: New Directions in Urban Architecture," Detroit Artist's Market, January 12 to February 16, 1996, Curators: Mark Nikita, AIA and Dorian Moore, AIA.

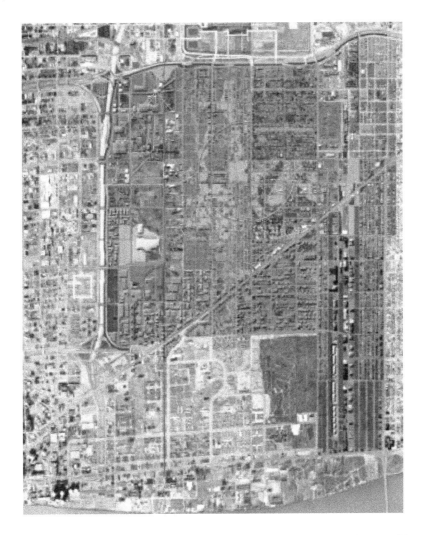

FIGURE 9.12 Plan for Adamah. Source: University of Detroit, Mercy School of Architecture. Photo: Studio of Kyong Park and Stephen Vogel.

Strategy in 1999 illustrated the daylighting of Bloody Run Creek as a part of the long-term strategy for the lower east side.[16] No process of implementation of this strategy has ever been put forward.

In 2000, it was decided that an architecture studio would be conducted at the University of Detroit Mercy that would attempt to combine the cultural and environmental factors outlined above into a new community that encompassed the watershed area of Bloody Run Creek. Stephen Vogel and Kyong Park were studio instructors and six selected students, Louis Farris, Stefan Lennon, Victoria Matous, Christian Pomodoro, Rebecca Raleigh, and Shane Terpening performed the work. It was decided early on that the community would have the following characteristics:

- Be rooted in the French farming tradition and the agrarian roots from the southern immigrants (Figure 9.13)
- Focus on "walk to work" job creation including, in addition to gardening and crop farming, fish and mollusk farming, arts and crafts studios, cottage industries, machine shops, and woodworking shops

---

[16] City of Detroit, Community Reinvestment Strategy, Cluster 4 Community Reinvestment Report, 1997.

FIGURE 9.13 Louis Farris's "Circle of Life," Adamah. University of Detroit, Mercy School of Architecture. Photo: Studio of Kyong Park and Stephen Vogel.

- Recognize the family unit and the single family home as a basic building block of the community and include intergenerational, eco-village and co-housing concepts
- Utilize sustainable systems, including the reuse of existing buildings and materials, decentralized and green utility and infrastructure systems including wind power (Figure 9.14)
- Propose settlement patterns that keep new buildings to the absolute minimum and either reuse existing buildings or move and/or combine existing buildings
- Allow for a form of government that is based on the "town meeting" model of democratic self-determination
- Daylight Bloody Run Creek not only for recreational and storm run-off purposes, but also for irrigation, gray water recycling, and power (Figures 9.15 and 9.16)
- Propose "low" technology for public transportation including bicycles, shared, community-owned automobiles, and the reuse of the former Detroit trolley system
- Propose "high" technology for communications and data transmission
- Utilize fuel systems that included ethanol refining and hydrogen fuel cells
- Create economic systems that included cooperatives and trading and bartering of food for service (Figure 9.17)
- Found school systems that included home schools, foster schools, and neighborhood-based charter schools

FIGURE 9.14 Louis Farris's "Bio-mass Generator and Wind Farm," Adamah. University of Detroit, Mercy School of Architecture. Photo: Studio of Kyong Park and Stephen Vogel.

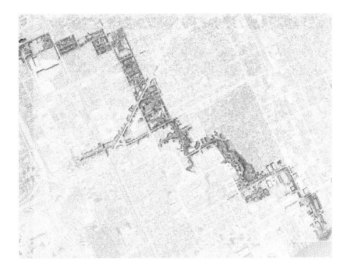

FIGURE 9.15 Stefan Lennon's "Parent Creek Linear Park." University of Detroit, Mercy School of Architecture.

FIGURE 9.16 Detail of Stefan Lennon's "Parent Creek Linear Park." University of Detroit, Mercy School of Architecture.

FIGURE 9.17 Stefan Lennon's "Chene Market Cooperative," Adamah. University of Detroit, Mercy School of Architecture Photo: Studio of Kyong Park and Stephen Vogel.

The research that created the *Adamah* concept included utopian communities of the 19th century, and there are certainly aspects of *Adamah* that are utopian in nature. However, concepts proposed for *Adamah* are either occurring currently in the community, albeit in a small scale, or have been proposed previously, such as the daylighting of Bloody Run Creek.

The work of the studio struck a cord with members of the local community directly affected by the *Adamah* proposal, It is clearly the most comprehensive and far-ranging approach to revitalizing a community that Detroit has seen—regardless of the antitraditional development approach to the project. The Boggs Center, a local community nongovernment organization, has been working with hundreds of individuals who are committed to seeing the *Adamah* vision implemented. Implementation strategies include the following:

- Creation of small scale land trusts to secure land and implementation of a co-housing and eco-village community
- Implementation of art installations highlighting community issues.
- Workshops that advance the ideas of *Adamah* sponsored by the University of Detroit, Mercy School of Architecture and the International Center for Urban Ecology and others
- Ongoing studios and research by the University of Detroit, Mercy School of Architecture
- Presentations and publications of the work, including national presentations by the Boggs Center and others[17]

The ultimate question that *Adamah* poses is whether it is an elaborate land-banking program that holds property until such time in the unknown future that the economical, social, and environmental conditions of Detroit can be overcome and shrinkage reversed, or is it truly a new middle landscape. Is *Adamah* the "brave new world" of postindustrial Detroit?

---

[17] Cuyette, Curt, Greening of Detroit, *Metrotimes*, Detroit, MI, October 31–November 6, 2001 includes the web site of the Boggs Center.

10 Rebuilding Salmon Relations
Participatory Ecological Restoration as Community Healing

René Senos

CONTENTS

FIRST SALMON CEREMONY

Northwest Coast Haida fisherman's prayer at first sight of migrating salmon:

> WELCOME, friend swimmer,
> We have met again in good health.
> Welcome, Supernatural One,
> You, Long-Life-Maker,
> For you come to set me right again
> As is always done by you.
> Now pray take my sickness
> And take it back to your rich country
> At the other side of the world,
> Supernatural One.

Salmon answers:

> HA, I will do so.

(Stuart, 1977: 163)

ABSTRACT

New models and creative strategies are urgently needed to reverse, repair, and prevent further depletion of life-sustaining natural systems, as well as to promote healthier interactions between humans and land. Ecological restoration has emerged in recent decades as a promising intervention in a range of settings. As it is currently framed, however, restoration generally focuses on the scientific and political parameters of a specific project but rarely considers broader social needs and benefits in the course of structuring, monitoring, and evaluating restorative processes and outcomes. Furthermore, restoration is often implemented on a project-by-project basis, rather than cast within a larger community or regional visioning process. A regenerative strategy that addresses both the intrinsic social and environmental complexities of a given community is crucial, particularly in urban settings where natural systems are marginalized or where city residents are disconnected from the myriad benefits of nature.

This chapter explores how ecological restoration may be a holistic model for healing both land and human community. It advances participatory restoration as an inclusive strategy that addresses a broader array of community needs, including the psychological, social, political, economic, and ecological needs of community members, and describes how participatory restoration may be conceived as a reciprocal process of healing for both the social and physical environment.

The uncertain future of the Pacific Northwest salmon is viewed as a strong catalyst and opportunity for engendering stronger relations between people and ecological systems. Keystone salmon are deprived of spawning streams, while watershed citizens are deprived of their biological and cultural connection with salmon, demonstrating the acute need for a bioregional restoration process framed by both social and ecological imperatives. Key projects that structure a more comprehensive approach to community health recovery are presented, and suggest strategies for rebuilding relations with salmon while restoring whole community health.

INTRODUCTION

New models and creative strategies are urgently needed to reverse, repair, and prevent further depletion of life-sustaining natural systems, as well as to promote healthier interactions between humans and land. Ecological restoration has emerged in recent decades as a promising intervention in a range of settings. As it is currently framed, however, restoration generally focuses on the scientific and political parameters of a specific project but rarely considers broader social needs and benefits in the course of structuring, monitoring, and evaluating restorative processes and outcomes. Furthermore, restoration is often implemented on a project-by-project basis, rather than cast within a larger community or regional visioning process. A regenerative strategy that addresses both the intrinsic social and environmental complexities of a given community is crucial, particularly in urban settings where natural systems are marginalized or where city residents are disconnected from the myriad benefits of nature.

This chapter explores how ecological restoration may be a holistic model for healing both land and human community. It advances participatory restoration as an inclusive strategy that addresses a broader array of community needs, including the psychological, social, political, economic, and ecological needs of community members, and describes how participatory

FIGURE 10.1 1899 Astoria, Oregon fisherman with salmon: Does restoration offer us an opportunity to redefine the measure of our ecological identity? *Source:* Alex Blendl.

restoration may be conceived as a reciprocal process of healing for both the social and physical environment (Figure 10.1).

RESTORATION: INTERVENTION OR INTERACTION?

Defining what we mean by ecological restoration is crucial in order to think, communicate, and act at the landscape scale, and by extension, if we wish to deliberately and thoughtfully chart a course of action for social and physical community change. Definitions of restoration, or "stewardship semantics," have co-evolved with our expanded understanding of restoration ecology. Restoration terms are often used interchangeably despite conveying very different motives and actions. While some restoration theorists claim that the distinction is not important because various activities fall along a "continuum of restoration" (Hobbs and Norton, 1996), other restoration philosophers contend that this is not a satisfactory explanation if one wishes to arrive at an ethical, consistent, and holistic view of ecological restoration (Throop 2000; Bazerman 1997; Pister 1997; Jackson et al. 1995; Baldwin et al. 1994). A science-based view of restoration, with a vocabulary that seemed to condone further environmental destruction with terms like "mitigation," or the R words of "rehabilitation," "reclamation," and "revegetation," raised the ire of restoration pioneers William Jordan and Frederick Turner. Turner vigorously challenged a limited definition of restoration devoid of human interaction, whereas Jordan pioneered the concept of restoration as ritual, asserting that restoration activity entails four dimensions of value, specifically; product, process, experience, and performance or ritual (Turner 1984; Jordan 2003, 1995, 1986).

Restoration is the most fitting word because it entails a high level of commitment to the recovery process. To honor this commitment, restorationists are compelled to ask critical questions that might otherwise be overlooked, which range from researching historical ecological and cultural conditions to defining the role of humans in shaping landscapes. As the scope of restoration is expanded to include a broader realm of concerns, including social, political, economic, cultural, and ecological aspects, a new definition of restoration activity is required. A definition that captures the dimensions of reciprocity, dynamism, mutual nurturing, creative learning, and relationship implicit in meaningful restoration is essential to appropriately cast humans as community members rather than resource managers.

Since Jordan first proclaimed his views on restoration as a model for a healthy relationship between humans and the environment, many other voices have joined in, and their revolutionary ideas have evolved into a rich body of theory. Critically, these ideas are being tested on the ground. Recent research has begun to advocate a broader approach to reciprocal restoration. For example, Michael McGinnis defines mimesis or miming as a process of "culturally and imaginatively constructing nature," and emphatically states that "community-based restoration—in the form of dance, poetry, theatre, other arts, and ritual—is a means to recover a wild sensibility so that we can learn lost social and community values. By fostering restorative relationships with nature, we can create a healthier community" (McGinnis 1999: 219–226).

To go one step further, restoration may be seen as an act of rebuilding our ecological identity. McGinnis, House, and Jordan suggest that as "disabled creatures working in a wounded landscape" restoration gives us a vital opportunity to renegotiate our relationship with nature and reclaim our ecology of shared identity (McGinnis et al. 1999: 206). By re-engaging the land with our minds, hearts, and hands we reactivate the collective memory of our biological and spiritual connection with the land that remained vital until just very recently in human history.

Paul Shepard asserts that humans are intrinsically connected to nature, and share a common ecological identity. He describes a rich tradition of ecological mimesis, where through ritual or dance, humans mime an animal or a part of a landscape in order to cross over to spiritual, and material realms of nature (Shepard 1996). Adopting a totemic animal such as the salmon to represent a clan, family, or group is a powerful mechanism for expressing this common identity.

Although restoration ecology is directed at mitigating dysfunctional natural systems, parallel currents in the field of landscape architecture are focused on improving or creating healthy habitats for people. The most exciting work happening today is where these fields converge, and cultural and ecological values are integrated in a cohesive plan to enhance overall community functioning. Randy Hester, Mark Francis, Ann Spirn, Clare Cooper Marcus, and others were early advocates of the concept of "environmental equity" with regards to a social and environmental valuing system (Marcus 1997; Hester 1990; Francis et al. 1984; Spirn 1984). Michael Hough is a renowned proponent of integrating natural systems in cities to create more humane environments by "making visible the processes which sustain life" (Hough 1995: 30). These and other design professionals established a conceptual framework for renovating diminished natural and built environments.

Elements of participatory restoration may be found in citizen-driven efforts like community garden projects, urban renewal projects, or social health projects. Urban neighborhoods create community gardens as a tool to promote social and environmental health. Derelict lots are converted into nurtured places, and neighbors collaborate while working the land. San Francisco League of Urban Gardeners (SLUG), Seattle Pea Patch gardens, and Philadelphia neighborhood gardens are examples of successful programs that have linked multiple values in a community restoration

FIGURE 10.2 Neighbors gather at their Oxbow Park community garden in South Seattle. *Source:* René Senos.

approach (Figure 10.2). Urban agriculture, community pride, stronger social ties, aesthetics, cultural expression, and improved ecosystem functioning are just a few benefits that result. Even deeper, however, is a level of exchange between residents and the land, and residents with each other. Participants invest their time and energy in caring for the land, and develop a sense of ownership and pride that is reflected back into the community. Even in San Francisco's roughest neighborhoods, urban gardens are not targeted by vandalism.

Other researchers have pursued a related vein of study by examining the human relationship to nature via perceptions or preferences, or by exploring the psychological benefits of human engagement in the plant world (Louv 2005; Marcus and Barnes 1999; Kaplan and Kaplan 1989; Lewis 1996; Hartig et al. 1994). The psychological and social dimensions of human living in a fragmented environment, and the subsequent need for renewal, are perhaps more critically examined in exploratory fields such as deep ecology, environmental ethics, or environmental psychology (Sessions 1995). "Green psychology" or "ecopsychology" examines our relationship to the earth, diagnoses systemic problems, and suggests paths of healing (Metzner 1999, 1995; Roszak et al 1995; Clinebell 1996). Ecopsychology explores the basic shifts in our patterns of identity and relationship that occur when we include our connection to the natural world as essential to human health. This unconventional approach offers many rich theoretical perspectives, diverse ideas, and unexpected connections. Aldo Leopold's work was a forerunner to this field, as he considered how to convert "collective amnesia" to "ecological conscience" (Leopold 1949). Increasingly, ideas emerging from these "deeper perspectives" are applied as problem-solving strategies in a range of settings, and ecological restoration is one such strategy.

BENEFITS OF PARTICIPATORY ECOLOGICAL RESTORATION

In the last decade or so, researchers have attempted to elucidate not only the ecological but social benefits of regenerating derelict places. Tools that measure social outcomes as well as ecological integrity are necessary to establish appropriate goals and measure progress toward a vision of whole community health. Balancing the needs of all watershed inhabitants requires an in-depth accounting, an informed awareness, active participation, and a strong social and political will.

Researchers are catching up to what communities are discovering on their own; that the benefits of participatory ecological restoration extend far beyond biophysical gains. Converting a culverted water channel to an open stream reinstates hydrologic and habitat functioning, and these benefits are conferred upon the human community. Flood control, improved water quality, aesthetics, increased property values, wildlife viewing, and recreational opportunities are desirable outcomes that improve quality of living for watershed residents (Gobster 2000; Riley 1998; Schroeder 1998; Hall 1997). Communities are also recognizing less tangible values gained through hands-on work, such as stewardship and environmental education (Figure 10.3).

Social research validates the intuitive notion that we depend on a connection with the natural world for our physical, aesthetic, or spiritual well-being. On a behavioral level however, our feelings toward nature are acted out in a conflicted way, and at best, reflect ambivalence toward the natural world. This ambivalence also shows up in surveys of public views of ecological restoration; in general, public opinion indicates strong support for restoration but is divided over how this gets carried out. The Snake River dam controversy or the Chicago Wilderness debates demonstrate that restoration may be either a uniting or divisive issue in communities (Gobster 2000; Barro et al. 1998; Barry 1998).

In restoration we make restitution, we return the gift given to us. In view of the time, commitment, and effort involved, conscious restoration may be an act of sacrifice or love. Engaging in the activity of voluntary restoration encourages the participant to suspend the ego, cross over the boundary of perceived self, and enter into a whole-world view of being. Not surprisingly, this transition is of mythical proportions; the manifestation of dreams, imagination, stories embedded in the land, performance, dance, art, celebration, and ritual.

Anthropologist Victor Turner suggests that in order to be effective, ritual must be a cultural activity in which community members feel obligated to participate, such as primitive cultures where

FIGURE 10.3 Mattole residents work hands-on to rebuild native salmon populations. *Source:* Mattole Salmon Group.

FIGURE 10.4 Acting for the good of the community: Volunteers undertake Portland, Oregon's largest scale fisheries restoration project at Crystal Springs. *Source:* Johnson Creek Watershed Council.

survival depends upon members' full involvement. Ritual in this context gives rise to a deep sense of *communitas* or feeling of oneness with the community (Turner 1969). Such rituals are not arbitrarily shaped, but are meaningfully generated by the participants who recognize the community's interdependence and feel a sense of obligation to act for the good of the community. The challenge is to reinterpret these ritualistic patterns of community building in new ways, make them representative of diverse groups, and as accessible to as many residents as possible (Figure 10.4).

MEASURING RESTORATION SUCCESS

The efficacy of restoration projects is often determined according to scientific parameters. Measuring the base level functionality and health of a physical ecosystem is a necessary starting point, and James Karr's Index of Biologic Integrity is the appropriate kind of yardstick to evaluate restoration efforts from a biophysical perspective (Karr and Chu, 1999). However, in a comprehensive model of restoration, the scope of success must be expanded to include measures of whole community functioning to meet the needs of all biotic members. Measures of success may be different with multiple viewpoints included. For example, a successful project may not just mean that salmon return to a particular stream but that school children have connected with their home place. Other possible goals might include building social equity, cultural continuity, environmental education, local food production, or economic improvement.

"Sustainability indicators" like those developed by Ecotrust and Fraser Basin Council are one way to gauge the level of community health, and provide a mechanism for assessing community functioning pre- and postrestoration (Ecotrust 2006; Fraser Basin Council 2003). These measures of sustainability capture quality of life issues missed by pure biologic monitoring. Researchers point out that the site-specific knowledge demonstrated in Oregon's Applegate River watershed, Rogue River basin, and other watersheds improves accountability of each watershed. Local residents want a "report card" or a social accounting of stewardship activities to show progress, and give credit for responsible citizenship (Figure 10.5). This form of social monitoring is seen as "indispensable to measuring the benefits of bio-social management and watershed restoration" (Preister and Kent 1997: 48).

FIGURE 10.5 Totem Salmon gives out stewardship awards to citizens at the annual Johnson Creek Watershed Summit in Portland, Oregon. *Source:* René Senos.

RESTORATION PRACTICE AND SALMON RECOVERY

Restoration definitions are made explicit in practice. First and foremost, restoration ecology is linked to scientific inquiry. The science of returning a natural system to an improved state of function is inexact and highly variable, and current literature is primarily concerned with setting restoration goals, technique, monitoring, and evaluation, and establishing a database. The literature is rife with "how-to" articles on reconstructing wetlands, estuaries, prairies, and forests—virtually every system impacted by human activity. Typically, these recipes focus on desired objectives and outcomes of a given project, and are designed and implemented by outside experts, then carried out in an isolated fashion. Restoration projects may be isolated in various ways: in terms of *spatial extent* in that there is little or no connectivity between the specific site and its surrounding systems; in terms of *project scope* (a single project is carried out on a single site); in terms of *temporal scale* (the restoration effort does not accommodate or connect to future conditions); or in terms of *social input* into the process.

In part, funding or political processes drive this isolated approach. Interventions tend to be single-issue driven and reactive, and often occur in the form of "regulatory-inspired" restoration. Even in larger-scale projects the frame of reference is often defined by political boundaries rather than watershed boundaries (Johnson et al. 1999; Williams et al. 1997). Yet a top-down approach to environmental problem-solving dramatically influences both landscape and community configurations. Solutions imposed on the landscape are also imposed on a given community of people. An explicit assumption in our political framework is that an outside authority can represent a community's values and be responsible for making decisions with community interests in mind. As more people become concerned about their livelihoods, quality of life, and their environmental health, they have begun to challenge this assumption, as in the case of Applegate watershed's "bottom-up" approach to watershed management (Lipschutz 1999; Johnson and Campbell 1999; Jones 1999; Apostol et al. 1998).

The failure of technocratic-based restoration models to affect, promote, and sustain restored health to diminished ecosystems is perhaps nowhere better exemplified than by the traditional approaches to salmon recovery. The hotly contested debate over decommissioning four lower Snake River dams blocking fish passage is a case in point. Initially, scientific and some political support mounted in favor of dam removal. Buckling under extreme political and economic duress, however, scientists produced a new study indicating that rather than removing the dams, the dams could remain in place to support barging and agriculture, and wide-scale restoration could be implemented, instead, to compensate for dwindling salmon runs. Subsequent studies demonstrated that this strategy failed, and in May 2003, a federal judge ruled that the current federal plan to recover wild salmon and steelhead in the Columbia and Snake Rivers is inadequate and uncertain, and thus illegal under the Endangered Species Act. This scenario illustrates how restoration may become a political justification touted by a scientific equation that calculates just how repair is needed in one system to allow continued damage in another system.

Fisheries biologist Jim Lichatowich traces the demise of the salmon and the Northwest's failed attempts to recover its plummeting fish populations in his book *Salmon Without Rivers: A History of the Pacific Salmon Crisis*. Lichatowich sets the Pacific salmon crisis in historic context by citing an evolutionary account of the salmon in its 40-million-year-old Pacific Northwest geologic home, as well as tracing 9,000 years of human interactions with the salmon. He presents a history of protection and restoration measures, and concludes that billions of dollars and various recovery efforts have failed because they neglect the root causes of the fish's extirpation (Lichatowich 1999).

As early as the 1870s, scientists recognized that overharvesting, damming, and river degradation were negatively impacting salmon runs, however, in the prevailing industrial climate these factors were largely ignored in favor of technological fixes such as hatcheries, fish introduction, and barging. Spencer Baird, head of the new U.S. Commission on Fish and Fisheries in 1875, identified overfishing, dams, and habitat loss as primary causes of salmon decline, and established the country's first hatchery program. Later these hatchery programs justified massive dam building projects throughout all the Pacific Northwest. Today, despite decades of evidence documenting the failure of artificial fish propagation, and exhaustive studies linking hatchery fish to disrupted native runs, hatcheries remain a primary tool in fish recovery programs.

One major reason for this flawed system is that the recovery itself has been based on the same economic, political, and environmental valuing system in which salmon exploitation for human consumption was deemed acceptable. In spite of increased understanding of salmon life cycles in the last century, decisions such as the construction of the Bonneville and Grand Coulee dams prevailed under the assumption that politics linked with science could fix subsequent problems. Freeman House, who has applied a participatory model of salmon recovery in California's Mattole Watershed, calls this approach "terrestrial, linear intelligence" and suggests that we need to subdue our quest for rational explanation and enter into a new relationship with salmon through empathy (Latin: "to suffer with gathered senses") and cumulative attentiveness. He asserts, "In the closed-mouth world of reciprocal perception there is no way to learn to live in the place but from the place itself" (House 1999: 5, 99).

However, this ritual passage of learning about a place is difficult when ecological and cultural diversity diminishes to the point of scarce nature and culture to derive from, to intuit and perceive, or to learn from and preserve. Michael McGinnis notes that our lack of discrepancy between real (nature) and illusion (artifice) is perpetuated by industrial mimesis, which places us in a state of hyper-reality. He warns us that restoration can easily become yet another product of the virtual industrial age, and cites salmon restoration as demonstration of this point. McGinnis queries, "The question is what has been restored—self-regulating wild systems or the *spectacle* of wildness?" (McGinnis et al. 1999).

A snapshot of Seattle's Ballard Locks on any given sunny late-summer day where tourists and residents watch through a concrete-encased glass panel as embattled Chinook salmon shoot the

FIGURE 10.6 Direct encounters with migrating Chum salmon at Vancouver Island's Goldstream Park. *Source:* René Senos.

chute of the hydro-engineered ladder would confirm the notion that we have been lulled by the spectacle of wildness. This scene inspires a sense of hopelessness to those of us who know that even if that brave fish can do the impossible and dash up the narrow tunnel through hurricane force water, the chances of her progeny making the return trip to sea are slim.

Contrast this scene to the one at Oxbow Park in the Portland, Oregon, region, where children and families gather streamside to celebrate the return of the salmon with music, food, stories, and direct encounters with salmon returning to natal spawning grounds. At Vancouver Island's Goldstream Provincial Park, the stench of decaying salmon greets visitors who step over carcasses tidally flung up on the trail while thousands of salmon thrash in the river in the midst of their sacrificial spawning rites (Figure 10.6). Tribes and First Nations peoples across the Pacific Northwest still practice "First Salmon" ceremonies, an annual celebration that honors the returning salmon and the cycle of life. Many northwest indigenous peoples are also very engaged in salmon restoration efforts (Senos et al. 2006; Hollenbach and Ory 1999). In the latter examples, no barrier exists for either the migrating fish or people seeking encounters with salmon. Rather, these rare scenes poignantly recall a vital ritual that formally occurred throughout thousands of Pacific Northwest streams.

TOTEM SALMON MODEL

Pacific salmon are extinct in 40% of the rivers they once inhabited, and 34 native salmon stocks, called "evolutionary significant units," have been or are in the process of being listed under the Endangered Species Act (ESA). Sweeping 1998 and 1999 ESA salmon listings and related court rulings were implemented to protect numerous salmon species throughout all Northwest rivers. For the first time, a keystone species was listed as threatened or endangered in regions that included major metropolitan areas such as Portland or Seattle. Strikingly, the problem was no longer a rural or public lands management problem; the anadromous salmon that swims everywhere had just landed in everyone's backyards.

A single-species approach to restoration is problematic from many perspectives. An ESA listing is a fail-safe measure of last resort, and many populations go extinct before the federal government's crisis intervention. Also, it is a system of "head counts" that relies on population numbers and does not calculate the full impact of damage to an ecosystem. Sheer numbers as a way to measure losses (extinction) or gains (species stabilization) in a complex ecological system is arguably a crude way to evaluate or resolve the originating problems. Some ecologists and wildlife biologists complain that a single-species approach ignores myriad other species and ecosystem functions at risk, and that subsequent restoration and management plans tend to be reactive (Frissell 1997).

A single-focused view of the problem leads to restoration approaches not grounded in ecological context. Agencies often focus restoration efforts on the most-degraded sites. These enhancements allow continued degradation of relatively intact habitat in other parts of the watershed, and the result of this strategy is that disturbance is distributed across the landscape and virtually all sites ultimately become degraded. In the course of funneling resources to degraded sites while failing to protect the most productive sites, the sensitive species will most likely have disappeared (Doppelt et al. 1993; Frissell and Bayles 1996; Frissell 1997).

A single-species strategy also potentially ignores the needs of other community members, whether that species is salmon or human, and may perpetuate the very imbalance that helped create the problem. It divides rather than unites, as seen in the Northern spotted owl controversy. As one Washington watershed council member noted, the salmon, beloved by many, is at risk of being scape-goated (Willapa Bay Alliance, personal interview, 2000). "Scape-goating" is an interesting word choice given its definition: "a person or group made to bear the blame for others or *to suffer in their place*."

As evidenced in the Northern spotted owl debacle, an artificial dichotomy is created between the endangered species and residents' livelihoods, and certainly Pacific Northwest residents are divided over the matter. Clearly, a whole community view and bioregional approach is needed to resolve these dilemmas. Additionally, a *partnership ethic* is required that asserts, "the greatest good for the human and nonhuman communities is to be found in their mutual living interdependence" (Merchant 1999).

Our cultural, economic, and ecological relations with salmon may serve as a guide for our ethical orientation and decisions acted out on the land. As a significant totemic icon in Pacific Northwest consciousness, our identification with the salmon offers an entry point for reconstructing our ecological identity. Numerous social groups, from schoolchildren to fishermen, have organized around salmon recovery (Figure 10.7). This outpouring of strong public response to the plight of these fish indicates strong recognition of salmon's critical role in the ecological and cultural community.

Identifying salmon as a significant cultural and ecological icon may serve as a directional, a symbol of the mysterious and intrinsic connection between all members of the community, and a critical measure of healthy relations. In other places, different icons derived from a given community's unique relationship to the land will function as indicators of health. On the Atlantic seaboard,

FIGURE 10.7 Kids plant trees to restore salmon habitat in Portland, OR. *Source:* Johnson Creek Watershed Council.

for example, Chesapeake Bay oysters are integral to the fisheries economy and bay water quality; massive oyster reefs in the 1950s filtered all the bay's waters in three days—a task that now takes an entire year. Pacific salmon and Chesapeake Bay oysters are both powerful cultural icons that signify the nexus between economic, social, and ecological dimensions. They are the "cultural keystone species" that community members depend upon for survival (Garibaldi and Turner 2004). In restoration practice, totemic symbols can help communities form a multidimensional perspective while crafting a new vision of their home place.

PARTICIPATORY RESTORATION: STRIVING TOWARD HEALTHY COMMUNITY

Citizen-based restoration projects offer lessons that may inform a participatory restoration approach to regenerating bioregional health, and suggest ways to renew our relationship with fellow species like salmon. In the Pacific Northwest, there are over 400 nonprofit organizations and watershed councils focused on rebuilding healthy communities. While the effectiveness of these organizations is not fully understood, several of these groups have been highly successful in building partnerships and attracting local citizen participation.

The following case studies highlight the potential of citizen-based restoration to revitalize whole community health. These projects vary in their focus, strategies, landscape scale, and social configuration; however, they hold in common the notion of redefining human interactions with the surrounding natural world. The proponents of these projects have adopted new ways of relating to ecosystems that diverge from a strictly utilitarian view of the land. Community members seek healthier interactions with their physical and social sphere, which include engaging in more deliberate action with the land and each other; reconceptualizing community health more holistically; and working cooperatively at a grassroots level to achieve consensus on appropriate economic, social, and ecological interactions with the land. These adaptive, long-term approaches have typically evolved in response to escalated environmental, social, or economic crises, such as depleted fisheries or forestry productions. Best science is integrated with community visioning to develop restoration strategies. Critically, these creative community-based solutions directly address the linkages between ecological vitality and quality of life for the human community.

MATTOLE WATERSHED

Located in the Coast Mountain Ranges of northwestern California, the 64-mile-long Mattole River drains a 304 square mile watershed into the Pacific Ocean 35 miles south of Eureka. Less than 3000 residents live in the watershed, yet 90% of the old growth forest is logged. In 1980 the near extinction of Chinook and Coho salmon prompted Mattole residents to form the Mattole Watershed Salmon Support Group (now called Mattole Salmon Group). The express goal of this nonprofit organization was to increase wild salmon runs through the vehicle of river restoration. The group's work combined comprehensive assessments with intensive hands-on labor. Members surveyed fish populations and habitat conditions to determine causes for the salmon's decline, and engaged in numerous projects to improve salmon habitat and numbers of returning fish. One major project involved the operation of homemade, in-stream hatching and rearing facilities. Volunteers and biologists trapped, spawned, and hatched native salmon, increasing the egg-to-fry survival rate from 15% to more than 80% (House 1999).

Mattole restorationists soon learned that it was not enough to reconstruct physical connections in the watershed; rather, new social connections were required to halt further destruction to salmon. Through their fisheries work, the group determined that erosion and sedimentation from road building, ranching, and logging activities posed severe threats to fish. The salmon coalition joined forces with loggers, ranchers, and other Mattole residents to establish the Mattole Restoration Council (MRC). The Council recognized "it was time to think like a watershed," and linked individual ecological recovery efforts to form a comprehensive, bioregional plan (Zuckerman 1997: 223).

Conflict among different resource user groups served as a catalyst in developing a consensus-based program, while collaboration toward a common goal helped dissolve tensions and factions within the watershed community. Greater numbers of participants engaged in widespread action across the watershed. Through community mapping, watershed citizens were able to describe their home place, track its condition, and help determine its future. This place-based knowledge, when combined with scientific data and methods, resulted in an adaptive, innovative, and extremely powerful approach to watershed restoration (Figure 10.8).

Mattole restorationists are cautiously optimistic about the results of their efforts as the watershed is still in an early recovery stage. During its 22-year history the MRC has successfully collaborated with private landowners and conservation agencies to protect critical salmon habitat and old-growth areas. Joint MRC and federal and local efforts have resulted in the designation of over 70,000 watershed acres in conservation status. MRC is currently implementing a 2005 comprehensive plan that sets forth a 30-year vision for building a healthy Mattole watershed community.

As an MRC partner organization, the Mattole Salmon Group has provided rearing, research, and in-stream restoration for native Mattole River salmon for 26 years. Fish surveys showed a dramatic downward spiral of Mattole salmon to an all-time low in 1990, followed by a gradual increase of Chinook and Coho salmon populations since 1990, which suggest that the native salmon populations appear to be gaining ground. Nearly ¾ million native King salmon have been released into the river to ensure the genetic integrity of the run, and the Mattole salmon rescue program costs a negligible $40,000 a year, a cost far below any high tech hatchery program.

Despite the gradual upward trend of salmon returns, however, the populations fluctuate on a yearly basis in response to changing oceanic, atmospheric, and local watershed conditions. Adequate water flow is a major issue; the Mattole River is fed by groundwater and not snowmelt. Mattole headwaters have discharged low or zero flows in recent years, resulting in an 8-mile dry stream stretch in 2001. This loss of water recharge has negatively impacted salmon runs, and in 2006 the Coho runs are not as strong as expected. The Mattole Salmon Group and MRC, finding that residential water use is a contributing factor, are working with community members to install rainwater collection tanks at their homes and to reduce water consumption (Reid Bryson, personal communication. 2006).

FIGURE 10.8 Salmon recovery became the impetus for community building as Mattole residents learned to "think like a watershed." *Source:* Mattole Salmon Group.

FIGURE 10.9 Schoolchildren release salmon fry incubated in their classrooms. An entire generation of Mattole children has grown up educated in watershed citizenship. *Source:* Mattole Salmon Group.

Nearly one-fourth of all watershed residents have contributed thousands of volunteer hours to the Mattole recovery effort, from school children who incubate salmon eggs in their classroom to out-of-work fishers who assist with salmon hatching, fish surveys, and stream monitoring (Bernard and Young 1997, Zuckerman 1997: 232). Since the Mattole Salmon Group and Mattole Restoration Coalition were formed, a full generation of children has grown into adults educated in Mattole watershed community ethics and behavior (Figure 10.9). Although activities like logging and fishing still occur in the Mattole watershed, remarkably, there are more jobs in restoration than in logging (Bernard and Young 1997).

McGinnis observes that the ongoing citizen work in the Mattole is at the level of "shared service," which is a crucial first step to building community (McGinnis et al. 1999: 218). Restoration becomes an active exchange of gifts, and in the exchange, participants gain a greater awareness of the instrumental role they perform in the watershed. As citizens renegotiate relations with each other and their home place, solutions that are more imaginative, flexible, inclusive, and sustainable emerge from the process. This shift away from self-determined individualism to a "watershed-based democracy" has strong economic, cultural, and political implications, in addition to positive environmental consequences. Community members with divergent perspectives came together around a common goal: to restore salmon and the related political economy of the Mattole region. Interestingly, what began as a single-species recovery approach evolved into a process of watershed community regeneration.

THE APPLEGATE WATERSHED

In the 500,000-acre Applegate River watershed in southern Oregon, controversial events coalesced to create an environment favorable to a community-driven approach to watershed design and planning. In response to a political gridlock, watershed members formed an alliance in 1992 called the Applegate Partnership, which consisted of residents, environmental proponents, and timber industry representatives who initially disagreed but opted to work together to develop a participatory watershed planning approach. U.S. Forest Service (USFS) and U.S. Bureau of Land Management (BLM) personnel later joined the partnership.

The Applegate Partnership is consensus-based and community driven at its core. It advances the use of "best science" and sound natural resource principles to promote ecosystem health and diversity. Significantly, the Partnership links forest health to overall community health (Figure 10.10). The group's early success in collaborative planning among diverse interests has drawn the attention of politicians and planners nationally who cite the Applegate experience as an intriguing example of participatory watershed planning and development (Jones 1999; Preister and Kent 1997; Clinton's Northwest Forest Plan 1994).

Progenitors of the Applegate experiment have experienced challenges and pitfalls along the way in spite of deliberate collaboration. A consensus-model of community building defies old ways of thinking and operating, and this is especially difficult to achieve in a politically charged, rural setting such as the Applegate where the debate of appropriate land management tends to be polarized. The relationship between public agencies and private interests was historically strained in the Applegate Valley, and the Partnership worked to overcome these differences. The Applegate experience illustrates that private–public partnerships are absolutely vital in places like the Applegate that are primarily in public ownership; where decisions about the land have traditionally been formed and imposed by government agencies without citizen input; and where the watershed atmosphere

FIGURE 10.10 Forest health on public lands was the uniting issue in Oregon's Applegate watershed. *Source:* Dean Apostol.

has been characterized by distrust between different interest groups. Applegate's success is rooted in its founding: it was a grassroots coalition formed by people living in the watershed and most impacted by land use decisions; in other words, a "bottom-up, inside-out" approach to community-building (Apostol 1998).

Public involvement shaped the Partnership's mission and research objectives, and continues to form the basis of the group. Early on, participants were able to break through stereotypes and misconceptions about "the other," and over time, build a diverse base of participants, including loggers, miners, farmers, ranchers, environmental groups, and public agency representatives, by repeatedly inviting the participation of all community interest groups. The question "Who else needs to be at the table?" was the frequent call and invitation of the group in its public outreach efforts (Johnson and Campbell 1999).

Using a bioregional approach, the Applegate Partnership "called for both physical and social ecosystem assessments with the ultimate aim being to integrate the two over time so that community and forest health would be synonymous within newly drawn human geographic boundaries" (Preister and Kent 1997: 37). Various reports documented the watershed's "state of health" in economic, ecological, and social terms. For example, Preister performed a community assessment of the Applegate watershed, and determined that industrial–economic practices such as clear-cutting dislodge local knowledge and practices, and cause polarized debate over land use. He surmised that resource-dependent residents do indeed exhibit a stewardship stance to the land that tends to be subsumed by such technocratic practices. For example, a stewardship ethic was demonstrated by the cooperative, self-regulatory practices of farmers who recognized the destructiveness of the trapezoidal ditch to fish, and worked out a cost-sharing arrangement with the state to switch to sprinkler systems and install fish screens.

Stewardship behavior needs to be recognized and rewarded while local knowledge and practices are incorporated into watershed planning and management. Part of the success of the Applegate model is the thoughtful integration of local knowledge into public agency protocol. On several occasions, BLM and USFS worked with watershed residents via the Partnership forum to develop solutions that addressed economic, ecological, and social issues. Drawing citizens together early in the process increased their understanding of their landscape and of the design process. Shared information, an open dialogue, and an inclusive process of decision-making resulted in increased trust and adaptive solutions that benefited the whole community.

WILLAPA BAY ALLIANCE

Willapa Bay, located on the southwestern Washington fringe, includes a 600,000-acre watershed and an 88,000-acre bay and estuary. Nearly 20,000 residents inhabit the watershed, and 88% of the land ownership is private. Willapa Bay is one of the cleanest, most productive estuaries in the lower 48 states, and supports 25% of the nation's productive shellfish habitat (Figure 10.11). Oyster aquaculture and other fisheries are the economic mainstay of the region, followed by forestry with over 80% of the watershed in timber production, as well as agriculture, and tourism (Sustainable Northwest 1997: 74; Willapa Bay Alliance, personal communication, 2000).

Like other natural-resource dependent regions, Willapa Bay is highly vulnerable to development and land use pressures. The tension between social economics and the ability of the land to support human uses is reflected in conditions such as depleted old growth forests, despoiled estuary habitat, declining salmon runs, and weakened oyster, clam, and shrimp populations. Other environmental problems include: water pollution from sewage and agricultural sources, habitat degradation caused by logging or diking, and spartina invasion (introduced grass species) displacing native eelgrass habitat.

In 1992, Willapa Bay watershed residents, with support and funding from Ecotrust and The Nature Conservancy, incorporated a nonprofit conservation organization called The Willapa

FIGURE 10.11 Willapa Bay watershed encompasses a highly productive estuary, bay, and coniferous forests. *Source:* Jones & Jones.

Alliance (TWA) to empower local people to address land management and community health issues. The group represented various stakeholder interests, including commercial, environmental, Native American, natural resource user groups, and community members who united out of a common concern about sustaining the environmental, economic, and social health of their region. The Willapa Alliance's stated mission was: "To maintain and restore a substantial ecosystem that produces a variety of resources, while providing the basis for a healthy human economy based upon these products." Organizers desired a participatory partnership process in favor of an external, regulatory approach to resolving the region's intractable social and environmental problems, and immediately began building constituencies toward common goals (Sustainable Northwest 1997).

The Willapa Alliance linked best science methods with place-based knowledge to develop a bioregional approach to sustaining the ecological health of the watershed as well as the livelihoods of its residents. The organization represented diverse interests of residents who depend on logging, forestry, farming, fishing, oystering, and tourism for income, and the Alliance drew upon local knowledge to supplement the scientific and technical expertise of groups like Ecotrust and its affiliate Interrain Pacific. For example, in the construction of its GIS database, the Alliance reviewed satellite imagery with oyster gatherers to confirm the location of the most productive oyster beds. Similarly, the conservation group's Salmonwalk Program trained volunteers as "citizen scientists" in streamside monitoring and assessment to generate salmonid data. The GIS database was then widely distributed to watershed residents, who could track the health of their basin.

The Willapa Alliance made great strides in its early years. Nearly 10% of Willapa Bay residents became involved in restoration efforts, including salmon habitat restoration. Several promising projects were initiated, such as the Bear River Restoration, which involved working with landowners to breach dikes and place large woody debris in stream; the Palix River restoration; and a 1,000-acre wetland restoration. Incentive programs were developed to encourage citizen stewardship. Ecotrust's affiliate Shorebank Trust provided "Green Economy" loans to watershed residents engaged in ecologically sustainable business practices. Willapa Bay Alliance and Ecotrust developed a set of community health indicators, called "Willapa Indicators for a Sustainable Community" to measure the state of the basin's health in three areas: natural wealth, economic wealth, and social wealth. These indicators provided residents with a way to track progress, and adjust the course of their actions.

Unfortunately, the Willapa Alliance disbanded after nearly a decade of conservation collaboration. Observers cite multiple causes, including nonprofit management struggles, financial problems, partial stakeholder participation, and unreasonable expectations held by both locals and outside supporters (Goebel et al. 2003). The Willapa Alliance was perhaps also unable to overcome the community's perception that the organization was strictly an environmental initiative. The Alliance's experience was that despite overall community support, many Willapa Bay residents were wary of restoration efforts or any intervention perceived as a threat to private property rights. Like the Applegate watershed, Willapa's rural constituency was concerned about what regulations an "outsider" might try to impose on their way of living. Residents disagreed on the identification of the problem, the prioritization of restoration goals, and the means to accomplish watershed health. In view of these differences, some people in the community questioned whether or not their prescriptions were being carried out on the land. (Mark Heckert, personal interview, April 2000.)

Willapa's successes and challenges can teach us valuable lessons about community-based conservation. The community must be fully represented and engaged in the effort to insure that their diverse interests are represented. Local residents need to define specific goals and objectives early in the process to track measurable progress, and create early successes. Building the organization's professional capabilities, forming key relationships with community members, and identifying local leadership are critical steps in the process. In a working community, it is imperative to tie restoration and conservation efforts to the goal of bolstering the community's financial viability so that the benefits are distributed across the social and environmental spectrums of the watershed (Goebel et al. 2003).

However, the Willapa experiment broke new ground as one of the Northwest region's first community conservation collaborative, and demonstrated the success of many innovative strategies to implement watershedwide restoration. The endeavor also caught fire in the imaginations of people both inside and outside the watershed. As a testament to Willapa watershed's high value, in March 2003, The Nature Conservancy purchased 5,000 acres at Ellsworth Creek near Willapa Bay, protecting a major salmon run, and the largest coastal stretch between the Canada border and central Oregon. The Conservancy will implement restoration work through the federal Jobs in the Woods Program, which assists displaced natural resource workers in finding new employment in the field of restoration. The Conservancy will hire local people who formerly worked in the timber or fishing industry.

These case studies and many other examples of "watershed citizenship" offer valuable tools and guidelines for developing a participatory restoration approach.

TOOLS AND GUIDELINES FOR PARTICIPATORY ECOLOGICAL RESTORATION

A bioregional approach to regenerating whole community health that interweaves ecological, economic, cultural, and political values:

- Tools are employed that enable community to form a shared vision and an ecological identity, such as community visioning or bioregional mapping exercises.
- A collaborative, community-based model of restoring watershed health is applied.
- Restoration is conceived as a way to reinhabit our natural surroundings, to gain full community membership standing, and to partner with nature.
- Place-based knowledge is linked with best science.
- Decentralized economic, social, and political structure is set within the supporting framework of particular biophysical characteristics of a region.
- Restoration is linked to local economic development and generates new markets, jobs, products.

Human interactions with the land are cast in terms of watershed or landscape scale implications:

- Human activities occur within the given ecological constraints of a particular ecosystem.
- Full accounting is made of "nature's services" and built into watershed economy. (Ecotrust and Rocky Mountain Institute are organizations that "do the math").
- Sustainable practices such as self-regulation are embedded in the political, cultural, and economic structure of a community.
- Ecological or bioregional identity is expressed through cultural devices of ritual, art, and celebration.

Restoration is undertaken for the primary purpose of restoring biological functionality, but may include social goals such as education, aesthetics, reconnection with the land, or community revitalization:

- Ecological restoration rises out of an ethics-based approach; it is not used to justify damage to a natural system.
- Restoration activity is framed within a comprehensive approach to restoring watershed or bioregional health.
- Scientific evaluation of ecosystem conditions; best-known restoration methods employed.
- Human interests and needs are identified and factored into the restoration process; projects benefit both people and environment.

Inclusive process that values all community members:

- Human relationship to natural surroundings reflects equal standing of all members.
- Empathy; no "they" or "other."
- Open invitation for people to enter the process from their particular point of reference.
- Varied opportunities for community members to engage in decision making about their home place, and to participate in restoration activities.
- Ongoing dialogue and consensus-based decision-making format.
- "Assets modeling" identifies the unique contributions each member brings to the table.

Outreach, education, and widely disseminated information:

- Extensive outreach to whole community through a variety of means (media, Internet, newsletters, meetings, door-to-door, word of mouth, demonstration sites, schools, summits, celebrations).
- Determined effort to reach all segments of population: schoolchildren, elders, blue collar workers, parents, diverse social and ethnic groups.
- Data and information about watershed or bioregion is collected, assembled, updated regularly, and broadcast to the public in accessible format.
- Visualization tools help citizens understand their community's state of health, make informed decisions about desired future conditions, and track progress.
- Ongoing "community education" occurs in myriad forms: workshops, summits, adopt-a-stream, students' curriculum, Internet, hands-on restoration.

MONITORING AND ADAPTIVE PROGRAMMING

- Long-term monitoring and evaluation conducted by citizen-scientist teams.
- Flexible programming; mistakes are embraced and incorporated.

- Accountability and benchmarks track progress and provide feedback to community about their efforts; critically, success is measured by both ecological and sociological transformations.
- Community health indicators depict the state of whole watershed health by including social and ecological variables.

SALMON COMMUNITY: BENEFITS OF RESTORING LOST CONNECTIVITY

While salmon are an economic asset to many Pacific Northwest communities, strong public interest in salmon extends beyond economics to cultural identification with the fish. Public attitudes toward salmon indicate that the essential values that provide meaning and depth to peoples' lives include cultural heritage or environmental integrity. Several studies support this lesson. A Washington State survey showed that three out of four voters believed that salmon are a key part of Northwest identity, and 77% agreed that salmon are an important indicator of overall environmental health (The Elway Poll 1992). In an *Oregonian* poll, 60% stated that salmon should be prioritized before commerce in managing the Columbia River system, and when asked why salmon should be recovered, 35% of respondents cited Northwest heritage, 36% believed salmon are a measure of the region's environmental health, and only 15% gave economic value as the rationale (Brinkman 1997). Other polls demonstrate that the public supports greater protection of salmon and wildlife on public lands such as federal forests and rangelands (OSU Survey 1991; WSU/USU Rangelands Survey 1993; Smith and Steel 1996).

Pacific salmon symbolize the unique ecological and cultural heritage of the Northwest, and their past and future lives are intertwined with our own evolution. Aldo Leopold described the cultural value of "wild rootage," and observed, "There are cultural values in … experiences that renew contacts with wild things," and these experiences "remind us of our distinctive national origins and evolution" (Leopold 1949: 177). As we aspire to be indigenous to place, we can feel a sense of belonging that binds us to our natural surroundings. Paul Shepard says: "Belonging is the pivot of life, the point at which selfhood becomes possible—not just belonging in general but in particular. One belongs to a universe of order and purpose that must initially be realized as a particular community of certain species in a terrain of unique geology" (Shepard 1996).

House observes that "too many of us come into a place and don't know how to be part of it, which is the same as saying we don't know where we are." He notes that each bioregion represents particular water and energy flows, flora and fauna, soil structure, micro-climates, "the very flesh and blood of a place, and human beings have evolved to experience their relationship to the biosphere through sensory perception of these specific variations" (House, 1999: 158). However, many humans have forgotten the skill of systematic attentiveness that opens us to the natural world's instruction.

As a direct teacher, ecological restoration allows us to reenter the natural world's classroom, and guides us to a state of belonging. Our memories and senses are reactivated in the process. We learn from the patterns of the place that originally shaped our biological and psychological responses. In order to understand the natural system in which we interact, we develop careful observation to the dynamic nuances of an ecological system, which in turn leads to thoughtful engagement with the land. Restoration work enlists all the senses; we are embodied in landscape as we apply our whole bodies and minds to the work of physical repair.

AN EXPANDED VIEW: PARTNERING WITH NATURE

A model of ecological restoration that encompasses all impacted systems, ecological and cultural, and recognizes their interdependency moves us away from isolated restoration concepts and applications. Participatory restoration calls for a full accounting of the community in terms of its ecological and social relationships, and a vision of watershed health that is shaped accordingly. Citizen-based restoration at a broad landscape scale is both the physical and the social construct for manifesting a sustainable community. Rather than a single-site, single-species point of view, the problem and its corresponding solution are cast in more inclusive, holistic terms. As a result, dislocated sites are placed into the social and ecological context of an entire watershed or bioregion. Dislocated species find their way home again, whether salmon or human.

Restorationists are learning that ecological restoration provides a paradigm for a healthy, mutually beneficial relationship between humans and nature. Ecological restoration may represent the next cultural leap in gardening or farming by modeling a nurturing relationship between human beings and the natural landscape based on reciprocity. The agro-ecology movement offers a compelling alternative to agribusiness by creating food production systems that build sustainable environments, economies, and communities (Imhoff 2003; Jackson 1994). Community gardens also illustrate this idea of symbiosis. By attempting to reconstruct ecosystems, we grow to understand these natural systems, and enter into a relationship with nature that engages the full range of human activities. A translation occurs between the methodical and the mythical.

Volunteers report positive feelings that relate to self-image and one's place in the community or even the perception of the community itself. On an individual level, people describe feeling greater harmony, responsibility, and connectedness with neighbors and nature as a consequence of direct work with the land. Restorationists, defined in the broadest sense as community members who make restitution, or who give back to their bioregional community, relate great satisfaction from their contributions. Homeowners, business operators, farmers, grandparents, and school children express enthusiasm and hopefulness that their activities will make a positive difference in their home places (Figure 10.12).

These intangible rewards are not limited to an individual sphere of influence but impact the general community as well. The built and social structures of a restored or cared-for place positively shape the perceptions of people who live in that place. Aesthetics are a crucial element in built environments, and urban conditions affect how one feels about his or her home place. "The physical condition of a community … plays a double role. For the community, it is a measure of itself; for outsiders it creates an impression of community quality and character" (Lewis 1996: 54). Similarly, participatory restoration moves from individual motives and rewards to communal benefits. The acceptance of locally created spaces by the larger community, as evidenced by San Francisco

FIGURE 10.12 Community members improve riparian habitat on the industrialized Duwamish River in Seattle's Georgetown neighborhood. *Source:* Nate Cormier.

gardens or Mattole hatchbox sites, indicates that the benefits of participatory restoration extend to all residents. In this way, restoration holds a communal role that, like public parks or city gardens, transcends private garden space.

Restoration can lead us out of nature–culture dichotomies by affirming our role as partners with nature. Gone are the old models that assume an equilibrium landscape state devoid of human intervention. As the field of ecology factors in dynamism, disturbance, unbalanced or unpredictable states of nature, a corollary movement toward a cultural interpretation of ecological restoration is gaining momentum. Restoration is conceived as a cultural interaction with nature, in a long line of human interventions with the landscape (Anderson 2005; Higgs 2003; McCann 1999).

This partnership model is consistent with recent discoveries about the biogeography of salmon species and early indigenous peoples in the Pacific Northwest. Biogeography refers to the spatial and temporal distribution of species as explained by current and historic events. Research suggests a strong correlation between the development of individual salmon strains along specific stream reaches and individual linguistic groups of Pacific Northwest American Indians inhabiting these watersheds over an extensive time period (Lake 2003). Other studies document sockeye salmon preference for certain migratory routes through the northern Georgia Straits as a result of indigenous peoples' reef net fishery technologies and long-term management of kelp beds (Barsh and Hansen 2002). These startling revelations from the arena of traditional ecological knowledge (TEK) illustrate three major points. One is that humans have shaped the landscape and its constituents for millennia. Two, humans coevolved with other species, and catalyzed the genetic adaptation of diverse salmon species. Finally, as TEK demonstrates, restoration may reinstate a mutually beneficial relationship between humans and a particular species and its shared habitat (Senos et al. 2006; Hollenbach and Ory 1999) (Figure 10.13).

A missing component in traditional restoration is watershed restoration planning, or the identification of an overall restoration plan within which a particular project fits. A comprehensive watershed or bioregional restoration plan is necessary to salvage damaged ecosystems with potential, and critically, to protect functioning ecosystems. Assessment, identification, and prioritization of current ecological values are necessary to prevent high functioning places from being destroyed while low functioning places are "restored." Structuring interventions hierarchically as scientist Chris Frissell proposes may maximize ecological recovery. Salmon recovery depends upon operating at appropriate temporal and spatial scales, and adopting a salmon's eye view of the watershed

FIGURE 10.13 Karuk tribal members decommission a road and restore habitat for salmon and other species in the Karuk watershed in Northern California. *Source:* Frank Lake.

(Lombard 2003). In the face of rapid population growth and urban development in the next 50 years, citizens, policy makers, and scientists will need to make difficult choices to prevent the extirpation of wild salmon in the Pacific Northwest. Connecting restoration efforts to an overall ecological and social context is essential to reprise whole watershed health.

The Willamette River Initiative (WRI) in Oregon and the Greenprint for Puget Sound are two innovative projects underway that tie large-scale basin planning with prioritization of local conservation and restoration strategies. WRI created spatially explicit "future landscapes" maps of the Willamette Valley based upon scientific data and local knowledge to provide watershed residents with a powerful tool to make informed decisions about a desired vision for their community (Hulse et al. 2000). In addition to demonstrating the likely effects of different scenarios acted out on the land, the WRI plan identifies sites with the best restoration potential. The Trust for Public Lands (TPL) is currently developing a Greenprint for Puget Sound by mapping the basin and its current political, ecological, and social state. Working with other conservation agencies, TPL will prioritize sensitive lands for conservation acquisition, and will develop funding and partnership opportunities to protect these places. The Willamette Basin and Puget Sound projects both link GIS technology (Geographical Information Systems) with community input to develop conservation and restoration strategies at the watershed scale before the basin is entirely subsumed by urban development.

These projects offer a promising alternative to restoration initiated by bureaucratic or regulatory agencies that does not necessarily represent a community's self-interest or values. Many projects have failed without community input and investment. As people are left out of the opportunity to participate in developing creative solutions that represent individual and community needs, this exclusion fuels debate over appropriate land use. Sides are taken up around resource-based uses like fishing, felling, farming, or framing, and the community is left even more fragmented and divided.

ETHICS, RELATIONSHIP, AND ECOLOGICAL IDENTITY

It is our intention, and the quality of our relationship with nature, that distinguishes true restoration from tinkering. To move beyond casual tinkering, an ecological ethic is required to guide restoration work so that it is a healing art. This requires a clear definition of restoration that literally means to return to a state of health or to give back.

Participatory restoration at a bioregional scale facilitates community transformation, evidenced by places like the Mattole, Applegate, or Willapa. This evolves out of the process of defining new relationship and behavior patterns toward the land to achieve Aldo Leopold's description of a "mutually beneficial relationship" between nature and culture, in which humans are full participants in the reciprocal exchange of goods and services. As Leopold observed, "ethics are possibly a kind of community instinct in the making" (1949: 203). Participatory restoration is grounded in an ecological ethic that recognizes the intrinsic rights of all community members. Inclusiveness and equal standing among differing social and ecological interests are built into social, economic, and political structures.

Empathy grows out of an increased understanding and connection with ecosystems and social systems, developed through careful attention and deliberate engagement. The restoration process itself calls upon the physical, emotional, social, and intellectual faculties of human experience, and enhances ways of knowing that go beyond rational understanding. It expands human awareness of the natural world and our place in it. As such, restoration is a hands-on dialogue with the landscape.

Our cultural and ecological self-definition rises out of the mutual interaction that occurs between humans and ecological systems. Participatory restoration allows us to move past being overwhelmed by the infinite ways we impact ecological systems to a point of active engagement. As ecological systems are subsumed by resource extraction or random urban development, it is not just the physical landscapes themselves that are lost, but also our longstanding cultural ties with these natural life systems. This is why we need to worry about disappearing landscapes and species to draw lessons from: as evolutionary creatures, we are losing the very basis to learn about our selves and our place in the natural world.

FIGURE 10.14 Environmental curriculum begins in the classroom and extends outdoors as children learn the salmon life-cycle. *Source:* Mattole Salmon Group.

Disconnection from nature in our daily lives has created a sense of grief and helplessness. Many people seek a way to bridge this artificial separation. The growing success of programs such as community gardens, wilderness retreats, therapeutic horticulture, and citizen-based restoration attests to this strong desire to reconnect. Children especially respond enthusiastically to programs that introduce them to their natural world and allow them to contribute in some fashion (Louv 2005; Lev 1995; Sutton 1996; Sivakumarun 1996; Tapsell 1997; The Wetlands Conservancy 1997; Bomar et al. 1999). On a broader level, environmental curriculum is taking hold in school districts across the country (Figures 10.14 and 10.15). The great success of the Conservation Youth Corp, particularly

FIGURE 10.15 Environmental curriculum begins in the classroom and extends outdoors as children learn the salmon life-cycle. *Source:* Mattole Salmon Group.

in the Pacific Northwest landscapes, evokes the tradition of the 1930s Conservation Civilian Corp (CCC) movement.

Community education is not just for children but involves all citizens. Adopt-a-Stream projects, Friends groups, Naturescaping workshops, community gardening, community mapping and visioning exercises, and hands-on restoration offer fertile ground for developing fuller awareness of ecological and social processes (Aberly 1994, 1993; Harrington 1996; Egan and Glass 1995; Andruss et al. 1990). The notion of "citizen-scientist" is an intriguing one in that it breaks away from a traditional concept of science for outside experts only. It also suggests that watershed residents have the capacity to hone powers of observation and knowingness that lead to "self-education" and informed interaction with local places.

Ecological restoration provides us with an opportunity to become indoctrinated in our natural surroundings. As we develop greater appreciation for our community assets, and adjust our relationships accordingly, a richer way of being in the world unfolds. Restoration may be a rite of passage to assume our place in nature as full community members. Participatory restoration framed by ecological and social imperatives is an opportunity to reinstate our ecological identities, or as Freeman House puts it, to unearth our "futures primitive." A meaningful, holistic process like bioregional restoration or the "totem salmon" approach may facilitate this discovery process. Critically, the driving mechanism for defining this approach is particularized to the place: it may be Chicago prairie landscapes, or Chesapeake Bay oyster reefs, or Pacific Northwest salmon.

This approach does not suggest that human interactions with the local landscape such as forestry or fishing are derided or scrapped entirely. Rather, these practices inform the context of our actions, reinforced by political, economic, and social institutions, so that we may translate this eco-cultural basis in a more sustainable way. Technologies, tools, and information gained in previous land interactions are adaptively incorporated into contemporary relations. An excellent demonstration of this concept is the U.S. Fish and Wildlife Service's Jobs in the Woods program that hires former loggers and forest workers to implement watershed restoration projects in northern California, Oregon, and Washington. Forestry skills are refocused on building healthy forests rather than wholesale timber extraction. Furthermore, the new physical forms and aesthetics of restoration are integrated with historic landscape structures and patterns, and provide interpretive and educational value.

COMMUNITY-BUILDING: NEGOTIATING RELATIONS WITH EACH OTHER IN THE ENVIRONMENT

Steve Packard, pioneer of the Chicago prairie restoration movement, believed that entire "congregations of people" were needed to restore prairie landscapes (Stevens 1995: 50). His hope was that if he could recruit cadres of volunteers to engage with prairie landscapes, they would develop an emotional bond with the land, and these remnant prairies would have dedicated constituents vested in their recovery. Indeed, the implementation of a large volunteer network through the Nature Conservancy resulted in the restoration of several thousand acres of prairie and oak savanna that otherwise would have been neglected or lost (Figure 10.16).

Restoration not only took hold in physical form on the land, but in the social form in Chicago's "restoration mindset," as the prairie restoration impetus grew into a much bigger forum called the Chicago Wilderness. Public debates related to the form and function of restoration, while controversial, demonstrate that restoration has become embedded in the social fabric of this major metropolitan city. The recent construction of a green roof on Chicago's city hall is the quintessential expression of a restoration paradigm.

A restoration effort of great magnitude requires the involvement of not just a handful of people but rather a large segment of the population. Restoration provides a forum for people of divergent cultural and ecological experiences to come together. A true bioregional approach offers varied

FIGURE 10.16 Steve Packard and citizen volunteers restore Chicago's vanishing prairie and oak savanna systems. *Source:* Dean Apostol.

opportunities for community members to engage in decision making about their home place and to participate in restoration activities. Multiple opportunities for entry into the process are needed to capture the interest of as many constituents as possible. Levels of potential engagement may range from low involvement to high in order to encourage greater participation.

A key concept is that restoration represents a level of *shared service*. In an inclusive restoration model all members are needed and valued. There is an open acknowledgment of the assets and unique contributions each community member brings. Given the critical role each member plays, there is a sense of "communitas," or an obligation to participate for the greater good of the community. Rather than passing by as listless observers, we arrive, discover, and attend to our home places. Restoration as ritual is a cultural activity whereby citizens are committed to participate because community survival depends upon it.

New cultural expressions of language, myth, art, ritual, and celebration emerge out of this transformation. Rituals and celebrations such as restoration workdays or first salmon ceremonies or watershed summits take place. Language and terms like "no *they*," "thinking like a watershed," "citizen scientist," and "just show up" incorporates restoration lexicon and the community's shift toward a collaborative dialogue. The craft of restoration and the aesthetic of "health" create new physical imprints. Public artworks and inspiring aesthetic forms are integrated into restored places and help reflect a distinct regional identity (Figure 10.17).

Restoration can be a powerful form of community building. Public participation in organizing, collaborating, deciding community futures, making new choices on the land, and mobilizing those actions in physical, social, economic, and political form is a process of community regeneration. This is not a quick fix approach by any means, but requires long-range vision, flexibility, persistence, immediate and long-term strategies, monitoring, and regular community health check-ups.

While an undertaking of this magnitude seems daunting, and undoubtedly requires courage and strong commitment, it may be worth considering the alternative scenario. For in the absence of a holistic community healing approach, it is unlikely that salmon will return to their home streams, that kids will be able to play in those streams, or that our home places will support inhabitation of our full physical, social, and spiritual selves. In the Mattole, Applegate, Willapa, and many other places, the absence of community-based restoration and involvement would have resulted in a very different physical and social landscape (Figure 10.18).

FIGURE 10.17 Reedroll furrows at riparian edge: new landscape aesthetics emerge out of the craft of restoration. *Source:* René Senos.

FIGURE 10.18 Communities organizing around the perspective of the salmon have catalyzed changes in their social and physical landscapes. *Source:* Mattole Salmon Group.

Participatory, bioregional restoration allows members to reconfigure their relationships with each other in the environment. As tenets of accountability and inclusiveness are instituted, watershed members construct appropriate economic, social, and political institutions that support the community's healing process. Through restoration, we make amends, learn anew about the place we live in, and develop a stronger bond with the land. In recrafting relations with our fellow species and our shared landscape, we rewrite our salmon story.

LITERATURE CITED

Aberly, D., ed. 1994. *Futures by Design: The Practice of Ecological Planning.* Gabriola Island, B.C.; Philadelphia, PA: New Society Publishers.

Aberly, D. 1993. *Boundaries of Home: Mapping for Local Empowerment.* Gabriola Island, BC; Philadelphia, PA: New Society Publishers.

Anderson, M. K. 2005. Tending the Wild: Native American Knowledge and Management of California's Natural Resources. Berkeley and Los Angeles, CA: University of California Press.

Andruss, V., C. Plant, J. Plant, and E. Wright. 1990. *Home! A Bioregional Reader.* Philadelphia, PA; Gabriola Island, BC: New Society Publishers.

Apostol, D., M. Sinclair, and B. Johnson. 1998. Design Your Own Watershed: Top-Down Meets Bottom Up in the Applegate Valley. *1998 ASLA Meeting Proceedings.* Portland, OR: American Society of Landscape Architects.

Baldwin, A. D., Jr., J. De Luce, and C. Pletsch, eds. 1994. *Beyond Preservation: Restoring and Inventing Landscapes.* Minneapolis, MN: University of Minnesota Press.

Barro, S. C., and A. Bright. Summer 1998. Public Views on Ecological Restoration: A Snapshot from the Chicago Area. *Restoration and Management Notes* 16(1): 59–65.

Barry, D. Winter 1998. Toward Reconciling the Cultures of Wilderness and Restoration. *Restoration and Management Notes* 16(2): 125–127.

Barsh, R. and K.C. Hansen. August 2002. A TEK-based Marine Habitat Restoration Program: The Samish Indian Tribe of Puget Sound from the Joint 2002 Conference of The Ecological Society of America and The Society of Ecological Restoration, Tucson, Arizona.

Bazerman, M. H., D. Messick, A. Tenbrunsel, and K. Wade-Benzoni, eds. 1997. *Environment, Ethics, and Behavior.* San Francisco: New Lexington Press.

Bernard, T. and J. Young. 1997. What We Have in Common is the Salmon: The Mattole Watershed, California, in *The Ecology of Hope: Communities Collaborate for Sustainability.* Gabriola Island, B.C.: New Society Publishers.

Bomar, C. R., P. Fitzgerald, and C. Geist. Summer 1999. Ritual in Restoration: A Model for Building Communities. *Ecological Restoration* 17(1&2): 62–74.

Brinkman, J. December 7, 1997. *Salmon Tops Environmental Worries.* The Oregonian. Portland, OR.

Bryson, R. (personal communication, 2006). Mattole Salmon Group.

Clinebell, H. 1996. *Ecotherapy: Healing Ourselves, Healing the Earth.* Minneapolis, MN: Augsburg Fortress.

Doppelt, B., M. Scurlock, C. Frissell, J. Karr. 1993. *Entering the Watershed: A New Approach to Save America's River Ecosystems.* The Pacific Rivers Council. Washington, D.C.: Island Press.

Ecotrust. April 25, 2006. http://www.ecotrust.org/. 721 NW Ninth Avenue, Portland, OR 97209.

Egan, D., and S. Glass. 1995. Watershed Volunteers: A New Approach to Dealing with Cross-Boundary Influences. In the 1995 Society of Ecological Restoration Conference Proceedings: Seattle, Washington.

Elway Poll. 1992. http://www.elwayresearch.com/ElwayPoll/. Elway Research.

Francis, M., L. Cashan, and L. Paxson. 1984. *Community Open Spaces.* Washington, D.C.: Island Press.

Fraser Basin Council. 2003. A Snapshop on Sustainability: State of the Fraser Basin Report <http://www.fraserbasin.bc.ca/>. Vancouver, BC.

Frissell, C. A. 1997. Ecological Principles. In *Watershed Restoration: Principles and Practices*, eds. Williams, Jack E., C. Wood, M. Dombeck. Bethesda, MD: American Fisheries Society.

Garibaldi, A. and N. Turner. 2004. Cultural Keystone Species: Implications for Ecological Conservation and Restoration. *Ecology and Society* 9(3): 1. [online] URL: http://www.ecologyandsociety.org/vol9/iss3/art1.

Gobster, P., and B. Hull, eds. 2000. *Restoring Nature: Perspectives from the Social Sciences and Humanities.* Washington D.C.: Island Press.

Goebel, J.M., C. Fox, and K. Wolniakowski. November 2003. "The Role of Grassroots Action for Biodiversity Conservation and the Transition to Sustainability: Practical Experiences from the Pacific Northwest United States." Case Study paper prepared for NATO Advanced Research Workshop, Krakow, Poland, November 2003.

Hall, M. 1997. Co-workers with Nature: The Deeper Roots of Restoration. *Restoration & Management Notes* 152: 173–178.

Harrington, S., ed. 1996. *Giving the Land a Voice: Mapping Our Home Places.* Salt Springs Island Community Services: Salt Spring Island, B.C.

Hartig, T., P. Bowler, and A. Wolf. 1994. Psychological Ecology: Restorative-environments Research Offers Important Conceptual Parallels to Ecological Restoration. *Restoration Management Notes* 122: 133–137.

Heckert, M. (personal interview). Willapa Alliance director. April 2000.

Hester, R. T. 1990. *Community Design Primer.* Mendocino, CA: Ridge Times Press.

Higgs, E. 2003. *Nature by Design: People, Natural Process and Ecological Restoration.* MIT Press, Cambridge, MA.

Hobbs, R. J., and D. A. Norton. June 1996. Towards a Conceptual Framework for Restoration Ecology. *Restoration Ecology* 42: 93–100.

Hollenbach, M. and J. Ory. 1999. *Protecting & Restoring Our Watersheds: A Tribal Approach to Salmon Recovery.* Handbook. Portland, OR: Columbia River Inter-Tribal Fish Commission.

Hough, M. 1995. *Cities and Natural Processes.* New York: Routledge.

House, F. 1999. *Totem Salmon: Life Lessons from Another Species.* Boston, MA: Beacon Press.

Hulse, D., J. Eilers, K. Freemark, D. White, C. Hummon. 2000. Planning Alternative Future Landscapes in Oregon: Evaluating Effects on Water Quality and Biodiversity. *Landscape Journal* 19(1&2).

Imhoff, D. 2003. *Farming with the Wild: Enhancing Biodiversity on Farms and Ranches.* Berkeley, CA: Sierra Club Books and University of California Press.

Jackson, L. L., N. Lopoukhine, and D. Hillyard. June 1995. Ecological Restoration: A Definition and Comments. *Restoration Ecology,* 72–75.

Jackson, W. 1994. *Becoming Native to This Place.* Lexington, KY: The University Press of Kentucky.

Johnson, B., and R. Campbell. 1999. Ecology and Participation in Landscape-based Planning Within the Pacific Northwest. *Policy Studies Journal* 27(3): 503–529.

Johnson, K. N. et al. 1999. *Bioregional Assessments: Science at the Crossroads of Management and Policy.* Washington, D.C.: Island Press.

Jones, S. Spring 1999. Participation and Community at the Landscape Scale: Current Issues and Future Possibilities. *Landscape Journal* 18(1): 502–529.

Jordan, W. R. III. 2003. *The Sunflower Forest: Ecological Restoration and the New Communion with Nature.* Berkeley and Los Angeles, CA: The University of California Press.

Jordan, W. R. III. 1995. Good Restoration. *Restoration & Management Notes* 13(1): 3–4.

Kaplan, R., and S. Kaplan. 1989. *The Experience of Nature: A Psychological Perspective.* New York: Cambridge University Press.

Karr, J. R and E. Chu. 1999. *Restoring Life in Running Waters: Better Biological Monitoring.* Washington, D.C.: Island Press.

Lake, F. K. 2003. *Biogeography of Salmon and Indigenous Peoples in the Pacific Northwest.* Publication pending. Portland, Oregon: Ecotrust.

Leopold, A. 1949. *The Sand County Almanac and Sketches Here and There.* Special commemorative edition, 1989. New York: Oxford University Press.

Lev, E. 1995. Youth Conservation Corps Carries Out Streambank Project. *Restoration & Management Notes* 13(1): 20–21.

Lewis, C. A. 1996. *Green Nature/Human Nature: The Meaning of Plants in Our Lives.* Urbana and Chicago: University of Illinois Press.

Lichatowich, J. 1999. *Salmon Without Rivers: A History of the Pacific Salmon Crisis.* Washington, D.C.: Island Press.

Lipschutz, R. D. 1999. Civil Society and Environmental Governance. In *Bioregionalism,* eds. McGinnis, Michael Vincent. London; New York: Routledge.

Lombard, J. 2003. The Politics of Salmon Recovery in Lake Washington, in Montgomery, Bolton, Booth and Wall, eds., *Restoration of Puget Sound Rivers.* Seattle & London: Center for Water and Watershed Studies in association with University of Washington Press.

Louv, R. 2005. *Last Child in the Woods: Saving Our Children From Nature-Deficit Disorder.* North Carolina: Algonquin Books of Chapel Hill.

Oregon State University Survey. Greater Protection Should Be Given to Fish and Wildlife Habitats on Federal Forest Lands. Oregon State University: Corvallis, OR, 1991.

McCann, J. M. Fall 1999. Before 1492: The Making of the Pre-Columbian Landscape. Part II: The Vegetation, and Implications for Restoration for 2000 and Beyond. *Ecological Restoration* 17(3): 108–119.

McGinnis, M. V. Winter 1999. Re-Wilding Imagination: Mimesis and Ecological Restoration. *Ecological Restoration* 17(4): 119–226.

McGinnis, M. V. 1999. Making the Watershed Connection. *Policy Studies Journal* 27(3): 497–501.

McGinnis, M. V., ed. 1999. *Bioregionalism.* London; New York: Routledge.

McGinnis, M., and J. T. Woolley. 1999. The Discourses of Restoration. *Restoration & Management Notes* 15:1: 74–77.

Marcus, C. C., and M. Barnes. 1999. *Healing Gardens: Therapeutic Benefits and Design Recommendations.* Hoboken, NJ: John Wiley and Sons.

Mattole Restoration Council. April 20, 2006. http://www.mattole.org/about_us/index.html.

Merchant, C. 1999. Fish First! The Changing Ethics of Ecosystem Management. In *Northwest Lands, Northwest Peoples: Readings in Environmental History*, eds. Gobel, Dale; and Hirt, Paul. Seattle, WA: University of Washington Press.

Metzner, R. 1999. *Green Psychology: Transforming Our Relationship to the Earth.* Vermont: Park Street Press.

Metzner, R. 1995. The Psychopathology of the Human-Nature Relationship. In *Ecopsychology: Restoring the Earth, Healing the Mind*, eds. Roszak, Theodore, Gomes, Mary E., and Kanner, Allen D. San Francisco, CA: Sierra Club Books.

Pister, E. P. 1997. Ethical Principles. In *Watershed Restoration: Principles and Practices.* Williams, J. E., C. Wood, M. Dombeck, eds. Bethesda, MD: American Fisheries Society, pp. 17–27.

Preister, K., and Kent, J.A. 1997. Social Ecology: A New Path to Watershed Restoration. In *Watershed Restoration: Principles and Practices*, eds. Williams, J. E., Wood, C. A., Dombeck, M. P. Bethesda, MD: American Fisheries Society, pp. 28–48.

Riley, A. L. 1998. *Restoring Streams in Cities: A Guide for Planners, Policy makers, and Citizens.* Washington D.C.: Island Press.

Roszak, T., M. Gomes, and A. Kanner, eds., 1995. *Ecopsychology: Restoring the Earth, Healing the Mind.* San Francisco, CA: Sierra Club Books.

Schroeder, H. W. 1998. Why People Volunteer. *Restoration & Management Notes* 16(1): 66–67.

Senos, R., F. Lake, N. Turner, and D. Martinez. June 2006. Traditional Ecological Knowledge and Practice in the Pacific Northwest. *In Restoring the Pacific Northwest: the Art and Science of Ecological Restoration in Cascadia*, Apostol, Dean and M. Sinclair, eds. Washington, D.C.: Island Press.

Sessions, G., ed. 1995. *Deep Ecology for the 21st Century.* Boston, MA: Shambala Publications.

Shepard, P. 1996. *The Others: How Animals Made Us Human.* Washington, D.C.: Island Press.

Sivakumarun, S. 1996. The Right to Play: Towards an Urban Environment with the Child in Mind. In *Public & Private Places*, eds. Nasar, Jack L. Nasar and Brown, Barbara. 27th Annual Conference Proceedings, The Environmental Design Research Association EDRA. Edmond, Oklahoma, pp. 130–135.

Society for Ecological Restoration International Science & Policy Working Group. October 2004. *The SER International Primer on Ecological Restoration.* www.ser.org and Tucson, Arizona: Society for Ecological Restoration International.

Spirn, A. W. 1984. *The Granite Garden: Urban Nature and Human Design.* New York: Basic Books.

Stevens, W. K. 1995. *Miracle Under the Oaks: The Revival of Nature in America*, New York: Pocket Books, Simon & Schuster.

Stuart, Hilary. 1977. *Indian Fishing: Early Methods on the Northwest Coast.* Seattle, WA: The University of Washington Press.

Sustainable Northwest. 1997. *Founders of a New Northwest: People Working Toward Solutions.* Portland, OR: Sustainable Northwest.

Tapsell, S. M. March 1997. Rivers and River Restoration: A Child's Eye View. *Landscape Research* 22(1): 45–66.

Throop, W., ed. 2000. *Environmental Restoration: Ethics, Theory, and Practice.* New York: Humanity Books.

Washington State University Survey. 1993. Greater Protection Should Be Given to Fish Such as Salmon on Rangelands. Pulliam, Washington: Washington State University.

The Wetlands Conservancy. 1997. *Tools, Trees, and Transformation: A Collection of Restoration Stories from Schools and Community Groups in and around Portland*. The Wetlands Conservancy: Tualatin, OR.

Williams, J. E., C. A. Wood, and M. P. Dombeck, eds. 1997. *Watershed Restoration: Principles and Practices*. Bethesda, MD: American Fisheries Society.

Zuckerman, S. 1997. Thinking Like a Watershed: Mattole River of California. In *Watershed Restoration: Principles and Practices*, eds. Williams, J. E., C. A. Wood, and M. P. Dombeck. Bethesda, MD: American Fisheries Society, pp. 216–234.

11 Renovation of Byzantine Qanats in Syria as a Water Source for Contemporary Settlements[1]

Joshka Wessels and Robert Hoogeveen

CONTENTS

ABSTRACT

In the summer of 2000, a small group of Syrian villagers living near the borders of the steppe southeast of Aleppo renovated and cleaned their own water supply with international help. The source of their water is an ancient Byzantine underground water supply system called the *qanat.*

---

[1] This chapter is a revised and updated version of a paper that has been published in Zafar Adeel (ed.), "Sustainable Management of Marginal Drylands, Application of Indigenous Knowledge for Coastal Drylands," proceedings of a joint UNU-UNESCO-ICARDA international workshop, Alexandria, Egypt, 21–25 September 2002, ISBN 92-808-8011-X.

The cleaning was a pilot project in an effort to contribute to the preservation of qanats in Syria. This chapter describes the qanat research undertaken by an international research team based at the International Centre for Agricultural Research in Dry Areas (ICARDA) during 1999 and 2001. Within the framework of a United Nations University project on traditional water management, some of the most important qanat sites were revisited in August 2002. Qanat renovation activities took place in the year 2000 until 2004. It was found that some qanats had considerably decreased in flow during the last decades and were at the verge of extinction. Ironically, the qanats have been flowing for 1500 years and over the last 15 years they are drying up. What is the benefit of this sustainable water supply system in this time of ecological farming, increasing environmental awareness and within a changing social and economic environment? Are informal, national, and international institutions able to maintain a traditional common water source like a qanat when groundwater abstraction through pumping gives much more short-term benefit for a larger population?

MODERN CHALLENGES FOR TRADITIONAL SYSTEMS

The research project described in this chapter looks at the use and values of a potentially sustainable qanat system in a changing modern environment. Lightfoot (1996) defines qanats as "a form of subterranean aqueduct—or subsurface canal—engineered to collect groundwater and direct it through a gently sloping underground conduit to surface canals which provide water to agricultural fields." In Syria, many ancient qanat irrigation systems have been abandoned due to falling water tables as a result of the increased use of modern electric and diesel-pumped wells. Lightfoot (1996) stated that "New and often rapacious water technologies have all but replaced traditional irrigation systems in the Middle East, aggravating an impending water crisis and further complicating regional water compacts [...] traditional, low-impact irrigation technologies can no longer support the region's rapidly burgeoning numbers of peoples."

In modern times, qanats are not able to provide enough water for large-scale agriculture and lose their importance. Traditionally, qanats should be cleaned on a regular basis to prevent silting, collapsing, and disfunctioning. This maintenance helps in keeping the qanat flowing even in dry seasons. But as soon as qanats are giving less water, young people lose interest and start looking for revenues in off-farm work. The urban environment is financially much more attractive than traditional qanat farming. This group of youngsters literally abandons qanats. With the abandonment of qanats, the indigenous knowledge and community cooperation critical for qanat upkeep also disappears, and more qanats collapse or dry up. A vicious circle is complete.

As a result a valuable cultural heritage is vanishing. Not only are qanats relics of a prosperous past, but also sustainable and environmentally friendly systems of extracting groundwater today. In Qarah, Syria, we have seen that combining ancient qanats and modern drip irrigation systems for fruit trees might prolong the life of some qanats and encourage younger generations to commit to their upkeep. Another option to think of is to encourage eco-tourism based around qanats to provide alternative income for the farmers.

QANATS IN SYRIA[2]

In 2001, our team explored qanat sites in Syria guided by a map published by Dale Lightfoot from Oklahoma State University in 1996. We documented geographical, socio-economic, and hydrological characteristics and interviewed local experts and officials from various institutions. We found a total

---

[2] Based on the extensive field collection of digital video footage, Joshka Wessels produced several films. The UNU film called *Qanats in Syria* covers the survey work undertaken during 2001 and 2001. *Little Waterfall* is an anthropological film about the pilot qanat renovation and the adapted television documentary *Tunnel Vision* was produced for the TVE Earth Report Series on BBC World. More information at www.sapiensproductions.com.

of 42 qanat sites containing 91 qanats, of which 30 were still in active use. Others were dry or only drizzling and almost abandoned. We tried to cover most of Syria; however, it is likely that Syria used to have many more qanats in the past. Today, they are difficult to locate, and many are beyond repair.

In Syria, running qanats are concentrated around Damascus, Homs, and in the steppe areas. The qanats used to provide the main water supply for drinking and agriculture. It is difficult to determine the age of qanats because of the small amount of artifacts that are found inside the tunnels. However, we can say through circumstantial evidence that Syrian qanats were already in use during the Roman period. The digging technique and type of the qanats varies considerably throughout the country.

The water of Syrian qanats has been used mainly for irrigation ever since the date they were dug. The division of the water is based on a local system of rights and regulations. The groups of users for each qanat we found is relatively steady, and each user household has an irrigation share measured in time, the so-called *dor* (turn). Irrigation shares can be traded among the users and are usually attached to land.

CASE STUDY IN NORTHERN SYRIA

As we have seen in countries like Oman, renovation of neglected qanats is viable. Successful renovation of qanats in Syria is technically possible but thorough social and hydrological assessment is required in advance of renovation. A pilot renovation was done in 2000 in a village east of Aleppo, and our team initiated a qanat cleaning based on the priorities and traditional knowledge of the community. The qanat was dated to the Byzantine period as a result of finding an oil lamp in a tunnel. It is the only source of water in the village. In collaboration with the museum of Aleppo, the scientists started up the cleaning of their own qanat.

METHODOLOGY AND APPROACH

The research and development methodology of the case study is based on one of the action models described by Chambers (1985): "Action anthropology begins with the premise that the anthropologist should operate within the framework of goals and activities initiated by groups seeking to direct the course of their development. The action anthropologist may use his or her technical skills to help a group clarify its goals, but generally avoids the temptation to direct the project." Action research is a subset of applied research. In this case the action is the actual cleaning and renovation of the qanat system in Shallalah Saghirah.

The project followed an integrated holistic approach led by the priorities and needs of the community. The anthropological action research was supported by other disciplines such as hydrogeology, archaeology, biology, agronomy, and soil science. An interdisciplinary team of scientists of both social and biophysical disciplines thus collected data on various topics. In general, the data collection can be divided into a social focus and a technical focus.

Initial contacts with the community were established in the second part of 1998, but the actual project started in October 1999. A good rapport developed with the local community. Overnight stays during the fieldwork enhanced and strengthened the relationship and mutual trust between researchers and respondents. Hydrological measurements were taken regularly, the social organization, history of the village, and water rights system in use investigated; and a genealogy of the households in the village finalized. Key informants, both male and female, have been interviewed on their sources of income.

AREA DESCRIPTION

The village of Shallalah Saghirah is located southeast of the city of Aleppo in the western part of the Khanasser Valley bordering the eastern slopes of the Jebl al Hoss. The Khanasser Valley is located between the 200-mm and the 250-mm rainfall *isohyets*, whereas Jebl al Hoss is located

FIGURE 11.1 Overview of Shallalah Saghirah, Syria.

between the 250-mm and 350-mm rainfall isohyets. The 200-mm isohyet demarcates the cultivated zone to the west and north and the steppe areas to the east. Shallalah Saghirah is a typical village because it finds itself in time and space in a transitional zone (Figure 11.1); spatially, because it is located between two different rainfall zones at the border of the steppe area, and in time because, as Lewis (1987) describes, this area has known rapid environmental, cultural, and economic changes from the 20th century to the present. At the time of this study, the village did not have electricity except from private generators and was not heavily influenced by modern developments elsewhere in the world. However, television has made its entrance, and the younger generation has been moving around and outside of Syria for off-farm migration work.

From 1998 until 1999 a groundwater and well survey was undertaken by Hoogeveen and Zöbisch (1999) in the Khanasser Valley to investigate the groundwater system and its use by farmers. They mention that part of the water pumped from the aquifer in the center of the valley is replaced by salt water from the isohyets in the North. Therefore, water tables in the valley are not falling as much as in those areas with comparable pumping activities. The limestone layer from which the qanat in Shallalah Saghirah derives its water is not very productive due to its low permeability and porosity. It is believed that the nearby pumping activities had little influence on the discharge of the qanat.

This ancient qanat system gave less and less water every year, according to the local inhabitants. They mentioned that many shafts of the qanat system were filled with debris collected over the years and that children have thrown stones in the shafts. Regarding the physical environment of the qanat, cleaning and renovation of the system could be beneficial for both the people and the environment. Elderly inhabitants and some of their sons expressed willingness for cleaning and renovation, but they did not have the financial means to do it. There was also a certain reluctant attitude towards cleaning the system by some of the local inhabitants. Birks (1984) mentions that changing socio-economic circumstances may be the main reason for this kind of reluctant attitude. Through applied anthropological research and community development this project aimed to overcome the various obstacles that prevented the sustainable use of an ancient qanat system.

HISTORY OF THE QANAT OF SHALLALAH SAGHIRAH

The water in the village is supplied by a 1500-year-old qanat. Figure 11.2 shows a cross-section of the qanat tunnel. Although it is the only working system in the vicinity, the qanat in Shallalah Saghirah is not isolated. An important site near Shallalah Saghirah is the old Byzantine site of

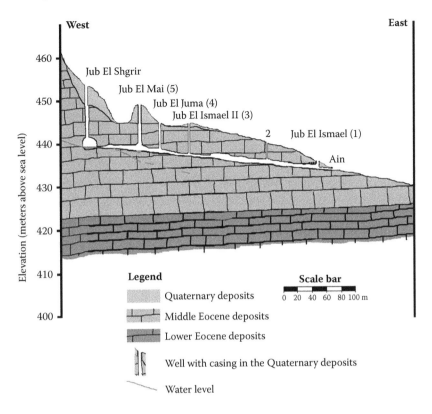

FIGURE 11.2 Cross-section of the qanat of Shallalah Saghirah.

Ras'm al Nafl. This village is strategically located northeast of Shallah Saghirah along the road between Sfeereh and Khanasser. The origin of Ras'm al Nafl is late Roman/early Byzantine era. The main evidence of Byzantine settlement is the concentration of church foundations west of the modern village towards the Jebl al Hoss. In their aerial research on Roman Syria, Mouterde and Poidebard (1945) observed the remains of several churches and an underground qanat that cut into the hills of Jebl al Hoss, at a site called Mu'allaq. He speculated that the sites were part of a Roman military range that later became Byzantine settlements. We have observed the qanat and, according to the nearby farmer, it has been dried up since the 1930s.

Looking at its geographical location, the village of Ras'm al Nafl is located at a strategic site on the fringe of a major salt lake. During the early Byzantine Empire it formed part of a second line of defense against nomadic raids and invasive troops trying to push back the borders of the Byzantine Empire. The first line of defense was situated more towards the Euphrates in the East (Haasse, 1983). In the fifth century A.D. Syria formed part of the Byzantine Empire after the dissolution of the Western Roman Empire. Looking at other Byzantine settlements in Khanasser Valley and Jebl As-Shbayth, one can assume that the outposts were positioned in two rows running parallel to each other from North to South: a line of outposts on the eastern side of Jebl As-Shbayth, with the plateau as buffer zone and a line on the eastern side of Jebl Al-Hoss. In between the rows of settlements was the Khanasser Valley, cut off in the south by the town of Khanasser (then called Anasartha) and few military outposts like El Hammam. This buffer zone of defense was suitable for agriculture and hence could provide supplies to the border troops. It protected the villages north and west of Jebl Al-Hoss and, most important, prevented invaders from pushing through to the city of Beroea (Aleppo) and the northern Byzantine cities like Antioch, the second most important city of the empire, where the main monasteries and philosophical schools of the eastern Byzantine Empire were located.

After Emperor Constantine had made Christianity the state religion, the Byzantine Empire was constantly threatened by attacks from outsiders. The defenses were necessary in the late Roman–Early Byzantine sixth century because the Sassanians in the East were a constant threat to the state. In the late sixth and early seventh century, two great powers, the Byzantine and the Sassanians, ruled the Near East (Bloom & Blair, 2000). The Byzantines had inherited the East Roman Empire that stretched from Byzantium or Constantinople (present day Istanbul) to south of Syria. Emperor Justinian (r. 527–565) brought relative peace in the empire. Justinian and his general Belisarius reconquered many of the further reaches of the empire. To celebrate his victories, Justinian spent a considerable amount of time building churches, monasteries, and schools. In 550 A.D. he rebuilt the famous Hagia Sofia in Constantinople.

Yet despite his successes, Justinian still faced a constant threat from the Sassanians in the East. These Persian-speaking Zoroastrians ruled a territory roughly covering modern day Iraq and Iran (Bloom & Blair, 2000). For centuries the Sassanians and Byzantines waged wars against each other, and the border between the Persian and Byzantine Empire was shifted periodically.[3] Procopius, historian and chronicler of Justinian and his general Belisarius, mentioned in his history of the wars that in April 531 A.D. the Byzantine army encamped in the city of Chalcis (modern Kinnesrin). The enemy troops were in a place called Gabboulon (near the Jabul Salt Lake, north of Khanasser valley) 110 stades[4] (approximately 20.2 km) away from Chalcis (Procupius, translated by Dewing, 1914). Once they learned of the proximity of the Byzantine army, the Persian army retreated to the city of Callinicus, on the bank of the Euphrates. The Byzantine army followed suit. A long battle followed that finally ended in the retreat of the Persians. The Syrian province had now been freed from the invaders. It gave way to a period of prosperity. Justinian made use of this period by rebuilding and restoring fortresses and towns like Antioch, Chalcis, and Sergiopolis (modern Rusafa). The emperor bestowed the same careful attention on all the towns and forts on the farthest borders of Euphratesia, namely Barbalissus,[5] Neocaesaria, and Gaboulon[6] (Procopius translated by Dewing, 1914). All these locations lie in the area around Shallalah Saghirah and Khanasser valley. It is most likely that the many churches and qanats in Khanasser valley were build during Justinian's reign.

In order to assure the safety of the Syrian province, military zones such as the ones in Khanasser valley were thus needed and very important. Obviously, the border outposts needed a water supply, hence the presence of qanats. During this period Justinian built many water supply works like dams, sluices, and aquaducts; although qanats are not literally mentioned, Procopius describes in his book *Buildings* how Justinian rebuild the water supply of the city of Cyrus (modern Chorres):

> The interior of this city had been destitute of water from ancient times; outside of it there had been a certain extraordinary spring that provided a great abundance of water fit for drinking … so he [Justinian] dug a channel outside the city all the way to the spring, not allowing it to be seen, but concealing it as carefully as possible, and thus he provided the inhabitants with a supply of water without toil or risk (Procopius translated by Dewing, 1914).

The likelihood of the origin of most qanats in Khanasser valley is therefore Byzantine, more specifically, Justinian. The fact that the present local population calls them "qanat *Romani*" does not necessarily point to a Roman origin; it instead refers to any ancient origin. In fact, the Byzantines, although they spoke and wrote Greek, considered themselves as the only true Romans, and Arab

---

[3] In 540 A.D. the Sassanians sacked Antioch in northern Syria and were controlling the region. The continuous fighting between Sassanians and Byzantines continued over almost a century until the mutually ignored threat from the South determined the future of the Syrian province; in 633 A.D. the Muslim Arabs conquered Syria and ultimately brought it to the centre of the Islamic empire when the capital was moved from Medina to Damascus by the Ummayads in 661 A.D.

[4] One stade is approximately 184.4 meters.

[5] Near modern Meskaneh (north east of Khanasser Valley).

[6] Modern Jaboul.

authors later acknowledged the Byzantines' claim by referring to them as *"Rum,"* meaning Romans in Arabic and calling the Europeans, including the pope of Rome, "the Franks" (Bloom & Blair, 2000). Although Safadi (1990) says that the appellation for *qanats* in official Syrian documents is *"foggara,"* we prefer to use a short form of the colloquial term "qanat *Romani.*" It is very well possible that the term "qanat *Romani*" in fact refers to "*Byzantine* qanats."

Further evidence of the likelihood of Byzantine origin of qanats in Khanasser valley has been found in Shallalah Saghirah. Underneath the *kubbeh* houses basalt remnants of foundations, most probably from Byzantine buildings, can be found. One of the main rooms contains a basalt doorpost. Also, in one of the tunnels of the qanat of Shallalah Saghirah, we found a Byzantine oil lamp during the renovation in the summer of 2000 (Figure 11.3). Ancient diggers or maintenance workers probably used the oil lamp. By finding this lamp, we can approximately date the age of the qanat. The precise dating of qanats is virtually impossible, unless their construction was accompanied by documentation or, occasionally, by inscriptions (Lightfoot, 1996). With the found artifact we have some idea about the age. We also found two Byzantine crosses carved in the sides of the walls; probably a digger died at these spots or water was found there. The lamp has been officially dated in the first half of the sixth century by archeologists. This period coincided with the reign of the Emperor Justinianus (r. 527–565).

We think the qanat of Shallalah Saghirah was a drinking water supply system for nearby military outposts guarding the plain between Jebl al-Hoss and Jebl As-Shbayth. We found remnants of buildings and pottery fragments that show the presence of either a small farm or permanent outpost. The presence of the churches in Khanasser valley indicates that the zone was relatively peaceful, more a place to rest and be fed for the soldiers who need to relax after the frontier battle duties.

Remnants of terraces or water harvesting systems can be observed on the hill slopes surrounding the *wadi* where the qanat draws its water. This suggests that relatively large-scale farming was practiced to feed the soldiers on the frontiers. Haasse (1983) suspects the terraces were used for vineyards due to the location towards the direction of the sun but the shape of the terraces do not suggest a particular crop. Some terraces run along vertical lines from top to bottom, and the width of the walls do not suggest they have been used for specifically grapes. Thorough archeological research is needed for clearer suggestions as to the agricultural use of these remnant terraces. During their transects in 1938, not far from Shallalah Saghirah, at the valley of Ruwayhib (*Wadi Boutma*), Mouterde and Poidebard (1945) found basalt remnants of a church foundation. One of the lintels bore the inscription dedicated to "the glorious Mother of God, the Virgin Mary" by "the famous leaseholder of the saltworks, Theodule" and was dated 553 A.D.

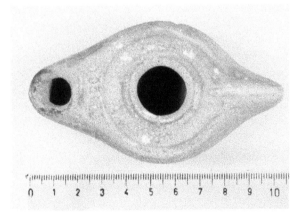

FIGURE 11.3 Byzantine oil lamp found during the renovation of the qanat of Shallalah Saghirah.

It assumes that during that time, the nearby Jabul Salt Lake was actively exploited and well protected as a commodity.

During Justinian's reign, it is assumed that the qanat of Shallalah Saghirah was regularly cleaned and maintained by the Byzantine state institutions. The construction of qanats in Byzantium was a state affair as Tate explains: *"La construction d'ouvrages d'irrigation, le creusement ou l'entretien des réseaux de* qanats *aux confines de la steppe, enfin, ont exigé des resources dépassant de loin celles des villageois: d'une manière ou d'une autre l'intervention de l'État a été nécessaire"* (Tate, 1992).

All but one of the qanats present in Khanasser valley have dried up or been neglected. The most well-known qanat is the system of the town of Anasartha (modern-day Khanasser). This qanat was estimated to be 12 km long. In 1959, Hamidé (1959) observed and described the qanat of Khanasser. At that time it irrigated 15 ha of gardens and had a discharge of 8 L/s. The qanat of Khanasser stopped flowing in 1975 after the introduction of pump wells in the area west of the Khanasser Valley. In the modern town of Khanasser the Byzantine remains are visible on the street, in newly built walls and inside residential houses. The amount of artifacts, foundations of houses, and churches indicate a flourishing Byzantine settlement that needed a substantial water supply.

Other antique qanats and water harvesting dams in the Khanasser valley can be found on the southern fringes of Jebl As-Shbayth, the settlement of Kleya and various other places. In the area around al-Hammam south of Jebl As-Shbayth, we found a line of airshafts with original casings. The water in the qanat was standing still but according to the local population, the qanat has flowed, and in times of heavy rain, the reservoir fills up with harvested rains. There has not been a systematic archeological survey or excavation of qanats in Khanasser valley. This study has not further considered dried up qanats. The likelihood of the presence of more underground water tunnels hidden underneath unexcavated Byzantine settlements is high.

HYDROGEOLOGY OF THE QANAT IN SHALLALAH SAGHIRAH

The qanat here has been dug in the limestone rock that is dated Middle Eocene. The limestone consists of chalk-like clayey limestone and marl. Flint has been observed in the limestone layers exposed to the surface. According to the inhabitants of Shallalah Saghirah, the water in the mother well is tapped from a layer where flint is present. After production and transport through the qanat, the water is directed through a small open canal (*saqieh*) running through the village (Figure 11.4). Figure 11.5 is an aerial photo taken with a kite and shows the open canal through the village. The water is collected in a reservoir (*birkeh*) at the end of the open canal.

This birkeh was built in the 1950s to collect the water used for irrigation of a community garden. The birkeh can be opened and closed for irrigation from an outlet closed off with stones and cloth. Several discharge measurements of the water entering the birkeh were taken in the winter of 1999–2000. The maximum observed discharge was 1.1 L/s. This was measured when the villagers did not use the water from the saqieh, and the full discharge of the water production section entered the birkeh.

In contrast, measurements that were taken when water was drawn from the *saqieh* gave an average discharge of 0.35 L/s. The water extracted by the qanat system in Shallalah Saghirah has been tested and proved to be of good chemical quality (Table 11.1). We have recorded chemical analyses of a sample taken on June 11, 1999. Values are compared with the standards for drinking water given by the World Health Organization. Excrement from bats present in the qanat cause biological contamination of the water that is harmful to human health. Therefore, regular biological testing of the water has been listed as a priority.

FIGURE 11.4 Open channel (saqieh) running through Shallalah Saghirah.

FIGURE 11.5 Aerial photo of center of Shallalah Saghirah.

TABLE 11.1

**Composition of the Water of the Qanat Compared
to the Drinking Water Standards of the WHO**

Parameters	Observed Values 1998	Maximum Values for Drinking Water
EC (at 25°C) (dS/m)	850	
PH	8.2	
Na^+ (mg/L)	91.9	175
K^+ (mg/L)	1.95	
Mg^{++} (mg/l)	27.34	50
Ca^{++} (mg/L)	58.12	
Cl^- (mg/L)	107.77	300
HCO_3^- (mg/L)	170.86	
SO_4^- (mg/L)	114.31	250
NO_3^- (mg/L)	26.04	50
IB_err	3.3	
Water type	Calcium chloride	

SOCIAL HISTORY, MOBILITY, AND INCOME
SOURCES OF SHALLALAH SAGHIRAH

The inhabitants of the village of Shallalah Saghirah are descendants of one ancestor who originated from the clan of Al-Hariri on the Hawran Plain in the south of Syria. Batatu mentioned that the clan was dominant in 18 villages on the Hawran Plain, its main seats being at Da'il and Shaykh Miskîn. The first inhabitant of Shallalah Saghirah was one of the two sons of who decided to migrate from the Hawran to the Khanasser Valley during the end of the 19th century. The Ottoman Sultan Abdul Hamid, who ruled from 1876–1909, owned estates northwest of Khanasser Valley. The first inhabitant worked on these estates to prevent his sons from being sent to the Ottoman army. Lewis describes the area of Khanasser Valley during that time as a frontier area with nomadic Bedouin tribes in the east and Ottoman landowners in the west.

Two years after he had bought the land, the first inhabitant started to clean the motherwell of the qanat (*ras el nebe'*). His five sons helped him with this. After the cleaning, the water returned. Hearing of this discovery, the former landowner wanted to have his sold share back. The first inhabitant refused and went to the powerful Bedouin *shaykh* Mujhim Ibn Muheid. The shaykh offered his protection, and from this day on, the protection and settlement of Shallalah Saghirah was established. The first inhabitant and his five sons lived prosperously on the benefits they gained from the water of the qanat.

After the land reform initiated by the Syrian government in 1958, the land of powerful landowners was divided among individual families and the property of Shallalah Saghirah became property of the family. However, the inheritance rules from before 1958 are virtually still in use among the villagers with regard to land ownership. The group of inhabitants in Shallalah Saghirah, is thus what is called a patronymic group, a group identified by a common surname (Mundy, 1995). The descendants in Shallalah Saghirah are divided into five groups. In Arabic, they are referred to as *bayt* (pl. *buyut*) or *hosh* (pl. *howaash*). We will refer to them as bayt. This is a unilineal descent lineage group in which the membership rests on patrilineal descent from one of the brothers of the first generation.

Until 1965, all descendants were resident in the village. The strain on land and water resources caused some households to leave the village in 1970; the group became too large for the land.

Economic pull drove some of the families to urban areas like Aleppo and Raqqa. In that period there was a considerable rural exodus happening In Syria. Data from 1970 on total migration flows in Syria show that 66% was rural to urban (Ashram, 1990). Tully reports that on very small farms, household members migrate so that subsistence farming is possible for those remaining behind (Tully, 1990). The migration of some in Shallalah Saghirah did not result in the self-subsistence of those who remained. The village became a residence from which households developed different types of income activities to survive, whether on-farm or off-farm.

There always has been a relatively high level of mobility of people in rural areas of Syria. Kin relations are a very important reason for travel. Of course, the villagers of Shallalah Saghirah have relatives in the Hawran plain from where they originally migrated, and they have regular contact with each other. Traditionally, the inhabitants of Shallalah Saghirah are used to traveling seasonally with their sheep to northern areas of Aleppo Province in late spring and summer to let them graze on areas with higher rainfall.

Birks & Sinclair (1980) mentioned that since the 1970s international migration in the Arab region altered the social organization of many villages. With respect to social mobility and relationships, the people of Shallalag Saghirah have connections with the cities of Aleppo and Raqqa, the town of Azzaz (60 km north of Aleppo), Sfeereh, Rasm El Nafl, Fijdan, and other villages in Khanasser Valley. Regarding international labor migration, the village has connections in Lebanon and Saudi Arabia. Jordan used to be a target for labor for sheep shaving but this work has shifted to Saudi Arabia. Lebanon is a preferred destination, especially with the younger generation, who find construction work in Beirut.

Marriage arrangements are changing in the village. Young men migrate to be able to earn their bride wealth to be paid to marry their parallel cousin or, recently, perhaps a nonrelated bride. Stevenson (1998) finds the same in Yemen: "Patterns of marriage arrangement are an indicator of changing father–son relations. Remittances so inflated the cost of marrying that most fathers were unable to finance their son's marriages. Young men migrated to be able to marry and provide most of their own marriage costs."

Until 1977, the main income source was agriculture. Sheep, rain-fed barley, irrigated fruit trees, and vegetables in the garden of the qanat (*bustan*) provided enough food and income for the people. After the evacuation in 1977, the sources of income changed radically. Along with the two-year evacuation, in the mid-1970s modernization and rural–urban migration patterns have altered the socio-economic landscape considerably as described by Stevenson. Selling sheep in the market is now usually practiced by the older generation. They own most of the sheep and have a long-term relationship with their seasonal contractors in the northern parts of Syria. The amount of income depends on the rainfall during the year. One of our key informants told us that in a good year he would receive an annual total of 170.000 Syrian *lira* (approximately $3,700) for selling his sheep. He estimated that in a dry year, he might not receive more than half of this amount. Table 11.2 shows the main categories of income and the respective locations where this work is found.

TABLE 11.2

Main Income Categories in Shallalah Saghirah

Category	Daily Income/Person (S.L.)	Location	Seasonal/Daily
Selling sheep on the market	400	Syria	S
Shaving sheep	750–1500	Saudi Arabia/Syria	S
Construction work	500–1000	Lebanon/Syria	S
Government	250–500	Syria	D

WATER USE AND RIGHTS IN SHALLALAH SAGHIRAH

The water taken straight from the tunnel outlet is, according to customary law, free to drink for everyone (Figure 11.6). The villagers use qanat water to irrigate a community garden (*bustan*) to grow food crops such as onions, cucumbers, tomatoes, and other vegetables for the additional nutrition of households (Figure 11.7). The garden also contains fruit trees such as mulberry, figs, and pomegranates. In addition to that, they grow irrigated barley to provide feed for the sheep. In the western part of the bustan can be found the orchards and in the eastern part, the arable land. Besides the irrigation of the bustan, elderly people in the village make use of the qanat water by irrigating small-scale private plots for growing vegetables and herbs.

The division of landownership of the bustan determines the rights for irrigation times. The five sons of the first inhabitant divided the bustan into five equal parts. They decided that each of them had the right to irrigate his land every 5 days. This order has not changed since then. The descendants of each of the five sons divided the land in mutual agreement according to inheritance laws. The ones who emigrated lost their rights to irrigation water. They can claim it back whenever they return, but only if they did not sell their land. Presently, the descendants who hold the right to irrigate and are resident in the village are seven elders from three lineages. These seven are called the *haquun* ("the holders of the right").

FIGURE 11.6 Women chatting at the outlet of the qanat.

FIGURE 11.7 Irrigating the bustan of Shallalah Saghirah.

Patrilinear descendants of the other lineages have either sold or rented their land to their cousins. Therefore, the descendants of one particular bayt hold the right to use the water from the birkeh 3 out of 5 days.

METHOD OF INTERVENTION

Together with the local village elders and their sons, the priorities of the community with regard to the use, repair, and maintenance of the qanat were discussed and determined. During the focus meetings, participatory tools like community maps were used to facilitate the discussion between qanats users. From these focus group meetings, and based on the local technical knowledge, a plan for the cleaning and renovation was developed and generally agreed upon. This cleaning and renovation took place in the summer of 2000 with financial support of the Dutch and German embassies in Damascus.

CONSIDERING CONSTRAINTS AND RELUCTANT ATTITUDES TOWARDS QANAT CLEANING IN FORMING AN AGREEMENT

Several focus group meetings have been held with the haquun. In the first group meeting, a community map was drawn of the construction of the qanat (Figure 11.8). This map was used in other group meetings. In the beginning it was impossible to get all seven haquun together due

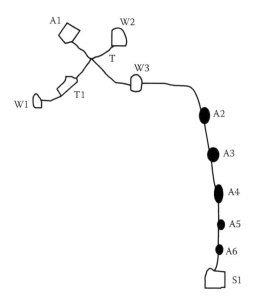

FIGURE 11.8 Community drawn map of the qanat of Shallalah Saghirah.

to an internal dispute between family members of different *biout*. Some family members have made attempts in the past to ease the tension and mediate between the different lineages through so-called "wedding alliances." In this case, a son of a particular bayt decided to marry a wife from "the other side" as a way of alliance. However, from time to time tensions arise and are expressed in little disputes.

With respect to the qanat, the use of rubber pipes for irrigating personal plots outside the rotation system was the subject of such a dispute. The villagers used rubber pipes as siphons to draw water from the *saqieh* for domestic water use. This was allowed throughout the daytime. However, when someone used this pipe for irrigation, the rotation system became the rule again. Villagers accused each other of the use of these pipes for irrigation without following the rotation system.

The village did not have a chief (*mukhtar*) and disputes were not solved immediately by the family themselves. Therefore, weak leadership formed a constraint for the regular maintenance of the qanat. Mainly, it was the haquun who were involved in these disputes, accusing each other, whereas the younger men were much more occupied with other things like migrant work to earn money for their weddings. Birks mentions that the pull of young men towards migrant work is another constraint and reason for a reluctant attitude towards qanat work. Despite the latent constraints and reluctant attitudes, a general willingness was felt for cleaning the qanat as was expressed in the group discussions. Also, some men of the younger generation became more and more interested through the discussions and fieldwork conducted. But after some time, it was realized that without getting the haquun together, the cleaning would not take place at all.

However, the presence of the scientists and the many group meetings had apparently stimulated the *haquun* to settle their differences during the feast (*'aid al fitr*) after the Islamic fasting, Ramadan, in January 2000. Therefore, another focus group meeting was planned, this time with all haquun present (Figure 11.9). It was felt there was a need to create an informal organization, and this was suggested to the haquun. They supported this idea and in the focus group meeting, the haquun made an informal written agreement among themselves to regulate the maintenance and renovation work of the qanat. They agreed upon regulations for the use of rubber pipes to extract water, and made a list of all the workers that would be available for the cleaning work, and at what times throughout the year. This last point is important because of the seasonal migrant work that many young men are doing. Also, it was decided that the haquun would be forming a committee that represents the village. With this agreement and a technical work plan/budget, the committee and the researchers would initiate a search for funds necessary for cleaning and renovation.

FIGURE 11.9 Group meeting among haquun members discussing the work plan.

PRIORITY ACTIVITIES

Before the informal agreement, several focus group meetings were held with the haquun who had a good relationship with each other. Because the ongoing dispute between them had little to do with the qanat itself, the haquun gradually came to an agreement on the technical work plan. First of all, the haquun decided that the villagers themselves should do the cleaning work. Also Birks (1984) mentioned that local communities should carry out repair and improvement instead of hiring outside labor.

The haquun put together priority activities for the renovation work using the indigenous terms to refer to the important parts of the qanat. Because the water production section of the qanat is of direct benefit to the haquun, they decided that this should be their first priority for cleaning. They stated correctly that if this dries up, the village will have to evacuate again. The priorities based on the different sections of the qanat system are summarized below according to activity:

1. It was suggested to start at air shaft A1 called the *sunduq* ("the box"), which is closed by debris and boulders from above. This airshaft connects the western well (W1), called the *jub el saghir* ("little well"), with the main tunnel. This airshaft provides oxygen for workers down in the qanat tunnel. Once this airshaft was cleaned from above, it became possible to observe the damage on the jub el saghir.
2. According to the haquun, water well W1 is filled with debris from above and the basalt walls are collapsed at certain places. After cleaning airshaft A1, this well had to be cleaned from above and below, and a wall constructed to enforce the well and prevent future collapsing.
3. Tunnel T1, which leads towards W1, is intersected by a low roofed reservoir of 3 m × 3 m. This reservoir, called *el ghurfah* ("the room"), was reported to be filled with both water and debris, needing major cleaning.
4. An unsuccessful attempt was made in the past to drill an airshaft in the motherwell (W2), called *ras el nebe'* ("head of the spring"). It was suggested to locate W2 from above and drill a shaft to make it more accessible for the workers. The well needed major cleaning. After that, the tunnel towards the motherwell (T) and two shafts (A3, A4) required cleaning.
5. Airshafts A2, A5, and A6 needed some cleaning but their construction was completely intact. Also, the first water production well (W3) needed cleaning of the walls.
6. S1 is the source of the qanat, where the water reaches the surface (*el a'yn*). This source needed extension of the walls if more water was to be collected. Also, the canal (*saqieh*) running from the source to the collection reservoir (*birkeh*) needed reconstruction and the reservoir had to be cleaned of debris.

The technical work plan developed by the haquun included the priority activities and the estimated number of working days for each activity.

CLEANING WORK

After developing a research proposal based on the outcome of the group meetings, local funds were granted by the Dutch and German embassies in Damascus, and cleaning work started on June 17, 2000 (Figure 11.10). The village committee compiled a group of workers, and they chose a supervisor from the village community itself. The community work plan was followed, and the supervisor made a weekly work program with names of the workers. The whole cleaning activity was officially regarded as an archaeological excavation as it concerned a Byzantine site. Therefore, on a daily basis a representative of the Aleppo Museum, who was very instrumental in keeping the work spirit high, attended the worksite. In case of difficulties between workers, he would always mediate.

The first step, cleaning the *sunduq*, went quite smoothly; the team spirit was high and work progressed well. After 6 weeks however, the initial enthusiasm was lessening, and following some trouble between the villagers' elected supervisor and some group members, the elected group supervisor thought that it was best if the government representative of the museum was put in charge of the workers' program. This was done, and everything was back on track again.

FIGURE 11.10 Renovating an airshaft of the qanat in Shallalah Saghirah.

The team developed more and more, and a certain community strengthening was noticeable—until some 4 weeks before the proposed final date. At that time, a cousin who had been doing migration work in Lebanon returned back to the village and decided to revive an old case of revenge. The workers group split up in two factions, and the work had to be halted for 22 days until the problems between the cousin and his family were solved. Eventually the situation was eased with the help of a Bedouin judge, and the work could continue. The spirit of before, however, had deteriorated significantly. But the final day of restoration did arrive on September 16, 2000, and was concluded with the slaughtering of three sheep for a communal meal.

IMPACT AND LESSONS LEARNED

After the cleaning work, we inspected the tunnel with our key local representative (Figure 11.11). The technical impact of the cleaning was measured by a flow meter, placed in the open channel running through the village, and we recorded an increase of water flow in winter, which means that the recharge from rainfall was being directly caught by the tunnel and the water had become free to flow. Another promising result was that 16 young men from the community had been trained for qanat cleaning and would able to maintain their qanat in the future. Whether that is socially sustainable or not can only be observed over the long term. When we returned in the summer of 2002, the village was still divided in different descendant groups, and social tension was still present, but the qanat was flowing and had given a substantial amount of water throughout the year. Figure 11.12 shows a renovated airshaft near the village and Figure 11.13, the amount of irrigation water collected in the birkeh in an afternoon.

FIGURE 11.11 Inspecting the tunnel of the qanat after the pilot renovation.

FIGURE 11.12 Renovated airshaft near the village center of Shallalah Saghirah.

FIGURE 11.13 Qanat water fills the irrigation reservoir (*birkeh*) of Shallalah Saghirah.

The cleaning drew much attention from Syrian and international officials, which benefits public awareness of these sustainable water supply systems. This is necessary to ensure a future for those working qanats still surviving in the modern world. From the experience we have had with the cleaning in Shalalah Saghirah, we have developed some feasibility criteria that can be used for any other qanat sites in the Middle East.

These criteria are:

1. A stable groundwater level: Pumping is a major threat to qanats. If there is a fast decrease in the groundwater level, it is impossible to reuse qanats for agriculture unless the pumping stops within a range of 3.5 km from the qanat tunnel.
2. Consistent underground tunnel construction: Many of the ancient qanat workers died because of the danger of the job and potential collapsing of tunnels. If there is any doubt about the consistency of the underground construction, care should be taken and renovation reconsidered due to safety reasons.
3. Strong social cohesion in a community: This is a condition for any management of qanats as a common water resource. It should be noted that social cohesion differs and that it therefore should be studied on a case-by-case basis. In the Arab rural areas, a strong village or family leader is usually a condition for good social cohesion.
4. Clear ownership of qanat: This is a requirement so that there won't be any problems or conflicts about claiming ownership when there is more water coming from the qanat.
5. Existing system of rights and regulations on water to be used when water increases.
6. Willingness of users, who are the ultimate beneficiaries. If they are not committed to regular maintenance, the work is not likely to be sustainable.

DEVELOPMENTS AFTER THE PILOT PROJECT

Lessons learned from the pilot renovation led to the development of renovation criteria that can be used to decide whether it is profitable to renovate in other cases. In 2001 we conducted a national survey of remaining qanat sites in Syria. We used a structured method of observation and reporting that brought together researchers from several disciplines to conduct interviews with knowledgeable farmers and prepare reports on hydrogeology, damage status, irrigated gardens, and gradient of the tunnels. From our survey data, we selected three possible sites for renovation: Dumayr, Qarah, and Arak.

In March 2002 renovation works finished in Dumayr with the generous support of the Swiss Development Cooperation Fund. The users community is well organized in a traditional system of "water committees" and "water guards" supervised by the farmers' cooperative. The cooperative also paid part of the renovation costs themselves from their credit system. Also actively involved were the General Directorate of Antiquities and the Regional Directorate of Irrigation of the Awaj/Barada Basin that is active in qanat renovation in Damascus Province. In this way both the formal and the informal institutions are participating. The ultimate responsibility and monitoring of the renovation is with the farmers' cooperative.

In April 2004, the renovation of the qanat Ain el Taybeh at the Monastery of Deir Mar Yaqoub was completed. Figure 11.14 shows the monastery, the main water shareholder of the qanat. The monastic community had requested the research team to compile a scientific advice for them based on survey work done in 2001 and 2002. This advice was duly developed in September 2002.[7] The advice concluded with social, institutional, and technical recommendations for the qanats. The advice stated that the qanats of Qarah have an active and well-established traditional user's community, but there is a tendency for the younger generation to be moving away from farming towards urban employment, specifically in Damascus. The government support is fairly well developed, but

---

[7] Wessels & Hoogeveen, 2002, Internal advice and report on Qanats of Deir Mar Yaqub-an extract of the scientific reports for ICARDA and the United Nations University prepared for the monastic Community of Deir Mar Yaqoub.

FIGURE 11.14 The Monastery of Deir Mar Yaqoub, Qarah, Syria.

farmers should be involved at an equal level in renovation efforts. The presence of the renovated monastery, its frescos, and the importance of the site for international heritage justifies the assistance to maintain and conserve this valuable ancient installation as a human ecosystem.

The discharge of Ain el Taybeh had been significant in the past, and the report from expert advice included several remarks on the prospect of the discharge of the qanat after renovation works. We inspected the tunnel of Ain el Taybeh with local key advisors (Figure 11.15). During our survey we could not identify any severe groundwater abstraction around the proximity of the qanats in Qarah, and the official irrigation laws prevent drilling of wells within 1 km of the sources of qanats. However, near Ain el Taybeh, two groundwater users are present; the monastery itself and the neighboring poultry farm. It was advised that both users would agree upon the amount of water to be used. The qanat showed a downward trend of water production, and it was recommended that the source of the water towards the qanat should be thoroughly researched and identified. It was advised that a groundwater balance should be drawn up by the Directorate of Irrigation to determine the exact influence of pumping activities. Another recommendation concerned the building of infiltration dams, as it was observed that near the qanat of Ain el Baidah, the infiltration dam had been beneficial for the discharge. Infiltration dams are thought to increase infiltration into the alluvial fan that feeds the qanats. Alternative water resources and ways to use them were also discussed in the advisory, such as the reuse of domestic wastewater for the monastery and the introduction of drip irrigation techniques. A precedent was already set by farmers in several of the gardens of the qanats in Qarah.

Institutionally, it was advised that the monastery renovate the qanat of Ain el Taybeh in cooperation with the other stakeholders such as the farmer's committee, the local government representatives. The position of the monastery was favorable because it already had received large sums of financial resources to renovate the ancient church and monastery itself, and had access to both ministerial representatives and foreign donors. Furthermore, the analysis included a recommendation that the Directorate of Antiquities and the Irrigation Directorate of Awaj/Barada be closely involved in the renovation. The Directorate of Awaj/Barada had experience of several years of government-subsidized qanat renovation and the Directorate of Antiquities of Damascus countryside had gained considerable experience of qanat renovation during spring 2002, with the execution of the renovation in Dmayr.[8]

---

[8] The Community Based Intervention at Dmayr had been carried out in Spring 2002 within the framework of the overall study on qanats. A detailed description can be found in the project report; Wessels, J.I., R. Hoogeveen, Aw-Hassan, A., Arab, G. (2003) the Potential for Renovating Qanat Systems in Syria through community action–final project report for NRMP, ICARDA, Syria, 110 pp.

FIGURE 11.15 Inspecting the damage to the tunnel of Ain el Taybeh, Qarah.

Together with the area maps and the recommendations, the Monastery of Deir Mar Yaqoub started negotiations with their fellow stakeholders and successfully approached international funders at the embassy level in Damascus. In August 2003, a draft agreement was developed between three European embassies and the monastery representing the farmer's community of the qanat of Ain el Taybeh. In the translated agreement that follows, the farmer's community had stated that they were prepared to invest in the material costs of the renovation:

… The farmer's community, along with the Mar Yaqub Monastery, has decided to restore the Qanat of Ain el Taybeh, following the survey of Joshka Wessels who had, last summer in a very successful speech, shown the possibility of rehabilitation of the qanats.[9] [Support of] the Mayor of Qarah has also been gained for this purpose. All three partners involved are willing to put their capabilities together to be able to make the best restoration possible, hoping to reinstate the important water supply that represents the future of the bustan. Mar Yaqoub monastery had intended to dig a well to try to overcome the water crisis, but this experience was not as successful as was expected. Water came out salty and could not be used for drinking or agriculture. The only hope for the Ain el Taybeh is to join efforts and restore the ancient qanat. All our studies, beginning with those of Joshka Wessels, indicate that a successful restoration of the qanat would supply at least 300 to 4000 m² a day, which is almost two-thirds of the last debit (amount of water used).

---

[9] This section refers to a video feedback session organized in Qarah in 2002.

[At this time] the farmers' community is transmitting its request [for assistance] to three embassies (in alphabetical order)—Germany, the Netherlands, and Switzerland. These countries have [previously] stated their readiness to participate in the material costs:

1. The farmers are ready to provide free workers up to three per day for 3 months.
2. The Mayor of Qarah is ready to provide a Bobcat and trucks as long as needed.
3. The monastery of Mar Yaqoub is ready to host and feed free of charge the two supervisors of the project.
4. Furthermore, the farmers' community is ready to support the necessary maintenance and cleaning in future.

After a survey of the situation, all three partners are transmitting their request to the above mentioned embassies asking for cofinancing in this important matter of survival.

Project general objective:

"To renovate and sustain the vital water resources represented by the Ain el Taybeh qanat. Restoration works will emphasize preserving the authenticity of the Qanats by using traditional local material and techniques …"

—Project agreement "Rehabilitation of the Qanat Water System Qanat Ain el Taybeh Qarah, Syria," compiled by the Monastery of Saint Jacques de Mutile or Deir Mar Yaqub)

FIGURE 11.16 Renovation of qanat Ain el Taybeh, Qarah.

With this agreement, international funds were allocated to the monastery, and logistics for the renovation work could be arranged. A meeting with the farmer's community was held to discuss the organization of the renovation. A specialist group of qanat cleaners both in Ma'aloula and Nebk was approached to carry out the daily work, a daily supervisor would be appointed from the Ma'aloula group (Figure 11.16). The Directorate of Antiquities was informed, and it was decided that a representative of the Irrigation Directorate of Awaj/Barada would visit the site regularly for governmental supervision. The Mother of the monastery would take care of the overall project coordination.

By March 11, 2004, the renovation work finished. In her final report the monastic community stated that the work lasted 30 days longer than predicted. More than 850 m of underground tunnels were explored and 25 tons of mud removed. The gradient of the tunnels have been restored to increase the water flow. More than eight karst springs were found and connected to the system. The casing, walls, and vaults have been rebuilt using the original material and building techniques to preserve the cultural heritage value. The monastic community organized an inauguration day on the April 29, 2004, with representatives of all main participants present, including European embassy delegations with the German and Swiss ambassadors and a representative of the Dutch ambassador. This inauguration gave high profile to the CBI and was covered on national television. A special plaque built by the municipality of Qarah was revealed to commemorate the renovation work in front of the qanat Ain el Taybeh.

We hope, with these renovation efforts, to encourage preservation of indigenous knowledge on the qanats that are still extant in Syria, and by starting with community needs and priorities, more qanats will be revived and sustained for continued use in the future.

CONCLUSIONS

Cleaning of an ancient qanat is not an easy exercise. Not only is the work itself technically difficult but also the social organization around a qanat has major implications for the sustainability of a qanat system. In the pilot case of Shallalah Saghirah, a good hydrological result from the qanat renovation was based on a community work plan. However, tensions between individuals and weak leadership may hamper the long-term progress and prevent maintenance of the qanat. Also, the changing economic circumstances that force the younger generation to look for other sources of income than agriculture, and the high social mobility that is found at the village level, influences the sustainable maintenance of qanats.

This project aimed to characterize and describe the social and physical world around a qanat in order to understand the different forces that affect the use of a qanat in a modern environment. The project showed that focused group meetings on community level help in developing successful project proposals. The approach used here started with the direct users of the qanat water. Individuals who express the need for renovation but do not have financial resources can serve as facilitators and key resources to motivate other inhabitants. Focus group meetings can help users to conceptualize their needs, rank their priorities, and formulate a work plan and budget, themselves. Also focus group meetings can enhance communication between qanat users when problems from the past need to be solved. The creation of an informal institution such as a committee of elders, when weak leadership is present in a qanat village can possibly help in enhancing the sustainable use and maintenance of qanat systems.

Lessons learned from the pilot renovation led to the development of renovation criteria based on an interdisciplinary approach. These criteria are:

- A stable groundwater level
- Consistent underground tunnel construction
- Social cohesion in community
- Clear ownership of qanat
- Existing system of rights and regulations on water
- Willingness of water users to contribute

On a national level, qanats are rapidly drying up in Syria. During a field survey in the summer of 2002, an acceleration of drought conditions among qanats was found. Based on our national survey conducted in 2001 and with the knowledge of the pilot study, three sites were chosen for possible renovation as they still provided a substantial supply of irrigation water. The Drasiah Qanat of Dmeir was renovated in the spring of 2002, and the Qanat Ain el Taibeh of Qarah was renovated between 2003 and 2004. Both renovations were very successful and prove that qanat renovation is not a futile exercise they can provide a promising future for a new era. A thorough plan should be developed on the national level where all stakeholders are represented. This should provide an institutional framework that is vital for the sustainability of the use of the ancient qanats of Syria.

ACKNOWLEDGMENTS

We would like to thank all those who have been involved in the qanat project. At ICARDA, Senior Socio-Economist Dr. Aden Aw-Hassan, Senior Hydrologist Dr. Adriana Bruggeman, research assistants George Arab and Nasr Hillali, GIS specialist Piero d'Altan, and Remote Sensing Specialist David Celis; Pierre Hayek for showing us our first qanat; and the many students and friends that assisted us during fieldwork. Furthermore, Abu Zakki of the Museum of Aleppo for his guidance during the renovation work; the Directorate of Antiquities; Khaled Sawan of the Irrigation Directorate of Awaj/Barada; at UNU, Project Manager–Traditional Water Management Project Adeel Zafar; and Senior Advisor Professor Iwao Kobori, for encouraging us to continue with the research. This project was financially supported by DGIS, Ministry of Foreign Affairs, the Netherlands; UNU, Tokyo, Japan; and the Dutch, German, and Swiss embassies in Damascus, Syria. Last but not least, we would like to thank the qanat communities of Shallalah Saghirah, Dmeir, Qarah, and Arak for welcoming us into their lives and teaching us.

LITERATURE CITED

Ashram, M. (1990). Agricultural Labour and Technological Change in the Syrian Arab Republic, in *Labour and rainfed agriculture in west Asia and north Africa*, Kluwer Publishers, The Netherlands.

Batatu, H. (1999) *Syria's peasantry, the descendants of its lesser rural notables and their politics*, Princeton Press, U.S.

Birks, J.S. (1984). The falaj: modern problems and some possible solutions, *Water Lines* 2(4), 28–31.

Birks, J.S. & C.A. Sinclair, (1980) *International migration and development in the Arab region*, ILO, Geneva, Suisse.

Bloom, J. & S. Blair, (2000) *Islam: empire of faith*, BBC Worldwide, London.

Chambers, E., (1985) *Applied anthropology, a practical guide*, Waveland Press, U.S.

Haasse, C.P. (1983) *Ein archäologischer survey im Gabal Sbet und im Gabal al-Ahass*, Damascus Mitteilungen I, Ghoete Institute, Damascus.

Hamidé, A. (1959) *La region d'Alep étude de géographie rurale*, Universite de Paris, France.

Hoogeveen, R.J.A. and M. Zöbisch. (1999). Decline of groundwater quality in the Khanasser Valley (Syria) due to salt-water intrusion. Paper presented at the 6th International Conference on the Development of Drylands, Cairo, Egypt, 22–27 August 1999. 16 p.

Hoogeveen R.J.A., and M. Zöbisch. (1999) Well inventory and groundwater in the Khanasser Valley, Syria. Report on a field study in northwestern Syria. International Center for Agricultural Research in the Dry Areas (ICARDA), Aleppo, Syria.

Lewis, N. (1987) *Nomads and settlers in Syria and Jordan 1800–1980*, Cambridge University Press, U.K.

Lightfoot, D. (1996) Syrian qanat Romani: history, ecology, abandonment, *Journal of Arid Environments*.

Mouterde. R. & A. Poidebard, (1945) *Le limes de Chalcis, organisation de la steppe en haute Syrie Romaine*, Librarie orientaliste Paul Geuthner, Paris, France.

Mundy, M. (1995) *Domestic government: kinship, community and polity in North Yemen*, Tauris Publishers, U.K.

Procopius (1914) *History of the wars Books I and II*, with an English translation by H.R. Dewing, Harvard University Press, William Heinemann, U.K.

Procopius (1914) *Buildings* with an English translation by H.R. Dewing, Harvard University Press, William Heinemann, U.K.

Safadi, C. (1990) La Foggara, systeme hydraulique antique, serait-elle toujours concevable dans la mise en valeur des eaux souterraines en Syrie? in Geyer, B. (ed.) *Techniques et practiques hydro-agricoles traditionelles en domaine irrigue, approche pluridisciplinaire des modes de culture avant la motorisation en Syrie*, Tome 2, Paris, Librairie orientaliste Paul Geuthner, IFAPO, Beyrouth-Damas-Amman.

Stevenson, T. (1998) Migration, family and household in Highland Yemen: the impact of socio-economic and political change and cultural ideals on domestic organization, *Journal of Comparative Family Studies*, Vol. 28–2, 14–53.

Tate, G. (1992) *Les campagnes de la Syrie du Nord du IIe au VIIe siècle*, IFAPO, Paris, France.

Tully, D. (1990) Household Labour Issues in West Asia and North Africa, pp. 67–92, in *Labour and rainfed agriculture in west Asia and north Africa*, Kluwer Publishers, The Netherlands.

12 Growing Green Infrastructure Along the Urban River
Duwamish Stories

Nathaniel S. Cormier

CONTENTS

ABSTRACT

The Duwamish River has been dramatically transformed by the urbanization of Seattle. Once life-giving and wild, the river is now straightened and controlled. This chapter describes three community-based efforts to restore physical and emotional connections to the river. The Duwamish Riverfront Revival is a salmon habitat restoration project in the South Park neighborhood. Oxbow Park is a small open space in the heart of the Georgetown neighborhood, across the river from South Park. The Duwamish River Field School was a landscape planning and design studio that immersed students in the challenges and achievements of these river communities.

INTRODUCTION

Can attraction to water grow with stream order? Headwaters springs and babbling brooks have their special charms, but where all those delicate tributaries are gathered into a river, I am at home. Usually, the most impressive rivers find their way through cities. The cities, of course, found the rivers. In most urban settings, rivers have been straightened and controlled to prevent flooding and make more land available for human use. Precisely where water should be reaching its fullest expression, it is rendered powerless, or worse, invisible. Where I live, most of the oxbow-sketched floodplain and estuarine marshes are gone and even the rip-rapped waterway that remains is obscured by industry and highways. I will share three stories of community-based efforts to reconnect functionally and emotionally to Seattle's wet heritage. But first, let me define green infrastructure and introduce you to the Duwamish River and its communities.

WHAT IS GREEN INFRASTRUCTURE?

A single definition for "green infrastructure" is hard to pin down because the phrase is used to describe relationships between natural systems and the urban environment at many scales (Brady et al. 2001; Benedict et al. 2002). In its broadest sense, it can refer simply to the tapestry of open space in and around a city or region, from isolated street trees to connected healthy forests. Some communities are recognizing the valuable ecological services that nature can provide, especially in terms of stormwater management, microclimate moderation, and carbon sequestering. This is adding a strong new argument for preservation to the traditional accounting of a landscape's highest and best use.

Thanks to a generation of pioneering work, the built environment of the Pacific Northwest is awash in all manner of pilot projects for bioretention swales, rain gardens, cisterns, constructed wetlands, green streets, and the like (Wulkan et al. 2003). This expanding palette of innovative landscape features could be considered a more intentional and engineered understanding of green infrastructure. In Seattle, neighborhood-scale natural drainage systems like the Broadview Green Grid and at the High Point housing redevelopment suggest that we're even moving beyond the pilot phase of this type of green infrastructure implementation. Regulators and developers are finally catching on to the simple fact that open space is multifunctional and should be appreciated and supported for its diverse contributions to the urban environment.

At the intimate scale of human experience, green infrastructure needs to be a meaningful part of people's lives so that it will be loved and valued by citizens. In this respect, green infrastructure has many potential advantages over the traditional gray infrastructure of pipes, pumps, and treatment plants. For starters, it is visible. You can't love what you can't see. It is also diverse and responsive to regional conditions like climate and cultural identity. Finally, it grows in value over time instead of decaying. Multifunctional open space projects could be the most enduring public works of our time if we can connect them to people. I use the phrase "green infrastructure" to help people understand open space as an essential part of urban life, not simply an amenity. The examples that follow illustrate a range of tools and techniques for connecting green infrastructure to people in the communities of the Duwamish River.

THE DUWAMISH RIVER

The Duwamish River carries surface waters from a watershed of almost 500 square mi. At Elliott Bay, the Duwamish opens onto Puget Sound—a deep glacial fjord with over 2,000 miles of shoreline and a rich marine environment. Much of the promise and potential of the Puget Sound region is based on natural resources and the industries, tourism, and quality of life these resources support. Habitat degradation and pollution around the Sound have led to alarming declines in some fish and wildlife populations, most notably those of salmon and orcas.

Where the Duwamish River flows through Seattle, it is actually an estuary—the part of the river course affected by the mixing of salt water and fresh. The estuary plays a particularly critical role in the life cycle of salmon, which need to rest, feed, and grow there as young "smolts" before heading out to sea and need to pause there and acclimate themselves to fresh water again when returning to spawn as adults.

The Duwamish has experienced significant degradation of its habitats, water quality, and ecosystem processes. About 97% of the Elliott Bay/Duwamish River estuary has been dredged or filled and over two-thirds of the flow that once fed the river has been diverted out of the basin (Green/ Duwamish and Central Puget Sound Watershed Water Resource Inventory Area 9 Steering Committee 2005). Over the last hundred years, the Duwamish has been transformed from a shallow,

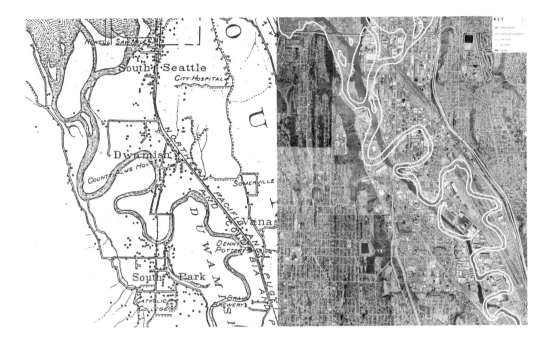

FIGURE 12.1 The Duwamish River in 1894 and today.

meandering river into a straightened, dredged waterway, reducing the length and increasing the depth of the channel. Figure 12.1 shows the path of the Duwamish River in 1894 superimposed over a recent aerial photograph. Where once a rich riparian forest of cedars and Douglas firs provided insects, shade, and logs to the river, vast stretches of impervious paving and rooftops now contribute increased and polluted run-off.

In September of 2001, the United States Environmental Protection Agency (US EPA) listed the entire Lower Duwamish River as a federal Superfund site, recognizing it as one of the most contaminated places in the country (US EPA 2001). Of particular concern are dangerous levels of polychlorinated bi-phenyls (PCBs) and mercury in the sediments of the river bed. Wild Puget Sound Chinook salmon (*Oncorhynchus tshawytscha*) still inhabit the Duwamish River and are listed as "threatened" by the federal government under the Endangered Species Act (ESA). Ecosystem restoration is key to the recovery of this and other fish species. Superfund and salmon recovery issues have ignited a number of efforts to improve the quality of quantity of aquatic and riparian habitat in the river corridor.

DUWAMISH RIVER COMMUNITIES—GEORGETOWN AND SOUTH PARK

The first river community residents were Native Americans of the Duwamish tribe. For thousands of years, they lived in large cedar longhouses and took fish from the river, grew potatoes, gathered bulbs and berries, and hunted game. Their story is a familiar one—violent displacement by arriving white settlers and their descendants (see also Chapter 15). Only a few Duwamish people remain, but they continue to tell their stories and practice their ancient rituals on the banks of the river that bears their name. Today's river residents inhabit the Seattle neighborhoods of Georgetown (Figure 12.2) and South Park (Figure 12.3), South Seattle places with their own histories of marginalization and resilience.

Early Georgetown was a brewery and railroad town. A large steam plant drew water from the Duwamish and provided power to the electrified Seattle–Tacoma Interurban Railway, breweries, and factories. Rainier Beer was born in Georgetown and grew to become the world's sixth largest

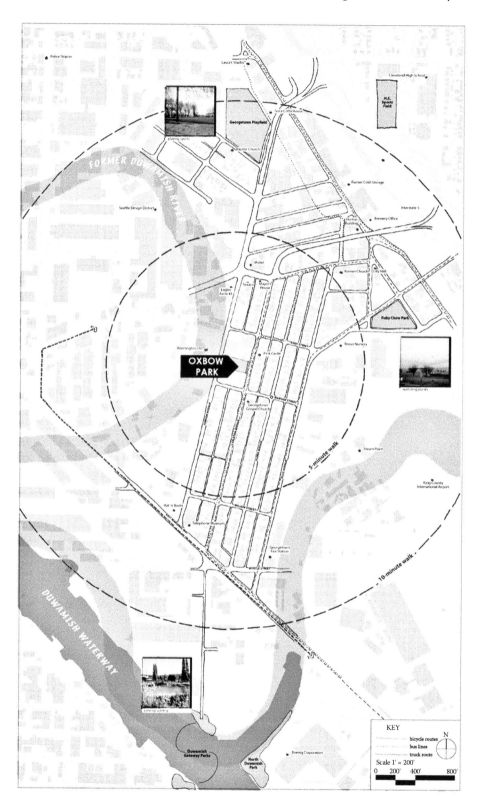

FIGURE 12.2 Map of Georgetown with historic river meander overlay.

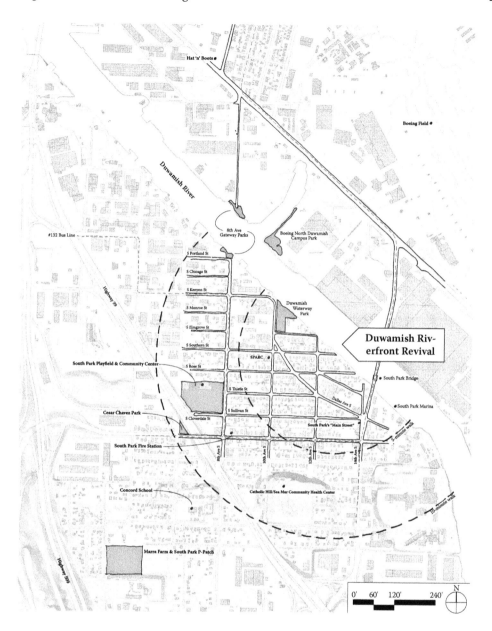

FIGURE 12.3 Map of South Park.

brewery for a time. At the turn of the 20th century, Georgetown, with its lax liquor laws, became a destination for revelers from Seattle. Full of road houses, beer gardens, and brothels, Georgetown developed a raucous reputation. Local voters eventually approved annexation by Seattle to help get things under control, but the neighborhood has hung on to its frontier spirit despite a century of challenges.

The straightening of the Duwamish had a major impact on Georgetown. Parts of the neighborhood that had once been beside the river ended up a half-mile away from the new channel. New land for industrial development rose from the mud flats on soil from the sluicing of Seatttle's surrounding hills. The Boeing Company began building airplanes for the World Wars on both sides of the Duwamish, and Seattle's first airport was built over the filled meanders south of the neighborhood.

In the 1940s, Georgetown was a thriving and prosperous neighborhood, but postwar planning would devastate the community. All residential zones were proposed to be eliminated in favor of industry. Construction of Interstate 5 severed the floodplain neighborhood from its hillside neighbors and decimated the once vibrant commercial core. As block after block of homes were demolished, the schools, markets, and small businesses left, too.

A small residential population has survived amid the industry and asphalt. They are down to 1200 residents, but determined to hang on to their heritage and find ways to reinvigorate their community. Georgetown's affordable workspace and tolerance for noise have begun to attract artists, craftspeople, and other pioneering hipsters. Bars, coffee, pizza, records, scones, and Vespas have sprung up amid the bustle of Georgetown's planes, trucks, and trains. Interdependence and ingenuity are the new ethos as a live–work hybrid is transcending old conflicts between industry and residents. The heading on its community Web site says it best "Georgetown: Seattle's Fiesty, Intensely Creative Neighborhood." Any visitor to their outstanding summer art and garden walk would definitely agree.

Rivaling Georgetown in feistiness, but with a flavor all its own, is South Park, the neighborhood directly across the Duwamish River. In the last half of the 19th century, development and industry favored Georgetown, which left South Park to the farmers, including many Italian and Japanese immigrants. Like Georgetown, South Park's character changed dramatically when the Duwamish River was straightened. Industry began to develop along the banks of the new waterway and with the rapid growth of World War II airplane and shipyard operations, the area experienced a critical housing shortage. The little farming community was flooded with newcomers. To make way for industrial expansion, most of the fertile bottomland that had attracted early settlers and, later, immigrant farmers, was paved.

South Park, like Georgetown, was also nearly erased by postwar zoning and highways. In the mid-1960s, when South Park was rezoned industrial, 4200 residents staged a protest at City Hall and got part of the neighborhood changed back to residential. One headline proclaimed, "South Park: A Square Mile of Defiance." Later, State Route 99 was rerouted through South Park, severing the only elementary school from the rest of the neighborhood. By the mid-1970s, crime was up and the area attracted mostly poor immigrants. The neighborhood hung on despite the pressures.

Over the last two decades, the City of Seattle and concerned organizations have reinvested in the neighborhood with a community center, an updated school, a health center, and a library. South Park's low-priced homes within city limits have begun to attract young families priced out of more expensive Seattle neighborhoods. This has invited retail businesses but also tended to drive up rents and housing prices. Today, about 3700 people live in South Park, and it is the youngest and most ethnically diverse neighborhood in Seattle. The Mexican-American influence is particularly visible in the restaurants and small businesses of the community.

Today, Georgetown and South Park are home to a healthy mix of artists, architects, activists, immigrants, and old-timers. Most residents don't have a lot of patience for ideas without action. They are busy enough with fundamental issues like getting rid of crack dealers and trying to attract a decent grocery store. Innovative projects can happen, but they require enthusiasm and endurance. The following stories are three efforts in which I have been involved.

DUWAMISH RIVERFRONT REVIVAL

From 2000 to the present, as a landscape architect at Jones & Jones and in close collaboration with Peter Hummel of Anchor Environmental, I have been leading the design of the Duwamish Riverfront Revival for the Environmental Coalition of South Seattle (ECOSS). ECOSS brings neighbors and businesses together to solve environmental problems. The Duwamish Riverfront Revival is a vision for salmon habitat restoration along a couple thousand feet of shoreline in South Park. The project showcases community participation and project visibility as fundamental goals of urban ecological restoration, and provides a model for public-private cooperation in salmon recovery.

Residents, neighbors, landowners, public agencies, and environmental organizations came together to learn about and design a stretch of mostly private riverfront as a habitat for salmon.

Over the last decade, there have been significant habitat restoration projects completed along the Duwamish. In particular, the Hamm Creek/Turning Basin area about one mile upriver from South Park and the Seaboard Lumber/Kellogg Island area about two-plus miles downriver from South Park have had large projects completed. However, these projects were done on government-acquired property with public funding. There are now fewer opportunities to buy large parcels of land along the Duwamish for restoration. Most of what remains is being used productively for homes and businesses, so new habitat projects will have to be developed in and around these land uses. This project is a prototype for restoration projects that complement the major government land acquisitions, serving as stepping stones that connect the larger habitat patches without displacing existing uses.

The riverfront addressed in this project, between Duwamish Waterway Park and the historic South Park Bridge, is the first view of the neighborhood that one sees when entering South Park (Figure 12.4). A restoration project in this highly visible location would have a significant impact on the perception of South Park by its residents and visitors. In spite of limited resources, land use pressures, and a diverse population this community could make a bold environmental statement. And, since the streetends near the South Park Bridge are within a very short walk of South Park's retail core, there is an opportunity to create an important civic place from which to observe the restoration and inspire further neighborhood activism.

We needed broad community participation to learn how South Park residents could best connect to the river and care for it. At the same time, more focused outreach was critical to build a dedicated coalition of riverfront landowners willing to contribute their own lands to the effort. We began the project by educating these landowners and their neighbors on the needs of juvenile salmon. We wanted everyone to start with a salmon's eye view of the river—what salmon are looking for, what the river once looked like to them, and what it looks like to them now. This salmonid perspective became the touchstone that everyone could return to during the inevitable conflicts that arose back on land (see Chapter 10 for another reflection on salmon and restoration).

The difference between high and low tide is nearly 12 ft in the Lower Duwamish. Juvenile salmon need suitable habitat throughout this range. Ideal habitat should provide shallow water for its slow currents and refuge from predators. It should have fine substrates (like sand, mud, branches, and marsh grasses) to host delicious invertebrates, and it should have overhanging vegetation to shade the river and drop insects (food) and woody debris (more substrate for more food). The Duwamish

FIGURE 12.4 View to Duwamish Riverfront Revival site from South Park Bridge.

still has a fair amount of mudflat available to salmon in the lower intertidal range, but the upper inter-tidal range, historically composed of emergent marsh vegetation, has been mostly replaced by rip-rap, bulkheads, and other hard vertical surfaces. Our primary goal was to lay back the banks of the river to create a band of emergent marsh vegetation in this critical upper intertidal range.

We assured the landowners that as a community-initiated process, everything we were discussing was voluntary. There would be no takings or condemnations. Quite the opposite, we were asking for them to allow us to "improve" their properties. Many of their banks, formed from waste materials and construction debris, were deteriorating so expanded salmon habitat with slope bioengineering could be a win-win for salmon and landowners. Figure 12.5 shows a typical riverbank condition. We offered to shape the project and process around their individual needs and the unique conditions of their sites. Full-scale mock-ups were used to discuss the extent of the project with riverfront landowners (Figure 12.6).

Reflecting specific conditions along the riverfront and the varying degrees of participation by adja-cent landowners, a series of eight distinct "zones" was defined for the project area. These are illustrated in Figure 12.7. A collection of design and engineering approaches was developed to improve the juvenile salmon habitat potential of these zones. Taken together, the zones represent a living laboratory where the effectiveness of the different techniques will be able to be evaluated and compared. The project will cre-ate or enhance approximately 2 acres, one half of which will be in the upper intertidal range, and will increase the length of the shoreline by 50%. Figure 12.8 illustrates the overall project, as viewed from the South Park Bridge.

FIGURE 12.5 Typical riverbank condition in South Park.

FIGURE 12.6 Full scale mock-up of project extents.

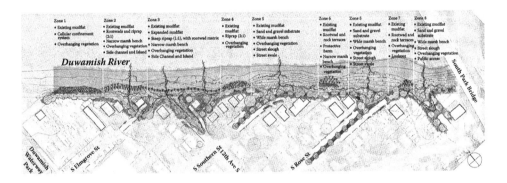

FIGURE 12.7 Duwamish Riverfront Revival Plan with 8 zones.

FIGURE 12.8 Illustration by Peter Hummel of Duwamish Riverfront Revival from the South Park Bridge.

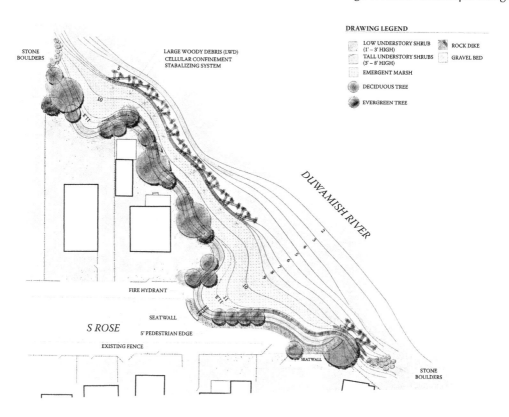

FIGURE 12.9 Duwamish Riverfront Revival Phase 1 Plan.

After the overall project planning was complete, we worked with ECOSS and the most supportive landowners to develop a first phase in greater detail. Figures 12.9 and 12.10 illustrate the design for this Phase 1. Government agencies donated the end of a street terminating at the river to be converted into a "street slough"—a benched cove of emergent marsh vegetation—where the entire neighborhood could experience and understand the restoration. Adjacent private landowners offered easements on their property to extend the marsh bench around to a second cove. Native plantings were designed to overhang the marsh, but openings were preserved for river views from the decks and windows of the homes. The project will be a great demonstration of the power of partnering with landowners for urban salmon habitat restoration.

Sadly, the Duwamish Riverfront Revival is currently on hold due to Superfund activities nearby. Our project actually began before the Lower Duwamish was declared a Superfund site. (At the time, it wasn't clear when, if ever, polluted river sediments, would be cleaned up.) As our project's design neared completion, we learned of plans to eventually dredge up contaminated sediments from around several Boeing facilities directly across the river. Based on some of the results from clean-ups that have already occurred, this dredging process can lead to resuspension and release of pollutants, including PCBs, in the surrounding river. Our project could be severely impacted by this pollution, so we'll have to wait until that clean-up is complete to implement this habitat restoration. The Superfund process is quite slow, so this was a major disappointment to the team and the community. Despite this setback, there were still some important lessons learned on this project, especially in terms of community participation.

First, citizens armed with knowledge made ambitious decisions and demanded more of designers, officials, and each other. Workshop participants were the neighborhood ambassadors for the project. It was a goal of the project that each participant could communicate key points from the events to

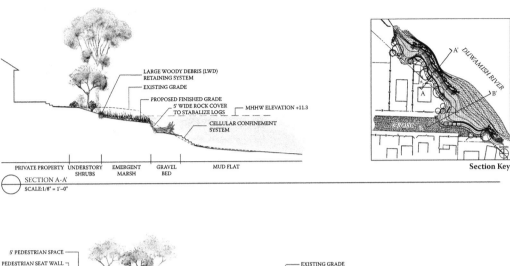

LARGE WOODY DEBRIS (LWD)
RETAINING SYSTEM
EXISTING GRADE

PROPOSED FINISHED GRADE
5' WIDE ROCK COVER
TO STABALIZE LOGS   MHHW ELEVATION +11.3

CELLULAR CONFINEMENT
SYSTEM

PRIVATE PROPERTY   UNDERSTORY   EMERGENT   GRAVEL   MUD FLAT
                  SHRUBS   MARSH   BED
SECTION A–A'
SCALE:1/8" = 1'–0"

A' DUWAMISH RIVER
A
B'
S ROSE B
NORTH
Section Key

5' PEDESTRIAN SPACE
PEDESTRIAN SEAT WALL

EXISTING GRADE
PROPOSED FINISHED GRADE
MHHW ELEVATION +11.3

S ROSE STREET   VISITOR LOW   EMERGENT MARSH   MUD FLAT
               EDGE SHRUBS
SECTION B–B'
SCALE:1/8" = 1'–0"

FIGURE 12.10 Duwamish Riverfront Revival Phase 1 sections.

his or her family and neighbors. Single page take-home-message flyers were handed out summarizing "Salmon 101" and other topics to make the project visible and clear beyond the workshops. The next step in cultivating stewardship was to turn understanding into authorship. Even project skeptics became champions when their markers hit the paper. Defensiveness gave way to creative problem solving. At our second public workshop the project really lifted off when it ceased to become about what ECOSS and the designers were proposing and instead became about what citizens wanted to do with their community and how this project could achieve those goals. Landowners began challenging each other to come up with more ways to help salmon. When one landowner volunteered to lay his riverbank back to create a marsh cove another responded "I'll dig a marsh channel all the way around my house!" When many people took ownership of the concept, a can-do attitude emerged.

Second, South Park neighbors and landowners had diverse individual hopes and fears for the project, but they were able to compromise for the sake of salmon. With complex jurisdiction, ownership, and physical conditions it was essential to recognize early and keep attention on values shared among the participants. With limited time and resources, all agreed to focus on the elements that received consensus support. There were a number of goals expressed that did not survive this test. For example, some neighbors suggested the need for greater public access along the river, perhaps in the form of a riverwalk. Most riverfront landowners, weary of unsavory characters that occasionally occupy the riverbank, took a position of no public access at all. Fortunately, all could agree to an approach of inviting public access at the streetends, especially near the South Park Bridge. These locations could receive enough activity to be safe open spaces and would be the best places to educate the public about the restoration. The remainder of the riverfront improvement would be focused on the needs of salmon. This kind of compromise was key to keeping broad support for the project.

Third, to get participation from people besides the usual suspects—designers, activists, and cranks—we needed to be good hosts. We planned meetings and events for nonworking hours and got our literature out in the languages of the people in the neighborhood. We served home-cooked meals and made sure everyone was enjoying themselves! The web of civic engagement woven at the public workshops was as important as the ideas generated there. Many participants have gone on to lead other community initiatives focused on salmon and the river. The community-building emerging around salmon recovery in the Duwamish River and around Puget Sound suggest that it is not we who will save salmon, but they/us.

OXBOW PARK

From 2001–2004, I led the design of the 1-acre Oxbow Park (Figure 12.11) for the Georgetown Community Council and Seattle Parks and Recreation. The park would become the new home of the neighborhood's most cherished icons, the Hat n' Boots (Figure 12.12). The Council wanted the park to integrate neighborhood history, the creative spirit of the current community, and the aforementioned roadside attractions. Laura Haddad, a community artist and collaborator, dubbed this mix "Industrial Artistic Cowboy." Rather than simply ape the "Western" theme of the Hat n' Boots, our challenge was to understand the historic and sculptural context of these overwhelming structures in order to root them in a park that expresses the community's identity. The result would be a place to showcase the Hat n' Boots as fascinating relics of an earlier era, but also to give them new life as an integral part of a robust place for visitors and residents.

The Hat n' Boots are an important part of Georgetown's history and a memorable landmark for most Seattleites. They were originally a gas station at the edge of the neighborhood on the former State Route 99, the main highway into Seattle from the south before Interstate 5 was built. You paid for your gas in the office under the hat, while the boots were actually a pair of restrooms. The Hat n' Boots were in the exuberant mid-century roadside style known as "googie." The best-known features of this genre are gigantic signs, eye-catching architecture, boomerang and crescent curves, and all manners of starbursts, sparkles, bubbles, amoebas, and ovals. Basic tenets of googie, a kind of Modernism for the masses, included combining design themes in an abstract way, making elements appear to defy gravity, making structural systems visible, and utilizing innovative building materials. Some of these principles would guide the forms and relationships of the landforms, terraces, and plantings of the park as we sought to create a complementary setting for the Hat n' Boots.

FIGURE 12.11 Oxbow Park in Georgetown.

FIGURE 12.12 Hat n' Boots in the 1950s and 1990s.

With so few public amenities in Georgetown, neighbors demanded an extraordinary number of programmatic elements for such a small park—a community garden, native habitat areas, shelter, children's play structure, amphitheater, plaza, picnic lawn, and an enormous vermillion cowboy hat and boots. The greatest challenge was developing a site strategy that could accommodate this broad range of elements and unify them into a coherent park experience.

Working with neighborhood historians to unearth some background materials about the park site and the neighborhood's relationship to the Duwamish River, we learned that before the Duwamish River was straightened, the parkland was on the right bank of a broad "oxbow" meander (Figure 12.2). In fact, the neighborhood street grid is cranked a bit off Seattle's typical north–south axis because this allowed most of the neighborhood homes to occupy higher land between two meanders. The site itself was a coal and wood storage yard where flat-bottomed scows could unload materials for the neighborhood. Nearby, beer gardens and roadhouses lined the avenue along the lively riverfront. This intriguing relationship to the river would ultimately give Oxbow Park its name, but more importantly it gave us a design concept to hold the park's many pieces together.

The "bones," as they would come to be known, are a web of low concrete walls inspired by the bunkers, or "cribs," of the former storage yard. The bones structure the park while creating places to sit, retaining garden terraces, disappearing into landforms, and reappearing in tall grass. The bones, only partially exposed, provide a sense of mystery in the park. What stories are children making up to explain their presence—some former industry, a new infrastructure? The bones in the middle of the site align with true north. The grid begins to come apart as it bends toward the streets at either end, expressing the different orientation of Georgetown to the rest of the city. The twisting grid suggests the former relationship to the Duwamish River while articulating distinct landscape rooms in the park. Figure 12.13 shows an early concept model of the park.

Community participation during the design, implementation, and stewardship of Oxbow Park has been critical to its success as a neighborhood gathering place and new civic landmark since its opening in 2004. As I briefly introduce the places of the park, I'll highlight some community-building achievements.

FIGURE 12.13 Early concept model of Oxbow Park.

FIGURE 12.14 Amphitheater landform at Oxbow Park

The Hat n' Boots anchor the industrial western edge of the park. An informal amphitheater is formed around the Hat n' Boots by a crescent dune landform (Figure 12.14). The landform and planting enclose and shield the park from road noise while allowing views in to the Hat n' Boots for passersby. A neighborhood committee has been focused on restoring the structures and sharing their history. Several hundred thousand dollars were raised to procure and move them to the park. In 2005, the Boots were completely restored. In the summer of 2006, Laura Haddad led a team of volunteers in covering the Hat in laurel and flowers (Figure 12.15) to raise attention and money for its restoration. Neighborhood businesses have since committed to providing materials and a nearby technical college training program will provide the labor to get the job done in 2007. The community has started a series of shows by neighborhood performing artists and bands for the Hat n' Boots Amphitheater. At the first event, an alt-country band played the park's first rendition (of many, I suspect) of "These Boots Are Made for Walking." Neighbors even peeled back the Hat's construction fencing for a day to allow two neighborhood park volunteers to have the first wedding under the Hat!

FIGURE 12.15 The Hat covered in laurel and flowers.

FIGURE 12.16 Community garden terraces at Oxbow Park.

The community garden terraces (Figure 12.16) are the dominant element on the residential eastern edge, the main entrance for neighbors walking to the park. Community gardeners provide critical eyes on the park during most daylight hours and throughout the year. The gardeners form a core stewardship group that tends the park plantings and organizes events at the park. The gardeners have also broadened their reach to tackle ecological restoration efforts in planting strips and vacant lots stretching all the way down to the Duwamish.

The garden plots are arrayed on a series of arcing terraces stepping up to maximize sunlight and views. The lowest level, actually a rain garden, receives piped run-off from the only impervious part of the park (a material drop-off area for the garden) and from the drain under the veggie and tool washing station. The rain garden was planted with native wetland plants in the park's first volunteer work party (Figure 12.17). Work parties during construction helped reduce costs while

FIGURE 12.17 Volunteer work party to construct the rain garden.

FIGURE 12.18 Volunteer work party to plant native shrub borders.

keeping neighbors engaged through the long wait between design and opening day. Volunteers also installed undulating native shrub borders to buffer the park from residences while creating habitat for wildlife and education (Figure 12.18). On Earth Day of 2005, neighbors and Jones & Jones staff teamed up to build a garden shed complete with South Seattle's first green roof (Figure 12.19).

At the heart of the park is an informal expanse of eco-lawn, a drought-tolerant and slow growing turf alternative. A series of depressions at the edges of the eco-lawn retain excess run-off, allowing the park to function independent of the city's conventional stormwater infrastructure. Along a central

FIGURE 12.19 Volunteers constructing the garden shed.

FIGURE 12.20 A big wrench embedded in the bones.

path, an orange metal children's climbing structure offers a family-friendly amenity with a nod to industrial scale and character. Conveniently-placed bones offer diverse places to sit and gather. Visitors sometimes notice steel and iron artifacts (Figure 12.20), collected from nearby foundries and scrap yards, embedded in the bones as a hint of the neighborhood's industrial heritage.

THE DUWAMISH RIVER FIELD SCHOOL

In 2003, I prepared a curriculum for the first of two innovative 1-month summer studios that Jones & Jones would teach for landscape architecture students of the Universities of Oregon and Washington. The Field School was an experiment in immersive education. It was also a way to share our approach to design practice as an ongoing process of learning about the places around us and working as activists with communities to make these places healthy. The first Field School was called, "Joining the River Communities." The students' mission was to "join" the Duwamish communities both in the sense of becoming a part of them and in the sense of bringing the two neighborhoods together around the river.

We wanted to explore how green infrastructure could help these neighborhoods realize their common relationship to the river and how a shared experience of the river could strengthen ties between them. We also wanted our students to weave themselves into the social and political fabric of these communities to become advocates of these places. While many design studios begin with the assignment of a specific program, we instead encouraged the students to find out for themselves what these neighborhoods really need and help discover what group or agency could become a client for this kind of change. In this way, the program and physical design become interacting, complementary aspects of a more propositional design process.

The studio was focused on the Gateway Parks, a pair of streetends on either side of the Duwamish that had been informally appropriated as open space by the Georgetown and South Park communities. The student design strategies could reach deeper into the communities or further up and down the river, but we wanted the projects to have these points of contact with the river in common. Designs could occupy the river, but had to allow for existing barge and boat traffic. The connection between the neighborhoods could be physical, or just implied, as long it made a powerful gesture of joining them.

"Otse"

The river bank
like lips, an edge
to trade words
across rocks
and teeth

saw sound
in two letters
cut in spit
and silt

Say, "Otse"
River, Language
Mar, Tseqw
and hear
water, Listen

to opposites
Formed by mouth
together on your tongue

Industry: Trout
Feathers: Lead
Progress: Culture

Meaning swims
in you like fish
sing salmon back
whole to taste
Say, "Otse"

write dreams, long
hand in clear ink, wet
your voice loud on air
Liquid letters. Shape shift ideas

And ask Questions
that dig up secrets
buried in the shore
Listen to the river

And say, "Otse"

FIGURE 12.21 "Otse," a poem by Iris Benson of the University of Oregon.

To appreciate the project's context, we began the studio way up in the headwaters of the Cascade mountain range, descended with the rivers to Puget Sound, surveyed the Duwamish River by boat, and trekked along the riverbanks and into the neighborhoods on foot. Along the way, students learned about issues facing the area from representatives of the Duwamish Tribe, environmental and community organizations, neighborhood councils, small businesses, city agencies, and countless neighbors. Students synthesized their initial impressions and research through mapping, essays, sculpture, and even poetry (Figure 12.21). In order to develop physical concepts and a program for their sites, we asked students to consider three important questions:

1. Where am I?
2. What can I do to strengthen the health of this place?
3. What can I do to connect people to this place so they will love and care for it?

The final designs, some examples of which are illustrated in Figures 12.22 to 12.25, were shared with neighborhood residents, agency representatives, and community leaders at an open house on the river, complete with a keg from a Georgetown brewery and a loaded taco truck from a South Park restaurant.

In 2004, we retooled the Field School curriculum to deepen the immersion in the river communities and add more opportunities for student initiative. The second Field School was called "Green Infrastructure for the Duwamish." Feedback from the first studio suggested that when students got back to the traditional confines of the design school studio space, they lost the sense of immediacy they felt along the river as they fussed with GIS and Photoshop layers and battled for plotter time. Although there is a valuable role for computers in professional practice and academia, in the fast-paced and hands-on environment of the Field School, we knew that we needed to pry the students

FIGURE 12.22 Student work by Travis Scrivner of the University of Washington.

away from the studio, get them to the river, and keep them there. Field School 1.0 had the fresh perspective of an experiment, but Field School 2.0 benefited from a series of radical revisions.

We encouraged the students to live by the river for the duration of the studio. Students rented out part of the Airlane Motel on East Marginal Way, the Georgetown strip that is about as down and out as Seattle gets—the same strip that was once SR 99, home to the Hat n' Boots. The rooms weren't comfortable or clean; there were more than a few shady characters around; and it was a real challenge to get decent groceries. However, the students' bravery gave them immediate street credibility with neighborhood residents. Neighbors recognized that the presence of these students was helping to deter crime, even if only for 1 month on one block, and they appreciated that. The motel also gave us a home base by the river instead of at the design school. From then on, all course meetings and "field crits" occurred *in situ* (Figure 12.26).

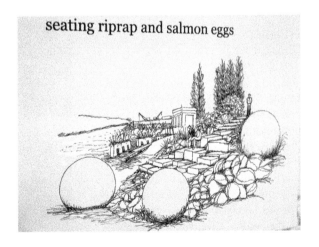

FIGURE 12.23 Student work by Becky Strickler of the University of Oregon.

FIGURE 12.24 Student work by Kitty Davis of the University of Oregon.

To facilitate this mobile studio approach, we developed a portable "field guide" that presented the research conducted by the previous year's students and structured all the activities and exercises of the studio. The field guide also contained space for students to record their impressions and design work. For presentations and reviews, students simply copied their field guides onto acetate and shared their work via an overhead projector.

We also added a team-based planning exercise to suggest programs and sites for the design work. A city can be understood as a web of dynamic systems. Each system can be examined independently, but the challenge is in balancing and integrating these forces. We described this effort with five related actions:

- Cleaning water
- Connecting habitat
- Growing urban villages
- Weaving transportation
- Developing eco-industry

Teams developed a planning framework and priority projects for each of the five actions, and then reconciled conflicts and identified opportunities for synergy among them. Overlaid together, the five integrated systems suggested a green infrastructure network for the corridor.

Individual students were then asked to site and design a project that could catalyze the shared green infrastructure vision. The first challenge was to decide where to intervene for the greatest effect on the health of the whole. Students used the intersections of the five planning actions as the basis of their designs, imagining projects that achieved

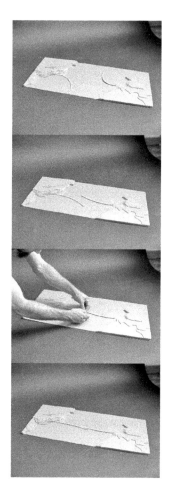

FIGURE 12.25 Student work by Dawn Sharpnack of the University of Oregon.

FIGURE 12.26 Mark Johnson giving a field crit to students.

multiple benefits. Students recorded their designs in three-dimensional models and drawings in their field guides. They also tested their designs at full scale with mock-ups and layouts using survey tape and flags (Figure 12.27). Their final designs, some examples of which are illustrated in Figures 12.28 to 12.30, were reviewed by professionals and community members at Georgetown's favorite pizza restaurant.

The Field School was a learning experience for neighborhood residents as well. For some folks, the sight of twenty "urbanauts" descending on their streets and leftover spaces with sketchbooks, water bottles, and digital cameras was surprising, even alarming, at first. But as the point of their expedition became clear, many residents seemed to appreciate the student presence and maybe see their own communities with new eyes because of the students' curiosity and enthusiasm. A documentary filmmaker, Patricia O'Brien, recognized that the brief experiences of the students were in microcosm similar to those of many Duwamish community residents as they first discover and then grow to care for the river, so she asked if she could accompany us for the second summer of the Field School. She recorded the poems, the motel frustrations, the wildlife and neighbor encounters, the field

FIGURE 12.27 A full scale mock-up by Phoebe Bogert of the University of Washington.

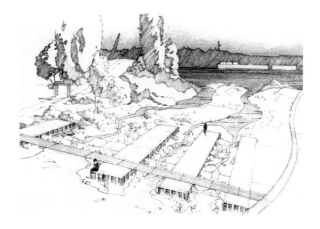

FIGURE 12.28 Student work by Nathan Hilmer of the University of Oregon.

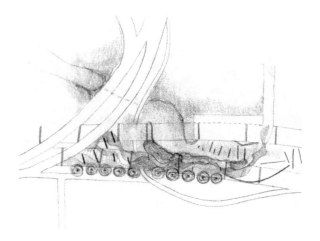

FIGURE 12.29 Student work by Dottie Faris of the University of Washington.

FIGURE 12.30 Student work by Nopporn Kichinan of the University of Oregon.

crits, the reviews, the beers, the pizza, all of it. Then she blended that footage with river scenes of industrial and natural beauty to create a one-hour film called "Duwamish: A River Transformed." She has been showing the film at festivals and gatherings to inspire environmental engagement in other communities.

THE RIVER ROLLS ON …

These were just three stories of community-based efforts along the Duwamish River and in its neighborhoods. There is a lot more happening—a park to share indigenous creation stories, a park dedicated to Cesar Chavez, a River City Skatepark, a Duwamish tribal longhouse, a Living Barge covered in native plants, habitat restorations, a river clean-up coalition, murals, trails, interpretive signs, and countless ephemeral events that demonstrate pride of place and an enduring connection to the river.

The community-building that accompanies these projects is as important as their physical outcomes. Green infrastructure can grow to accommodate all kinds of needs and dreams, but only with committed stewards to care for and animate it. In neighborhoods with many competing priorities, stewardship is cultivated by creating opportunities for diverse contributions toward a shared environment. When you do that, anything is possible with passion and perseverance. One final note—don't forget to really celebrate each achievement. This is when you can recognize the contributions of the team, attract new recruits, and scheme about the next project!

LITERATURE CITED

Benedict, M. and E. McMahon. 2002. *Green Infrastructure: Smart Conservation for the 21st Century.* Washington, D.C.: Sprawl Watch Clearinghouse, Monograph Series. pp. 5–6. Website: http://www .sprawlwatch.org/greeninfrastructure.pdf.

Brady, A., D. Brake and C. Starks. 2001. *The Green Infrastructure Guide: Planning for a Healthy Urban and Community Forest.* Princeton, NJ: The Regional Planning Partnership. pp. 11–12. Website: http://www .planningpartners.org/projects/gig.

Green/Duwamish and Central Puget Sound Watershed Water Resource Inventory Area 9 (WRIA 9) Steering Committee. 2005. Salmon Habitat Plan—Making Our Watershed Fit for a King. Prepared for the WRIA 9 Forum. Website: http://dnr.metrokc.gov/Wrias/9/HabitatPlan.htm.

United States Environmental Protection Agency. 2001. National Priorities List. Website: http://www.epa.gov/ superfund/sites/npl/nar1622.htm.

Wulkan, B., S. Tilley and T. Droscher. 2003. *Natural Approaches to Stormwater Management: Low Impact Development in Puget Sound.* Olympia, Washington: Puget Sound Action Team. Web site: http://www.psat.wa.gov/Publications/LID_studies/lid_natural_approaches.pdf.

13 Residential Street Design with Watersheds in Mind
Toward Ecological Streets

Stephanie Hurley and Megan Wilson Stromberg

CONTENTS

ABSTRACT

In the United States today, most city centers and the outer fringes of nearly all cities have been organized and developed to prioritize cars. Within cities, this pattern of urbanization, typically a matrix of impervious surfaces, generates diffuse sources of water pollution, flowing off the land in the form of stormwater runoff. The associated and well-documented consequences of these runoff patterns include: intensified runoff volumes and flooding, water resource contamination, overloading of combined sewer systems, reductions in groundwater recharge, and destruction of aquatic habitats.

Recognizing that streetscapes are corridors for stormwater, a new cohort of designers has embraced the challenge of water on the streets in exploration of ecological and aesthetic solutions for the stormwater problem. In this chapter, we review street design literature, discussing myriad design features that have been historically associated with "great streets." With this foundation, we explore the neighborhood context and elucidate the benefits of bringing stormwater to the surface to convert conventional streets to "ecological streets." Using precedents from residential Seattle, we propose reconceiving of the roles residential streets can have in shaping urban ecologies and communities. We argue that by departing from standardized street design guidelines, designers can provide attractive, community-focused neighborhood street design alternatives that also begin to restore natural hydrologic functions in urbanized watersheds, while increasing people's understanding of natural processes. We have found that—in addition to satisfying street design and stormwater goals—ecological street design can promote increased ecological understanding, build community identity, and celebrate an array of alternative urban ecological aesthetics.

INTRODUCTION

Street networks significantly shape economic, social, and natural processes in cities. Streets are ubiquitous in urban landscapes. They comprise up to one-quarter of the land in most cities (Richman 1999) and provide familiar backdrops for the activities of urban dwellers across the globe. In the United States today, most city centers and the outer fringes of nearly *all* cities have been planned and developed for cars. City planners have arranged public and private properties and rights-of-way under the assumption that social and economic growth will continue in the presence of automobiles. Although car-dominated cities have produced a spate of negative social and ecological consequences, cars persist at the top of the urban design hierarchy. Given that the automobile lifestyle is likely to remain prevalent throughout the coming century, we believe that it is the duty of contemporary planners and designers to acknowledge the environmental and social downfalls of cars and seek to mollify them through design.

For those who are interested in sustainable development and landscape restoration, urban streets—in their abundance and environmental manifestations—present formidable challenges. As such, we believe they are ideal places to begin the effort toward urban watershed restoration. Among the environmental concerns elicited from street ubiquity is the fact that streets—due to their largely impervious makeup—generate stormwater runoff. The authors of much of the street design literature of the past have been able to focus on terrestrial design issues, such as symmetry, access, and sight distance (Moughtin 2003, Jacobs 1996, Lynch 1971), but avoid the design issues pertaining to aquatic systems to which streets have now been linked by underground pipe networks. With the aid of conventional engineering, rainfall has been directed swiftly out of sight and into the gutters (street designers have effectively kept their minds out of the gutters). However, it is now widely known that traditional stormwater engineering has given rise to problems in terms of environmental quality. In response, a new cohort of designers has embraced the challenge of water on the streets in exploration of new ecological and aesthetic solutions for the stormwater problem. Among the leaders of this movement, designers in Seattle, Washington, have undertaken the goal of reducing the water quality and hydrologic effects of runoff to improve the ecological integrity of the waters in urban areas—starting in the streets.

Street rights-of-way offer design prospects that exceed traditional goals for transportation and infrastructure and become canvases for place-making and environmental education. The incorporation of stormwater management systems into the streetscape is an opportunity to merge the goals of street design and stormwater management and to rejuvenate the ecological processes, social fabric, and beauty of neighborhoods. Rethinking the linear landscapes of streets, which not only house myriad infrastructural elements upon which quality of life depends, but structure our experiences of cities and towns, is a necessary step toward watershed-supportive landscape restoration. In this chapter, we discuss the many design features that comprise great streets, and propose bringing stormwater to the surface to convert them to "ecological streets." We have found that—beyond street design and stormwater goals—ecological street design can promote increased ecological understanding, build community identity, and celebrate an array of alternative urban ecological aesthetics. Using precedents from residential Seattle, we propose reconceiving of the roles residential streets can have in shaping urban ecologies and communities. In doing so, we tackle design issues pertaining to street design, stormwater management, and livability, exploring the synergies between them within the public right-of-way.

STREET DESIGN + WATER = NATURAL DRAINAGE SYSTEMS

The notion of ecological street design is seemingly paradoxical. There is a wealth of urban planning literature on street design, but potential ecological benefits of these public right-of-way manipulations have rarely been identified. Yet, in practice and in theory, the following concept has recently emerged: streets can be designed to ameliorate the hydrologic problems they produce. Here, we briefly survey the common themes of the prevailing street design literature. We next review the basics of design for solving stormwater problems in urban residential areas and describe the arena where street design and stormwater management intersect in Seattle, Washington.

On Street Design

Street design literature calls for numerous criteria to be met in order to create desirable streetscapes. Numerous authors have offered up comprehensive recommendations for urban streets (Burden 1999, Jacobs 1996, Kostof 1992, Moudon & Unterman 1991, Anderson 1978). Although these authors itemize their respective criteria in detail, essentially, their recommendations fall into the following categories: mobility, safety, aesthetics, and community interaction.

Mobility refers to multimodal travel needs; the best streets are designed to allow access for all types of people to the street itself and to adjacent land uses. "Walkability" or pedestrian-friendliness has been highlighted in much of contemporary literature as a prerequisite for a street design success. *Safety* on streets goes beyond avoiding risks of traffic accidents; safety issues also include visibility, traffic calming, accessibility for emergency vehicles, and concerns for criminal activity. *Aesthetics* in the streetscape encompass numerous subcategories of design, including scale, definition, transparency, materials, and "qualities that engage the eyes" (Jacobs 1996). *Community interaction* refers to the function of streetscapes as backdrops for social interaction and commercial activity; again, the literature refers to designing for convenient and comfortable pedestrian movement. Long before they came to be auto-dominated, streets have fostered human existence in cities: "Streets are networked extensions of our brains; our use of them leaves traces of the social fabric woven through impressions of daily lives" (Lyndon 1997). Design and planning decisions shape our experiences of urban streetscapes. The potential influence of designers and planners upon the complex intersections of infrastructural networks, movement corridors, and community members' lives should not be overlooked.

Together, the four design components described above (mobility, safety, aesthetics, and community interaction) comprise a nearly complete list of essential goals for any urban street design agenda. But we believe one design consideration is missing: water. Highlighting streets as places for achieving multifaceted design goals, we add *stormwater management* to the list of necessary goals for successful street design.

On Water in City Landscapes

Water has been a central element in urban design throughout history. Incorporating stormwater into the streetscape, through beautiful and deliberate design, may be the latest trend in making water features a part of public life. Storytelling and imagery related to water invokes associations of cleanliness, rebirth, washing away of sins, bounty and life, fear and dread. It is also a unifying element; as a design tool water features can be used to draw literal and figurative connections within the landscape. "Water is the mirror of the world ... it disturbs present certainties and extends the possibilities of a brighter and better future" (Moughtin 2003). As evidenced by trips through city streets on rainy days, *stormwater,* via urbanization, has evolved as the contemporary urban "water feature." The interplay of stormwater and streetscapes (whether the result is positive or negative) mirrors cultural values about the role of natural processes play in cities.

ON STORMWATER MANAGEMENT ISSUES, GOALS, AND PRACTICES

Throughout centuries of urbanization, water has taken on meanings beyond beauty and symbolism in cities. Urban form is now characterized by impervious surfaces; the natural tendency of water to circulate from atmosphere to soils and back has been thwarted by sealing permeable landscapes with impermeable materials. Water is thus perceived as an element that requires conveyance and control. Accordingly, the disciplines of civil engineering and city planning have evolved methods for redistributing rainfall from the source where it falls to other parts of the landscape. They have effectively turned hydrology into hydraulics. Although these methods have solved some conveyance, flooding, and public health problems, urban pipe networks have also detrimentally altered natural water flows in the city: runoff no longer recharges the groundwater table, but courses at high velocities in pipes to receiving waters, often causing end-of-pipe erosion. In addition to compromised hydrology, stormwater runoff, laden with a toxic slew of contaminants—from heavy metals to oils to herbicides—threatens the waterways into which it flows. In response to these hydrology and water quality problems, many civil engineers and landscape architects are now exploring alternatives to conventional engineering based upon the use of soils and vegetation to attenuate and filter stormwater flows. Well-known contributors to this body of knowledge are the Low Impact Development (LID 2002) Center in Maryland, "Start at the Source" guidelines in the Bay Area of California (BASMAA 1999), Portland Oregon's Bureau of Environmental Services and METRO (2002) "Green Streets" programs, and Seattle's Natural Drainage Systems program (COS 2001), the latter of which will be discussed in detail throughout this chapter.

The techniques employed by these and other groups are often categorized as nonstructural and structural "Best Management Practices" (BMPs); they can be applied to both new and retrofit developments. Nonstructural methods for dealing with urban stormwater include policies, programs, and plans to control development practices and reduce impervious surfaces. In terms of structural BMPs, the common physical design elements are detention and retention ponds, constructed wetland systems, grass bioswales, sand filters, infiltration trenches, filter strips, and porous and pervious materials (Horner et al. 2001). In addition, "bioretention" swales can be used to soak up stormwater close to where it hits the pavement. Bioretention swales, also known as "raingardens" are essentially small ecosystems; they can be comprised of engineered soils or amended natural soils and planted with assemblages of water-tolerant plant species. Because of their flexibility in terms of shape and components, bioretention swales have emerged for designers as a promising functional landscape form.

THE NATURAL DRAINAGE SYSTEMS CONCEPT

Today, by merging street design and stormwater management practices, numerous municipalities have begun to explore ways in which water can be brought to the surface to enhance the ecological function of urban watersheds. The City of Seattle offers particularly promising examples of the movement toward ecological street networks. Within the confines of residential public street rights-of-way, stormwater treatment mechanisms have come to be artistic urban forms, demonstrating innovative street design. Starting with a year 2000 pilot project, Seattle Public Utilities (SPU) has adopted the term Natural Drainage Systems (NDS) to describe their program to implement bioretention in public rights-of-way. The SEA Street project (SEA = Street Edge Alternatives) has

FIGURE 13.1 SEA Street (Street Edge Alternatives) in Seattle has a narrowed meandering roadway, sheet flow over flat curbs, and bioretention swales within the public right-of-way. Photo by S. Hurley.

dimensions of approximately 1200×60 ft, spanning one long block of residential street in Seattle's Pipers Creek watershed (Figure 13.1). The combination of roadway narrowing, topography, soils, and vegetation attenuates virtually 100% of the runoff falling in the street's catchment (Horner et al. 2002, 2004). It is a public place where street design and stormwater goals are met simultaneously.

PRECEDENTS

Six years after launching the NDS program at SEA Street, two larger projects demonstrate the evolution of Seattle's NDS designs. The latest projects illustrate how the city has achieved success in melding street design goals—such as walkability, traffic calming, community revitalization, and aesthetic improvement—with ecological goals, including improved hydrology, infiltration, habitat, and water quality improvements. The first project is Broadview Green Grid, a retrofit of existing neighborhood streets in the Pipers Creek Watershed in northwest Seattle. Building on the success of SEA Street, construction of Broadview Green Grid was completed in 2004. Next, Seattle Public Utilities (SPU) followed Broadview Green Grid with the High Point housing project in the Longfellow Creek Watershed of West Seattle. With Phase I completed in 2006 and phase II underway, this project serves as a model for the incorporation of natural drainage principles from the inception of a new community development.

Both projects utilize innovative drainage systems to enhance traditional street models. Both projects target important salmon-bearing watersheds and aim to improve hydrology and water quality and, ultimately, to reinstate watershed conditions that more closely resemble predevelopment conditions. Both serve street design functions, attenuate stormwater, reveal ecological processes, and enhance place identity. Yet, these two projects aim to satisfy ecological and community goals by employing design forms that differ significantly from each other. The Broadview Green Grid and High Point projects demonstrate that there is no single template for creating an ecological street and underscore opportunities for design innovation. These residential Seattle street projects illustrate that multiple goals—e.g., for transportation, safety, hydrology, habitat, and community building—can be met within the streetscape; they offer hope for reshaping the future of urban watersheds.

Broadview Green Grid

With a catchment area of 32 acres (i.e., the area that drains to a single outfall pipe on the creek), Broadview Green Grid comprises about 2% of the Pipers Creek watershed and nearly an entire sub-basin. One SPU engineer describes the project as an opportunity to do a science experiment, a rare chance to monitor a large sub-basin both before and after retrofit (Tackett 2006). Baseline data of flow and pollutant levels were taken before project construction, and the University of Washington is scheduled to begin formal project monitoring in the rain season beginning in fall 2006 (Horner 2006).

Prior to construction, the drainage infrastructure at Broadview Green Grid was an informal ditch and culvert system. Many of the ditches were paved in the past to speed conveyance of storm-water, which resulted in flooding problems as well as environmental impacts including greater conveyance of pollutants and heavy scouring and erosion in the creek downstream. Over the years, as homeowners converted ditches to culverts and overlaid them with various paving materials, each block became a patchwork of pavement, gravel, bare dirt, ditches, and driveways at the road edge (Figure 13.2). Parking was also disorganized and pedestrians were relegated to walking in the street. In response to these conditions, the retrofitted drainage network utilizes a suite of natural drainage strategies including cascading swales with check dams on steep streets and vegetated bioretention swales along more shallow-sloped roadways.

Seattle Public Utilities teamed with the Seattle Department of Transportation to incorporate traffic-calming elements and pedestrian improvements with modifications to enhance drainage function. For this project, shallow sloped streets were retrofitted with meandering road surfaces with swales on both side and sidewalk on one side. The curving lines of the roadway allow for maximization of the size of the swales, which can be fit into the curves (Figure 13.3). The curving form also allows for a reduction of impervious surfaces and traffic calming. All but one of the retrofitted blocks saw a reduction in impervious area between 15 and 34% (Tackett 2006). Designers opted to narrow the paved widths of streets to as little as 18 ft in some places. To continue to provide emergency vehicle access, the street edges adjacent to the swale include a 2-ft wide flat curb on one side and a mountable 2-in. high curb at the edge of the sidewalk, totaling a minimum 25 ft of drivable surface. Streets are canted away from the mountable curb to sheet flow across the flat curb and into

FIGURE 13.2 A residential street in northwest Seattle is comprised of a patchwork of impervious surfaces with informal ditch and culvert drainage. Photo by M. Wilson Stromberg.

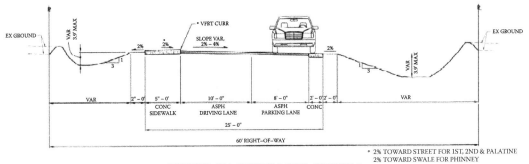

TYPICAL SEA STREET R/W X–SECTION
(1ST AVE NW, 2ND AVE NW, PALATINE AVE N & PHINNEY AVE N)
NTS

FIGURE 13.3 A typical section view of north–south streets in Broadview Green Grid. Bioretention swales flank roadways with widths as narrow as 18 ft across. Courtesy of Seattle Public Utilities.

the adjacent vegetated swale. The flat curb also serves to demark the roadway and organize parking along the road edge.

The vegetated bioretention swales (at a maximum of 3.9 ft deep) are designed to infiltrate within 72 h to eliminate the possibility of a mosquito vector. Underlying soils are either compost-enhanced topsoil or an engineered soil with high porosity, increasing the soil water storage capacity. As stormwater fills a swale and it reaches capacity, water can enter the culvert and flow into the next swale in the system. The plantings are mostly native species, selected and sited according to tolerance to water level fluctuation, micro-topographical variations, and input from residents (Figure 13.4).

Cascading swales (maximum of 4 ft deep) are designed to slow the stormwater as it flows down the steep east-to-west streets, maximizing infiltration along the way. To overcome the steep longitudinal gradient, designers included check dams (Figure 13.5). Two different methods were employed to address the limited space in the right-of-way by allowing for steeper side-slopes along

FIGURE 13.4 Broadview Green Grid plants were selected with inputs from residents. Photo by S. Hurley.

FIGURE 13.5 Check Dams on NW 107th Street in Broadview Green Grid increase retention time of storm-water on steep slopes (shown during construction). Photo by S. Hurley.

the cascades. Designers used "soil wrap" walls on NW 110th Street, set within the bioretention cell's banks to soften the edges, control erosion, and create a creek-like aesthetic (Figure 13.6). These are clay blocks, wrapped in UV-resistant fabric, within which native wetland and riparian plants are planted. Alternatively, rockery walls were used on NW 107th St to reinforce side-slopes, while creating deeper swale cells (Figure 13.7). Ironically, the willows planted along the soil wrap walls have grown so vigorously that they have become a safety concern. According to project manager Jim Johnson, the willows disguised the cascade pools so much that the city and residents became

FIGURE 13.6 Willows were planted into "soil-wrap" walls that line the swale systems along NW 110th Street. Photo by S. Hurley.

FIGURE 13.7 Rockery walls were used to reinforce the swale side-slopes along NW 107th Street in Broadview Green Grid. Photo by S. Hurley.

concerned that someone might unknowingly stumble into the swale, thinking it was much shallower than it actually is. The decision was made to cut out the willows to enhance the visibility of the grade change. Johnson ultimately felt that the cascades with rockery walls rather than soil wrap walls received greater acceptance due to their neater appearance and visibility (Johnson 2006).

HIGH POINT

High Point in West Seattle is a joint project between Seattle Public Utilities and the Seattle Housing Authority, funded in part by Hope VI low-income housing dollars. The funding source has specific implications for both ecology and aesthetics. Proponents seized this redevelopment opportunity as an opportunity for natural drainage systems designs by using a Hope VI clause that "mandates that public housing projects must resemble their surrounding residences so that those living in public housing are not physically or perceptually isolated from their neighbors" (Hill 2003, p. 40). This clause allowed for the High Point design to include narrower roadways which resembled the older Seattle residential street grid in the adjacent neighborhood. Overall, High Point contains a 120-acre housing project, with stormwater drainage consisting of gravel-lined pocket parks, compost-amended

FIGURE 13.8 Plan for the High Point Project. Courtesy of Seattle Public Utilities, Seattle Housing Authority, SVR Design Company.

soils beneath lawns, a stormwater pond, and a street network of vegetated bioretention and grass swales (Figure 13.8). The project comprises approximately 10% (or 140 acres) of the Longfellow Creek watershed. Like Pipers Creek, Longfellow is a salmon-bearing stream; it sees more adult salmon returning from the ocean than any other Seattle creek (Hill 2003). The project could be

FIGURE 13.9 High Point "deep swales." Photo by S. Hurley.

classified as a retrofit because this is not the first time the land is being developed: the project's 1600 mixed-income housing units plus a grocery store, library, neighborhood center, and public parks, was previously a 716-unit residential tract (Kirschbaum 2004). Yet, in terms of re-use of the former street pattern, designers and planners essentially started from scratch.

In part due to the Hope VI clause, designers chose to design streets with a typical residential aesthetic. Streets are aligned to be straight with curbs, sidewalks, and planting strips with grass and trees, yet slight modifications from the norm allow for a higher standard of hydrologic function. Two types of natural drainage systems have been applied in the right-of-way for different zones of the project: so-called "deep swales" and "shallow swales." The deep swales (12 in. maximum) are vegetated with native plants (Figure 13.9), similar in species type to those at Broadview Green Grid. Shallow swales (2 in. maximum) are vegetated with grass turf and street trees, resembling conventional street planting strips (Figure 13.10). Though curbs will separate the roadway from the roadside vegetation and sidewalks for both swale types, curb cuts conduct street runoff through the

FIGURE 13.10 High Point "shallow swales." Photo by S. Hurley.

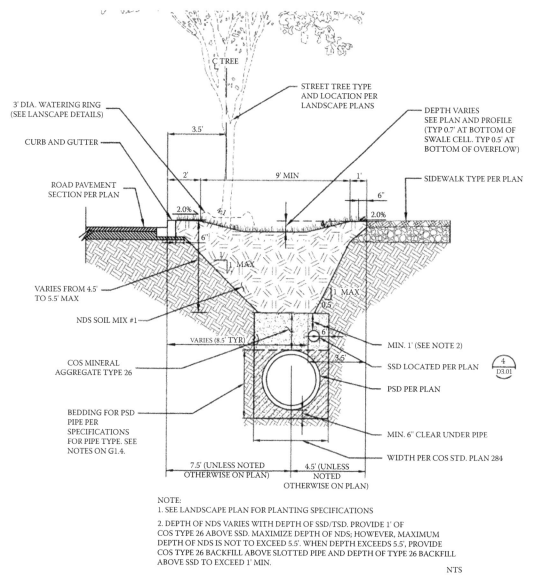

Shallow Grasslined Retention Swale Section W/NDS Soil Mix #1

FIGURE 13.11 Typical section drawing of Shallow Retention Swale with subsurface stormwater storage. Courtesy of Seattle Public Utilities, Seattle Housing Authority, SVR Design Company.

curb and into the natural drainage elements. Sidewalks are made from porous materials through which runoff will enter the swale system.

A key difference from the Broadview Green Grid project is that the shallow swale system at High Point offers a comparatively small amount of surface storage for stormwater. The shallow grass swales allow for only up to 2 in. of ponding water, whereas much of the hydrologic and water quality objectives are met beneath the surface via structural soils (Figure 13.11). The grass surfaces that cover the subsurface infiltration areas also serve as walkable and playable surfaces, a high priority in this relatively dense neighborhood (Maupin 2004). The engineered soil absorbs and treats stormwater; in high flows, overflow drains into a perforated pipe to be conveyed away. However, an

TABLE 13.1
Projects At a Glance

Broadview Green Grid	High Point
WATERSHED: 32-acre catchment in Pipers Creek Watershed.	WATERSHED: 140-acre catchment in Longfellow Creek Watershed.
LAND USE: Single family residential homes on 15 residential blocks covering approximately 13 acres.	LAND USE: 1600 mixed income homes including town homes and apartments in 120-acre development.
DRAINAGE PROGRAM: Retrofit of existing street network and informal ditch and culvert drainage to include vegetated bioretention swales, a narrowed roadway, on-street street parking, and sidewalks on one side.	DRAINAGE PROGRAM: Redeveloped parcel with new street layout incorporating porous pavement and pavers in parking areas and sidewalks, vegetated and grass swales, detention pond, and infiltration parks, rooftop rainwater dispersal.
MAINTENANCE: City provides one year of maintenance. Residents agree to maintain vegetation thereafter.	MAINTENANCE: Homeowner's association is responsible for planting strips including swales and the city is responsible for hardscape.
ART AND EDUCATION: Public meetings during the design and construction. Access to designers. Series of newsletters. Homeowner maintenance guide.	ART AND EDUCATION: Community workshops, graphic handouts, signage, interpretive public art.

explicit objective of the project's planners was to avoid using the perforated pipe system altogether (Kirschbaum 2004). In other words, drainage functions will be met through a combination of infiltration into the engineered soils of the subsurface natural systems for storage, exfiltration into the adjacent native soils, and absorption by compost-amended surface soils throughout the project's landscaping, rendering the overflow perforated pipes nearly obsolete. Current models predict that the High Point design will meet Seattle's performance goal of creek protection, which is accomplished by targeting frequent low-intensity storms that "recent science has identified as critical to fish health downstream" (Hill 2003).

Table 13.1 offers a summary comparison of the Broadview Green Grid and High Point projects.

BEYOND TREATMENT: WHY REVEAL RUNOFF IN THE RIGHT-OF-WAY?

Bringing stormwater to the surface within the right-of-way has functional and cultural implications. Revealing runoff can reinforce the cultural significance of natural processes, enhance ecological functions, promote environmental stewardship, and foster neighborhood identity and community participation.

Well-known requirements for great urban streets include: accessibility, bringing people together, *publicness*, livability, safety, comfort, participation, responsibility, and "magic" (Jacobs 1996). These are not precluded when ecological function is added to the list. Rather, these criteria are reconceived and incorporated into the ecological street concept. The introduction of vegetation and waterways into the streetscape is a way to target environmental goals, but also enriches the human experience and augments livability. Ecological streets that add opportunities for personal expression and environmental stewardship enliven neighborhoods, distinguishing them from the potentially repetitive street grid. In cities (such as Seattle) where citizens exhibit awareness and concern about environmental problems, the collective consciousness can be translated into the physical environment of daily lives via the streetscape. Philosophically, the idea that humans can take responsibility for the negative consequences of development practices by adopting natural drainage design strategies resonates with environmentally-minded citizenry.

Although spatial constraints may initially seem prohibitive, stormwater management can be accommodated within the right-of-way. The High Point and Broadview Green Grid projects were constrained to rights-of-way widths of 60 ft or less. Cleverly diverging from street standards without compromising access or safety, designers placed stormwater infrastructure into the streetscape. As streets comprise up to one-quarter of the land area in many cities, the potential cumulative benefits of retrofitting streets is far-reaching. As mentioned earlier, designers reduced impervious surfaces on all but one of the Broadview Green Grid streets by 15 to 34% (a figure which accounts for the area devoted to new sidewalks).

With the addition of green space in the form of vegetation-based stormwater treatment, pressure on strained infrastructure can be relieved and neighborhoods can be enlivened. The Broadview Green Grid project was catalyzed by flooding problems in an infrastructure system that was stretched beyond its limits (Tackett 2006). SEA Street, by contrast, was spurred by enthusiastic residents desiring sidewalks and formalized street architecture. SEA Street, now over 5 years old, has become a local destination for pedestrians. Although it is only one (1200 ft.) block long, local residents have noted increased foot-traffic. The neighborhood residents are likely attracted to its park-like character, comfortable pedestrian environment, and calm traffic. Just as designers and planners make decisions to allocate space in the right-of-way for pedestrian amenities, a similar process should be considered to create space for water in the streetscape.

As part of the city fabric, retrofits of urban streets can provide the necessary momentum for connecting green open spaces, introducing recreational spaces, and controlling noise and air pollution (Moudon and Untermann 1991). Retrofits and upgrades are opportunities to re-think the streetscape and apply stormwater friendly design alternatives. When modifications to existing transportation and drainage infrastructure—such as construction of new sidewalks or bike lanes or maintenance of storm sewer networks—are required, there are opportunities to consider an ecological streets approach.

Ecological Understanding

The current state of the urban environment is one that illustrates and promotes a disconnection between human action and ecological consequences. A closer look at the modern city reveals that nature is largely invisible or at most, confined to small pockets. We may even believe that ecological processes do not occur in the city. Yet, hidden beneath the façade of concrete and metal, ecological processes are present even in the most urban of places. The processes look different—water moves in pipes instead of streams, trees are dispersed by humans instead of wind or birds, constructed materials decay in the place of plant and animal debris—but these ecological processes are working within urban grids, just as in the preserved or pastoral open spaces where nature is more readily apparent. Therefore, we call upon designers to join the ranks of the aforementioned local government leaders in natural drainage street design and work to strategically educate the public, to raise cognizance about the ecology of landscapes in which we dwell. (Although not explicitly street designers, landscape architects such as Robert Murase, Herbert Dreiseitl, and William Wenk have already designed numerous projects in public spaces that aim to achieve these types of goals.)

It is noteworthy that many urban places that are perceived to be "natural areas" exhibit dysfunctional ecology. That which we may conceive as natural in cities may be only superficial or symbolic in terms of ecological functions, e.g. water quality, healthy soils, persistence of native flora, and wildlife habitat. Structurally and functionally, patches of urban nature do not cumulatively comprise complete ecosystems. Although living closer to nature may be the theory underlying the expansion of development in cities and suburbs, such developments frequently incorporate only the "trappings of nature like trees, gardens, and lakes, but are built with little regard for the processes of nature" (Spirn 1984). The "pedigreed landscape of mown turf, flower beds, trees, fountains, and planned spaces" (Hough 1995) have now become the predominant forms of urban green spaces, manifested as plazas, parks, streets, parking lots, and yards. These pedigreed landscapes generally lack the ecological functions of native soils and vegetation, and further compromise local ecology by consuming more water and chemicals than natural ecosystems would require.

Meanwhile, in the cases where ecological functions *are* incorporated into urban landscapes—e.g., in the typical mandated stormwater management schemes of suburban developments—the "environmentally-friendly" designs are often expressed as unappealing forms such as fenced-in trapezoidal detention basins. Despite the application of natural processes to serve hydrological and water quality purposes, the engineered geometries of these man-made systems appear intrinsically "unnatural." Thus, both the acceptance of the dysfunctional pedigreed landscapes as "natural" and the rejection of (admittedly unattractive) water treatment projects as "unnatural" perpetuate a societal lack of ecological understanding. Ultimately, to promote ecological understanding, designers must create places that are both attractive and functional, appealing to our senses as well as our social and environmental ethics. In this way, ecoliteracy may increase. "Being *ecoliterate* means understanding the principles of organization of ecological communities (ecosystems) and using those principles for creating sustainable human communities" (Capra 1996).

A near-synonym to ecoliteracy is the concept of "ecological understanding." Ecological understanding is "the expectation and awareness that human actions have consequences and that an intricate web of relationships connects patterns and processes in the physical, biological, and social environments" (Hill et al. 2002). To increase ecological understanding, "environmental education is essential in all its forms" (Nassauer 1997). One strategy for catalyzing ecological understanding is to design landscapes that protect or reveal ecological function (Nassauer 1997). Kevin Lynch's argument for spatial legibility of form is related to ecological legibility of forms: "It can be a means of extending one's knowledge of the world," he explains, and adds that legibility "confers the esthetic pleasure of sensing the relatedness of a complex thing" (1971).

Further, a recent movement in landscape architecture, called "eco-revelatory design," promotes the enhancement of ecological understanding through design. "Design has the capacity to make the invisible visible … our task is to comprehend patterns, divine meaning, and communicate understanding" (Helphand and Melnick 1998). True revelations are "moments of extreme clarity and insight with the potential to transform our consciousness and guide future action" (Helphand and Melnick 1998). Thus, in theory, promoting ecological understanding can motivate future actions by people who are better informed of the ecological consequences they set in motion.

The easiest approach to achieve legibility of urban ecological functions is to "expose infrastructure processes commonly hidden" (Brown et al. 1998). For instance, natural drainage systems convey stormwater on the surface of the landscape where we can see it flow from pavement into a swale. Broadview Green Grid achieves this transparency of ecological function. The water travels across the surface into bioretention swales (analogous to natural wetlands) before disappearing into the soil, the air, or a pipe. We see that the stormwater runoff is supporting the plants that inhabit the drainage system. We may also observe that the substances we deposit on the surface of our landscape like the oil from our cars and eroded soil from our yards, will be carried via the runoff into natural drainage system. We may begin to understand the connection between our actions and environmental consequences.

The incorporation of visual clues or "signifying features" of an underlying function is another strategy to provide ecological understanding (Brown et al. 1998). Visual clues and symbols can also include artistic treatment of infrastructural elements like the grates on curbs, sedimentation structures, and French drains. Necessary splash blocks from roof downspouts at the High Point project were shaped by a local artist to suggest the flow of the creek that lies at the bottom of the watershed (Figure 13.12). Artistic paving patterns are woven into intersections to indicate the flow path of stormwater. The High Point project also incorporates other sculptural elements, such as the "orca fence" around the detention basin and boulders sculpted with riparian insects suggesting the connection between the road, the roadside swale, and the creek or shoreline that could be miles away but is still intrinsically connected to the streetscape. Although it is subtle, the meandering pattern of the north-south Broadview Green Grid roadways hint at the connection between pavement and aquatic systems; the street design causes passersby to pause and observe the unique drainage of the neighborhood.

Direct education also plays an important role in enhancing ecological understanding. At High Point, interpretive signs explain the functions of the natural drainage systems with words

FIGURE 13.12 Downspouts on residences at High Point: Water splashes into sculptures that reference the creek downstream. Photo by Eric Higbee.

and symbols. After residents moved in, the city of Seattle provided educational outreach to the diverse community at High Point. With over 50 languages spoken in the neighborhood, the city's creek steward utilized graphic handouts to communicate information about the natural drainage systems (Maupin 2006). Public meetings played a role at High Point and at Broadview Green Grid. The project manager explains that during the design and planning process for the Broadview Green Grid city designers made themselves accessible to residents outside of meetings: phone interviews, parcel by parcel walk-throughs, and face-to-face meetings allowed designers to interact with each resident (Johnson 2006).

Inviting community members to take an active role in developing this partnership is an opportunity to teach a deeper understanding of ecological issues. In *Language of Landscape,* Spirn (1998) argues that isolation from nature has resulted in a loss of language and knowledge of the landscape limiting the "celebration of landscape as a partnership between people, place, and other life" and reducing "the capacity to understand and imagine possible human relationships with nonhuman nature." Project managers of the Broadview Green Grid project saw their efforts to engage residents pay off. All but one resident chose to go with a completely native planting palette in the natural drainage systems (Johnson 2006). Reaching out to the community there has resulted in resident satisfaction and comprehension of the overall ecological purpose behind the retrofit project. Participation can begin to restore understanding by involving residents in the development of the stories of the landscape and creation of metaphors about the processes occurring in the natural world.

The public streetscapes we share, ubiquitous in our daily lives, can become a backdrop for education. Environmental education can take place "in a neighborhood, a farm or a forest" … in any setting "where people are directly involved with a landscape …. Looking at the landscape as we go about our everyday travels, we constantly judge what we see and learn from it" (Nassauer 1997). Adding legibility to "everyday landscapes" contributes to the richness of experiences and raises ecological understanding within communities. Ecological street designs go beyond functionless symbolic imagery of nature. Done well, these streets have the potential to reconnect us with ecological functions in an aesthetically pleasing way and therefore to increase our ecological understanding and desire to sustain function and link us with the watersheds in which we live.

REGIONAL AND COMMUNITY IDENTITY

Ecological street design has implications beyond ecological enhancement. Natural drainage systems have the potential to alter the way in which we relate to the built environment. They provide the opportunity to increase ecological understanding and to enhance local and regional identity. By considering the cultural implications of infrastructure design and promoting ecological understanding, designers can help reveal connections between humans and the urban ecosystems they inhabit.

It is now common to hear community members express dismay that their towns and cities are taking on the look of "Anywhere, U.S.A." or "Generica." All too often, as Spirn explained in *The Granite Garden* (1984), "the potential of the natural environment to contribute to a distinctive, memorable, and symbolic urban form is unrecognized and forfeited." A culturally influenced character has replaced qualities associated with ecological integrity in the natural landscape.

A common example of a "culturally influenced" landscape that is strewn throughout many cities is the public park. Urban parks largely reflect the singular concept of the pastoral landscape, a cultural remnant of the industrial revolution when cities became unhealthful and parks were integrated as an "escape to the countryside from the smoke and stink" (Spirn, *in* Meyer 2001). Though elegant and ecological exceptions exist, many parks are designed more to achieve a particular standardized look rather than in response to the natural environment. These landscapes are generally groomed to remain static over time, inherently in conflict with plant growth patterns. Accordingly biodiversity is rare within them. Such places ignore natural processes and do little to raise ecological awareness. The appearance of these generic landscapes also reflects little variety from place to place, even though the regional ecology and cultural dynamics may vary greatly. Common urban streetscapes—with curb and gutter drainage, perhaps a grass strip, and row of street trees—are linear analogues to the "grass–tree–bench ecosystem" (Forman 2006) of standardized parks; their designs similarly lack reference to regional ecology and community dynamics.

Ecological streets may be the key to re-envisioning our urban public spaces and designing with the idiosyncrasies of the site in mind. The High Point project offers examples of designers logically straying from restrictive street standards, fire codes, and cultural expectations, while still achieving all the goals the standards had been developed to target. For instance, according to the Seattle Public Utilities planner Maupin (2006) planting strip bulbs allowed designers to save all but one tree at High Point. The curb bulbs (or chicanes) and narrow street widths were a result of negotiations between the Department of Transportation and Seattle Public Utilities, demonstrating the creativity of designers. Designers argued to reduce impervious street pavement widths to 25 ft, rather than the Seattle standard of 32 ft for this type of residential street. This was permitted on the basis that Hope VI projects should blend in with the surrounding neighborhood aesthetic. The early 20th century neighborhoods in the vicinity of High Point were designed with narrow paving. They function in this century as "queuing streets," a single lane, two-direction street with parking on both sides, where drivers pull over to let each other pass. After similar negotiations, the Broadview Green Grid project designers specified mountable and flat curbs to keep the roadway narrow but the access wide. Ultimately, the performance goal of emergency access was met for each street without the use of conventional standards. The meandering streets at Broadview Green Grid and the large trees that were saved at High Point also visually distinguish the neighborhoods from standardized developments.

Regional icons can also be the motivating factors around which communities rally. For example, the listing of several salmon populations that Seattle hosts as "threatened" under the Endangered Species Act has both ecological and cultural implications. There are also political implications, as this makes Seattle "and its region (along with Portland, Oregon) the first urban test case for that federal law" (Hill 2003, p. 39). Salmon are, and have been since those times when native tribes were the predominant human residents, Seattle's mascot species. Countless constituents, from tribes to government agencies to "Friends of" groups and nonprofit organizations, advocate for them. Although natural systems street designs "seem likely to provide the improvements in both water quality and stream flow levels needed to keep salmon swimming in urban creeks" (Hill 2003, p. 39), environmental education, particularly with regard to standard urbanization practices, is still essential for ecosystem survival.

Native plants can be elevated to the status of regional icons on ecological streets; likewise, they can embody a neighborhood's commitment to environmental stewardship. At Broadview Green Grid, where all but one resident chose a native plant palette, residents band together once a year at block parties to help each other maintain the natural drainage systems. Assigning a deeper significance to the plant palette, the Broadview project has not only improved hydrologic function and habitat but has created a community of creek stewards, strengthening residents' ties to the creek.

Raising our ecological understanding of places through design of ecological streets, designers can enrich streets to reconnect human communities with their local ecological communities. Such connections serve to foster a unique sense of place and identity based in native species and local ecological patterns. Recognizing patterns in the landscape and assigning meaning to them (e.g., by incorporating local native species and patterns that reflect ecological processes into urban streetscapes) can provide a basis for sense of place and regional identity in a largely homogenous urban landscape.

WHEN ECOLOGICAL DESIGN IS TAKEN TO THE STREETS: THE ROLE OF DESIGNERS AND AESTHETIC EXPLORATIONS

Sustainable landscapes need conspicuous expression and visible interpretation, and that is where the creative and artistic skills of the landscape architect are most critically needed.

—Robert Thayer (*in* France 2003)

It is now clear that designers can play an active role in bringing ecological streets to fruition. To connect people with the watersheds in which they live and with the human communities to which they belong, designers can utilize strategies that expose the processes of infrastructure, highlight regional icons, and provide visual clues to underlying functions. Through spatial arrangements, designers can also cultivate community participation as a means to provide education and experience and implant locally-based ecological signatures to foster a rich sense of place and identity.

For streetscapes, there is a spectrum of aesthetic options for natural drainage design. Seattle's Broadview Green Grid utilizes a naturalistic aesthetic that emulates a Pacific Northwest creek with rockery and dense plantings of native vegetation (Figure 13.13). The High Point project's plant palette and grading more closely resemble a typical residential neighborhood, though designers use art, interpretive elements, and symbolism to reveal ecological processes. High Point emulates the orderly aesthetic of conventional residential streets with straight alignment, vertical curbs, grass and trees; yet the streetscapes still accommodate for stormwater beneath the surface within the right-of-way (Figure 13.14). Presenting another aesthetic, the Broadview Green Grid streetscape varies from the norm of a conventional urban–suburban street corridor. Whether or not they immediately recognize the hydrologic function of the street, passersby are aware that the green grid is unique.

Public acceptance is crucial for ecological street design to gain momentum as a movement. Designers play an important role in this process. "By first being palatable, landscape aesthetics ultimately can go beyond the merely acceptable to evoke intelligent tending of the land so that aesthetic decisions become intrinsically ecological decisions" (Nassauer 1997). Designers can begin by simply asking residents what they like, as the city of Seattle did with the Broadview Green Grid planting palette. Mills' (2002) research in "Alternative Stormwater Design within the Public Right-of-Way: A Residential Preferences Study" indicates that residents in Seattle are accepting of a broad array of design alternatives. Residents in her study perceive the aesthetic and mobility-related benefits of ecological street improvements to be equal with those of conventional street improvements. Residents in all communities may not react in the same way, but Mills' findings—along with the noteworthy benefits to ecological function and affordability—justify the exploration of the aesthetic preferences of residents when undertaking alternative street design.

FIGURE 13.13 Broadview Green Grid's naturalistic streetscape with flat curbs, meandering roadway, rockery walls, and plants. Photo by S. Hurley.

Designers can generate stylish solutions for streetscapes that are in accordance with neighborhood preferences and the goals of the decision makers involved with each project. We must not think of natural drainage streets as having only one possible "look," but stretch our imaginations and plant palettes to explore and apply design options that are well suited to the sites in which they are installed. The curb–grass–swale design at the High Point project has proven that streets that incorporate natural drainage systems strategies do not have to change the overarching aesthetics of the urban environment. Natural drainage goals can still be achieved under the guise of orderly

FIGURE 13.14 High Point's orderly streetscape with vertical curbs, rectangular planting beds with subsurface storage, and a narrowed roadway that conforms to the existing street grid. Photo by S. Hurley.

FIGURE 13.15 Curb cuts allow stormwater to flow from roadway into vegetated swales at High Point. Photo by Eric Higbee.

or conventional aesthetics of symmetry with neat curbed edges, manicured grass planting strips, regularly spaced trees, and the wide straight roadway alignment that has become the norm for urban residential streets.

At High Point, the Hope VI clause that allowed designers to argue for narrow street widths (i.e., to match the context of the local neighborhood) hindered design innovation with regard to "alternative" aesthetics. Developers of the High Point project lobbied for conformance to the conventional aesthetic out of fears that alternative streets would be less marketable. Despite buyer and realtor focus group research showing that "buying green" was seen as an asset to the High Point project, the conventional aesthetic was seen as less risky (Herbert Research Group 2003). Thus, employing what Nassauer would call "cues to care," many swales are lined with clipped grass and framed by conventional curbs. The natural drainage modifications are aesthetically more subtle but still achieve stormwater goals. Maupin (2004) explains that curbs have cuts (Figure 13.15), planting strips are wider, and a portion of the swales are deeper and densely vegetated to improve overall hydrologic performance.

However, we also believe that a shift in aesthetic paradigms can take place through creatively and strategically taking advantage of opportunities to promote environmental education in the streetscape and by using design to enhance cultural meanings of ecological systems. There are many reasons to consider a new aesthetic for ecological streets. As we discussed earlier, breaking the mold of the conventional street has the potential to foster ecological understanding by making hydrologic functions more visible and forging symbolic connections with local habitats.

THREE AESTHETIC FRAMEWORKS

Ecological streets—accounting for community and ecology inasmuch as they account for economics and mobility—can be expressed in an infinite number of ways. Design choices depend on overarching project goals, spatial constraints, funding, and context. We therefore consider three very different types of urban residential street design aesthetics that might be used as templates for designers of natural drainage systems streets: the orderly, the naturalistic, and the expressive.

The "orderly" street design framework, inspired by the design for Seattle's High Point project, is one that might be seen in a typical urban setting (Figure 13.16). Street trees and grass are the

FIGURE 13.16 Orderly aesthetic Concept: Photo collage.

predominant vegetation, framing the streetscape and only subtly alluding to the fact that natural drainage strategies are being employed in the right-of-way. This aesthetic takes advantage of space available beneath the ground, relying on subsurface systems for stormwater detention and water quality improvements. The formal aesthetic—turf and trees—does not attempt to break the mold, hiding natural processes, and resembling the forms of conventional urban landscapes. Art and interpretive signs, however, enhance understanding of the underlying processes. We believe that this aesthetic has the potential to be broadly applied and accepted.

The "naturalistic" street design framework was inspired by Seattle's SEA Street and Broadview Green Grid (Figure 13.17). This "look" is characterized by native vegetation, arranged within bioretention swales according to microtopography. Plants are selected and placed based on their tolerances

FIGURE 13.17 Naturalistic aesthetic Concept: Photo collage.

FIGURE 13.18 The naturalistic aesthetic with edges kept trim, a sign the project is cared for. SEA Street. Photo by S. Hurley.

to water level fluctuation and local seasonal variations in climatic conditions. Depending on the plants selected and the proximity to patches of habitat within watersheds, this aesthetic also has the potential to form important habitat linkages within the street network. It is also important to note that the naturalistic model does not necessarily convey a wild (messy) appearance. Trimmed edges of vegetation clusters signify that the streetscape is cared for (Figure 13.18). Grading and drainage for curbless sheet flow of runoff—from hardscape to pervious areas—is optional for this aesthetic.

The third aesthetic framework we offer designers is "expressive" street design. It is inspired by several Natural Drainage Street prototypes including the Broadview Green Grid project and "rain-gardens" programs in Maplewood, Minnesota, in Kansas City, Missouri (2006), and elsewhere. In this case, designers establish an overarching spatial layout but neighbors are highly involved in the plant selection process and maintenance (Figure 13.19). In this concept, property lines are visually extended to the pavement edge, such that the public right-of-way edges take on the appearance of each privately owned parcel adjacent to the street. Much like a public community garden, or p-patch, neighbors are responsible for the care of the right-of-way plants nearest to their properties. In exchange for these maintenance contributions, neighbors have more control over the plant selection in their front yards. Employing the expressive aesthetic, designers can either provide a selected set of vegetation communities for neighbors to choose from (e.g., Maplewood), or allow virtually unrestricted plant choices but include caveats with regards to water tolerance, invasiveness, hardiness, and exclusion of fertilizers, herbicides, and pesticides. Over time, this highly interactive design approach may be the best option for achieving the additional goals of raising ecological awareness and fostering community identity with place.

EYES ON THE STREAM: FROM MAINTENANCE TO STEWARDSHIP

No matter what the aesthetic framework employed by designers may be, upkeep of the projects is essential to their hydrologic function and public perception. Though maintenance is a seemingly mundane issue, funding and ability and willingness to maintain the landscaped portions of ecological streets is a determinant of their success. From the outset, the ability to achieve proposed maintenance practices should inform the degree of complexity of the design. "Nearly all landscapes are judged and enjoyed according to the degree that they clearly exhibit care" (Nassauer 1997).

FIGURE 13.19 Expressive aesthetic concept: Photo collage.

Maintenance regimes have a direct impact on the attractiveness of both traditional street designs and engineered stormwater management operations; where these two merge on ecological streets, care is likewise imperative.

In some cases it is possible to shift the maintenance responsibility for the public space to private hands. This ensures that natural drainage street designs can be both functionally and aesthetically successful, while simultaneously lowering public maintenance costs. When local landowners are engaged with the public right-of-way projects they may be willing to help maintain them; residents (or owners of commercial spaces) may be asked to assist with plant care. Seattle's Broadview Green Grid project follows this model. Of course, neighbors are not responsible for maintenance activities that require special equipment or training. However, common practices such as pruning, weeding, and mulching can be deemed the responsibility of the neighborhood, just as on conventional Seattle streets neighbors mow the grass on public median strips outside of their yards. At Broadview Green Grid, the city has encouraged seasonal block parties for street gardening and supports the groups by providing a staff person to organize the parties and offering free yard waste pickup (Johnson 2006). Because residents have the opportunity to observe natural drainage projects throughout the year, and take note of seasonal changes, they are often the best candidates to be informal inspectors and monitors of the system. Accordingly, there are more "eyes on the street" (Jacobs 1961); as environmental goals are met at the watershed scale, there are also more eyes on the *stream*.

STREET DESIGN WITH THE WATERSHED IN MIND

Moving away from the standardized street sections, designers can provide attractive alternatives that also aim to restore natural hydrologic functions in urbanized watershed while increasing people's understanding of natural processes. Local successes in Seattle highlight the potential for replication of natural drainage projects and programs abroad. "If a prototype is cost-effective and socially useful, politically palatable, and gives aesthetic pleasure, it can be propagated through urban space—just like the American lawn or the ubiquitous stormwater pipe and catch basin system" (Hill 2003).

Although dramatically distinct designs may not be appropriate for every street (street designs should reflect the desires of the community, especially those of adjacent residents and business owners), we believe that the restoration of human relationships with ecological processes are more readily attainable when streets are released from the standard aesthetic. As such, designers can advocate and innovate for alternatives to traditional street functions and forms. The next steps will take us outside the temperate residential streets of Seattle to explore other types of topographies, climates, and land uses, integrating ecological streets with both their local waterways and the aesthetic character of their surrounding cities and regions.

ACKNOWLEDGMENTS

Material for this chapter was derived in part from our co-authored 2004 Master's thesis in architecture at the University of Washington entitled: Great (Wet) Streets: Merging Street Design and Stormwater Management to Improve Neighborhood Streets. We are greatly indebted to Richard Horner and Kristina Hill of the University of Washington and Miranda Maupin of Seattle Public Utilities for their inspiration, advice, and consultation on this work. We would also like to thank Tracy Tackett and Jim Johnson for insights offered for this chapter, as well as Denise Andrews, Darla Inglis, and all the other wonderful folks at Seattle Public Utilities for their work on natural drainage systems. Thank you to Eric Higbee for the High Point photos.

REFERENCES

Anderson, Stanford. (1978). *On Streets*. Institute for Architecture and Urban Studies. MIT Press, Cambridge.
Bay Area Stormwater Management Agencies Association (BASMAA). (1999). *Start at the Source: Design Guidance Manual for Stormwater Quality Protection*. New York: Forbes Publishing.
Brown, B., T. Harkness, and D. Johnston. (1998). Guest editors' introduction. *Landscape Journal*, special issue xii–xvi.
Burden, Dan. (1999). *Street Design Guidelines for Healthy Neighborhoods*. Local Government Commission and Center for Livable Communities. Sacramento, CA.
Capra, Fritjof. (1996). *The Web of Life*. New York: Anchor Books.
City of Maplewood. (2003). Creating a Rain Garden in Your Yard. Retrieved on August 24, 2003, from http://www.ci.maplewood.mn.us/index.asp?Type=SEC&SEC=%7B6D539312-B870-467B-8255-26B8FDB D5E5B%7D&DE=%7BB7D50FB1-1C85-415E-98CD-101B0D73F19C%7D.
France, Robert. (2003). Green world, gray heart. *Harvard Design Magazine*, (18): 30–36. Spring-Summer 2003.
Forman, Richard T.T. Personal communication with Stephanie Hurley on March 22, 2006.
Helphand, Kenneth and Robert Melnick. (1998). Editors introduction. *Landscape Journal*, special issue ix.
Herbert Research Group. (2003). Buyer and realtor focus groups, executive summary. Seattle Housing Authority. Seattle, WA.
Hill, Kristina. (2003). Green good, better, and best. *Harvard Design Magazine*, (18): 37–40. Spring-Summer 2003.
Hill, Kristina, Denis White, Miranda Maupin, Barbara Ryder, James, Kathryn Freekmark, Rebecca Taylor, and Sally Schauman (2002). In expectation of relationships: centering theories around ecological understanding. In Hill and Johnson (Eds.), *Ecology and Design* (pp. 271–304). Washington, D.C.: Island Press.
Horner, Richard R. (2006). Personal communication by email with Stephanie Hurley on April 8, 2006.
Horner, Richard R., Heungkook Lim, Stephen J. Burges. (2002). Hydrologic monitoring of the Seattle ultra-urban stormwater management projects. Water Resources Series Technical Report No. 170. November 2002.
Horner, Richard R., Heungkook Lim, Stephen J. Burges. (2004). Hydrologic monitoring of the Seattle ultra-urban stormwater management projects: summary of the 2000–2003 water years. Water Resources Series Technical Report, August 2004.
Horner, R.R., C.W. May, E. Livingston, D. Blaha, M. Scoggins, J. Rims, and J. Maxted. (2001). Structural and non-structural BMPs for protecting streams. In Ben R. Urbonas (Ed.), *Linking Stormwater BMP Designs and Performance to Receiving Water Impact Mitigation*.
Hough, Michael. (1995). *Cities and Natural Processes*. New York: Routledge.
Jacobs, Allan. (1996). *Great Streets*. MIT Press, Cambridge.
Jacobs, Jane. (1961). *The Death and Life of Great American Cities*. New York: Random House.
Johnson, Jim, (2006) Personal communication via telephone with Megan Stromberg on April 7, 2006.
Kansas City Missouri. "10,000 Rain Gardens" program. http://www.rainkc.com/home/index.asp, viewed on April 17, 2006.

Kirschbaum, Robin. (2004). American Society of Civil Engineers Lecture Series [Powerpoint Presentation]. May 20, 2004.

Kostof, Spiro. (1992). *The City Assembled: The Elements of Urban Form Throughout History.* London: Bulfinch Press. Little, Brown & Company.

Low Impact Development (LID) Integrated Management Practices Guidebook (2002). Programs & Planning Division, Department of Environmental Resources, Prince George's County, Maryland.

Lynch, Kevin. (1971). *Site Planning.* Cambridge: MIT Press.

Lyndon, Donlyn. (1997). Caring about places. *Places,* 11(2): 2–3.

Maupin, Miranda. (2006) Personal communication via telephone with Megan Stromberg on March 16, 2006.

Maupin, Miranda. (2004) Personal communication with Megan Wilson and Stephanie Hurley on May 7, 2004.

METRO (2002). *Green Streets Innovative Solutions for Stormwater and Stream Crossings.* Portland, OR.

Meyer, Beth. (2001). Territories: contemporary European landscape design. *Landscape Journal,* 20(2): 201–204.

Mills, Melanie. (2002). Alternative stormwater design within the public right-of-way: A residential preferences study. M.L.A. thesis. Department of Landscape Architecture. University of Washington. Seattle, WA.

Moudon and Untermann. (1991). Grids Revisited. Chapter 9 in Anne Vernez Moudon (Ed.), *Public Streets for Public Use.* New York: Columbia University Press.

Moughtin, Cliff. (2003). *Urban Design: Street and Square.* 3rd edition. Architectural Press.

Nassauer, Joan Iverson. (1997). *Placing Nature: Culture and Landscape Ecology.* Washington, D.C.: Island Press.

Seattle Public Utilities. (2003). Frequently Asked Questions about Natural Drainage Systems. Retrieved on August 19, 2003, from http://www.cityofseattle.net/util/naturalsystems/drainagesystems.htm.

Spirn, Anne Whiston. (1984). *The Granite Garden: Urban Nature and Human Design.* New York: Basic Books. Harper Collins Publishers.

Spirn, Anne Whiston. (1998). *The Language of Landscape.* New Haven, CT: Yale University Press.

Tackett, Tracy, (2006) Personal communication via telephone with Megan Stromberg on April 7, 2006.

Part 5

Heritage Sites

14 Cultural and Environmental Restoration Design in Northern California Indian Country

Laura Kadlecik and Mike Wilson

With contributions from

Paula Allen, Eric Johnson, Andrea Davis, and Helene Rouvier

CONTENTS

ABSTRACT

As elsewhere in the world, Euro-American settlement and development along the northwest coast of California has resulted in the dramatic disruption of the traditional culture and lifestyle of American Indian communities. These changes in lifestyle have caused trauma to the northwest California Indian population and as a result, a higher incidence of social, mental, and physical diseases such as diabetes and heart disease than the general non-Indian population. Two tribal organizations located in northwest California utilize a combination of traditional knowledge and contemporary landscape restoration technologies in order to restore balance and health, support culture, and provide venues that offer a sense of community to the local American Indian population.

The first organization is United Indian Health Services, Inc. (UIHS), a consortium of nine federally recognized tribes that operates to provide comprehensive ambulatory health care to 15,000 American Indian tribal members and their families. The second is the Wiyot Tribe, whose ancestors originally inhabited the shores and foothills surrounding Humboldt Bay. Together the UIHS Potawot Health Village in Arcata, California, and the Wiyot Tribe's Tuluwat Village Restoration project in Eureka, California, have contributed to an ongoing renaissance of local American Indian art and ceremony, and an increase in cultural awareness within the broader non-Indian community. Tools and technologies used in the projects and discussed in this chapter include vegetated swales, stormwater treatment wetlands, wetland restoration, edible and native landscape design, organic gardening, bio-engineering, hazardous materials and Brownfields remediation, geotextile materials, and salt marsh restoration.

INTRODUCTION

No amount of medical treatment can offer the healing that will take place when the Wiyot dance again on Tuluwat.

—Jerome J. Simone, Chief Executive Officer, United Indian Health Services, Arcata, California

Similar to most indigenous groups throughout the world, the standard of living among northwest California Indian tribes traditionally depended on the availability of natural resources. When settlers first arrived in North America, the landscape, and its diversity of ecosystems were being intensively managed to meet subsistence, economic, and spiritual needs. The management occurred within a value system and worldview wherein people were inseparable from the land. Not only did northern California Indians live in nature but they also saw, and continue to see themselves, as an integral part of it (Anderson, 2005; Blackburn and Anderson 1993; Margolin 1981; Heizer 1980; Powers 1976).

The natural resources and cultural traditions of northern California Indians who populated the area now recognized as Humboldt, Del Norte, and Trinity counties were fragmented under the impacts of modern industrialized civilization. This occurred in northern California much like it did throughout North America and as it continues today throughout the world. The California and

Oregon Gold rush occurred in the late 1840s. Pioneers flocked to remote, northern coastal mountain, river, and bay regions of the state seeking fortunes. Although most failed at finding riches in gold, they did discover vast and valuable giant redwood tree forests. The magnitude of the resource extraction caused boomtowns to occur overnight around Humboldt Bay. By 1930, there were dozens of timber mills along the perimeter of the bay and a network of railroads and shipping lines throughout the region supporting the construction of West Coast cities.

The timber industry brought agrarian settlements to the wetland prairies surrounding Humboldt Bay and the six river basins in its region, including the Van Duzen, Eel, Mad, Trinity, Klamath, and Smith rivers. Tribes consisting of what are now known as the Wiyot, Chilula, Whilkut, Yurok, Hupa, Karuk, and Tolowa Indians that populated the coastal portion of these rivers were decimated by the settlement period. While those that survived initial contact were being forcibly relocated to reservations far from their ancestral territory, the local rivers were dredged for gold, streams were straightened to facilitate the flow of fallen logs, and coastal islands and prairies in and around Humboldt Bay were cleared, leveled, and plowed (Loud, 1918). The native people that survived contact, and the settlement, reservation, and boarding school periods that followed in the first half of the 20th century, began returning to the rivers and urban centers of their ancestral land to find that everything had changed. After decades of struggle to gain a position in modern society, there are more than 10 federally recognized tribal groups in northwest California and an active local American Indian presence. However, the changes that occurred to their land and their lifestyle have caused trauma to the local California Indian community and as a result, a higher incidence of social, mental, and physical diseases, such as diabetes and heart disease, than the general non-Indian population. Several of the federally recognized tribal groups and associated tribal organizations are working to restore balance and health to their population by providing culturally appropriate places that offer a sense of community.

This chapter describes the planning and implementation of two cultural restoration projects that integrate traditional knowledge, contemporary landscape planning concepts, environmental restoration tools, engineering technologies, and permaculture strategies to achieve their specific restoration goals: the United Indian Health Services, Inc., Potawot Health Village, and the Wiyot Tribe's Tuluwat Village on Indian Island. They are separate projects that exist on the shores of Humboldt Bay, but both have implications for the American Indian communities throughout the Pacific Northwest (Figure 14.1). Both have been designed and are being implemented on the premise that cultural restoration can be achieved by utilizing environmental restoration techniques when natural resources and culture are intrinsically linked. Although the momentum of one has benefited the other, combined they have played an important role in supporting and contributing to an ongoing renaissance of local American Indian art, dance, and expression, and an increase in cultural awareness among the broader non-Indian community.

UNITED INDIAN HEALTH SERVICES' POTAWOT HEALTH VILLAGE

BACKGROUND

United Indian Health Services, Inc. (UIHS), is an intertribal, nonprofit organization that provides ambulatory health services to more than 15,000 American Indians and their families living in Humboldt and Del Norte counties of northwestern California, which is a service area of 5,000 square miles. UIHS clientele consists primarily of American Indians with Yurok, Wiyot, and Tolowa ancestry affiliated with nine federally recognized tribes but also includes other California and American Indians living in the area. Operating from a series of remote clinics along the coastal rivers since 1970, UIHS has recently built a new expanded facility, known as the Potawot Health Village. The facility is located 260 miles north of San Francisco on 40 acres in Arcata, and was designed to honor the traditional pole and split plank architecture of northwestern California Indians (Figure 14.2). The new facility functions as the central feeder clinic and supports five outlaying satellite clinics.

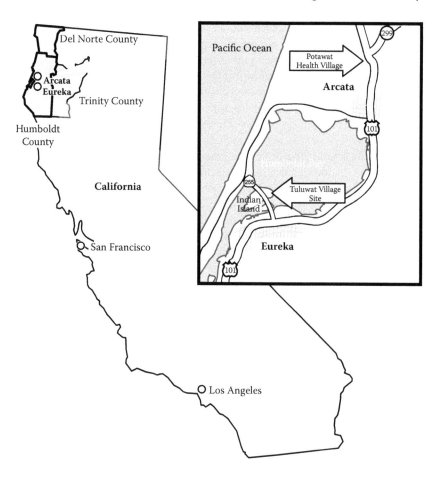

FIGURE 14.1 Location map, United Indian Health Services, Potawot Health Village, Arcata, California and Tuluwat Village Restoration project, Eureka, California.

Like many other indigenous populations throughout the world, the northwestern California Indians suffer from chronic, lifestyle-related diseases. Recent research regarding diabetes has shown its incidence is almost three times higher among American Indians than other groups in North America. These illnesses are in part due to inadequate diets, lack of exercise, and low self-esteem that are a direct result of trauma and dramatic changes in lifestyle experienced by the population within the past 150 years (Ferreira, 1996).

With the mission to provide comprehensive health care services that respect local California Indian traditions in a contemporary and culturally sensitive environment, UIHS set out to develop a clinical facility that was "more than just a building and a parking lot." After a 5-year search for an appropriate location, a project site in Arcata was chosen due to its centrality to the UIHS service area and proximity to the local hospital. In addition to meeting the goal of being able to provide ambulatory care, the site offered many opportunities to accommodate a range of community events in a culturally appropriate setting.

The intention to provide a culturally relevant landscape adjacent to the UIHS Potawot Health Village clinic site has its roots in a traditional worldview common to indigenous populations around the world, which is based on the concept that humans not only live in nature but also are themselves a part of it. It maintains that everything in nature is interconnected, such that anything that humans do affects the rest of the natural and spiritual world (Margolin, 1981: Heizer 1980; Powers, 1976). Underlying this view, and in a more practical sense, the traditional society and culture historically

FIGURE 14.2 Traditional plank house at Sumeg Village, Patrick's Point State Park, Humboldt County, California. Photo by M. Wilson (2005).

utilized and relied on the natural environment for their physical, mental, spiritual, communal, and artistic sustenance (Anderson, 2005; Blackburn and Anderson, 1993; Heizer, 1980). The concept of restoring the land adjacent to the Potawot Health Village is an extension of these traditional and cultural connections of people to their land.

Wendell Berry, the environmentalist, novelist, and poet, was quoted many times during the planning and development stages of the project in order to express this philosophy. Regarding public health, Berry (1996) states: "I believe that the community—in the fullest sense: a place and all its creatures—is the smallest unit of health and that to speak of the health of an isolated individual is a contradiction in terms."

The architect of the clinical facility was Robert Weissenbach of Mulvanny/G2, Seattle, Washington. The project landscape architects were J.A. Brennan and Associates, Seattle, Washington. Project management, stormwater designs, and environmental restoration designs were provided by Laura Kadlecik and Mike Wilson of HWR Engineering and Science, Arcata.

COMMUNITY-BASED APPROACH

In late 1997 during the initial stages of the project, the Potawot Health Village design team asked UIHS to convene a committee of local California Indian community members that had knowledge of traditional culture to design the environmental and cultural components of the facility. This committee comprised of elders, regalia makers, basketweavers, herbalists, and home gardeners met monthly for over 4 years to plan the restoration of the property surrounding the clinic facility so that it would complement the ambulatory programs located there and meet the local governments project conditions of approval (City of Arcata, 1999; Winzler and Kelly, 1997).

An active volunteer program was developed in conjunction with nearby Humboldt State University shortly after the community committee began meeting and several months before construction of the clinic began. The volunteer program carried out many of the plans developed by the committee and was able to utilize the strong desire of both the American Indian and non-Indian community to participate in restoration activities. This program draws upon students, community members, and clinical staff to implement many of the restoration elements and is successful in allowing many segments of the American Indian and non-Indian community to benefit from the Health Village site.

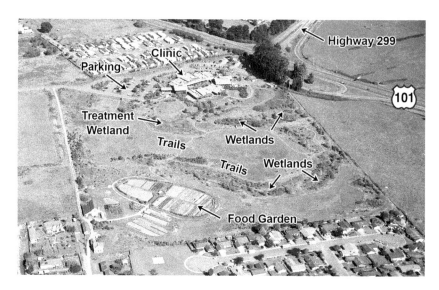

FIGURE 14.3 Aerial photo of 40-acre UIHS Potawot Health Village Project Area, Arcata, California. Photo by R. Fenton (2004).

POTAWOT HEALTH VILLAGE DESIGN

The 44,000-square-ft clinical facility was opened in September of 2001 and provides primary medical and dental care, obstetrics, diabetic, nutrition, and laboratory services, mental health counseling, pharmaceutical dispensing, special services for children and teens, tobacco cessation, and community health and substance abuse services. Each department of the health clinic is housed in its own building and the buildings in turn are arranged in a circle to create a village-like setting (Figure 14.3). The buildings are of a concrete tilt-up type construction; however, every one of the exterior walls was uniquely formed and stained to resemble the traditional pole and split redwood plank architecture. Even the metal roofing was designed to emulate plank roofing, including traditional "smoke hole" and "wind guard" details. (Figure 14.4)

FIGURE 14.4 UIHS Potawot Health Village clinic building entrance, Arcata, California. Photo by M. Wilson (2006).

FIGURE 14.5 Wellness Garden in the interior of the UIHS Potawot Health Village, Arcata, California. Photo by E. Johnson (2005).

It was important that the landscaping immediately surrounding the clinic blend seamlessly with the building as well as the restored surroundings. As a result, native plants, including those with cultural, medical, spiritual, and culinary significance, are planted throughout the facility grounds. Both the availability of fresh, flowing water and the opportunity for spiritual reflection are key elements to the local American Indian concept of health. As a result, a "wellness garden" exists in the center of the clinic featuring a recirculating creek and serves as a place for quiet reflection. From the wellness garden, the creek flows beneath the facility and empties into a series of wetland ponds that dominate the landscape outside (Figure 14.5).

KU'-WA-DA-WILTH RESTORATION AREA

Of the 40 acres that make up the Health Village site, 40 are set aside as a conservation easement held in trust by the City of Arcata. The 20-acre easement is aptly named the Ku'-wa-da-wilth Restoration Area, meaning "comes back to life" in the Wiyot Indian language. Before settlement by Europeans in the 1850s, the area was a wet coastal meadow dominated by native grasses, native berries, and sitka spruce. From 1876 to 1995, the area was farmed and grazed by cattle. When UIHS purchased the property in 1997, the seasonal wetlands that existed within the conservation easement area were much degraded due to changes in local hydrology, site topography, and the presence of hardy, introduced pasture grasses (Roscoe et al. 1996; Waller et al. 1996; Winzler and Kelly, 1996). Since that time work at the Ku'-wa-da-wilth Restoration Area has been conducted by UIHS under the guidelines of the UIHS Health Village Conservation Easement Management Plan (HWR, 2000) and has resulted in:

- Sixteen acres of seasonal wetland and meadow areas restored with culturally significant plant species; maturing stands of wetland vegetation and upland trees, shrubs, and native prairie grasses now dominate the landscape.
- Interpretive signage, viewing decks, and a basketry garden that are a venue for environmental and cultural training and education.
- Walking trails extending 1½ mi for physical exercise, psychological healing, and spiritual reflection.
- An exceptionally productive 1½-acre organic food and flower garden, greenhouses, and a 1-acre fruit orchard that are central to UIHS's nutrition education program.

With the understanding that all things can contribute to physical, mental, and spiritual healing of an individual, the family, and the community, the Ku'wah-dah-wilth Restoration Area and all of the activities that occur there are meant to compliment the existing clinical programs for both mental and physical health.

STORMWATER MANAGEMENT AND CONSTRUCTED WETLANDS

Appropriate stormwater management for both volume control and water quality treatment was identified early in the project as a cornerstone of project site design and is a critical element of the Health Village design concept because of the importance of the environment to UIHS but also due to flooding concerns by the City of Arcata. Situated on the north end of Humboldt Bay, Arcata is crossed by several creeks that cause flooding during winter months. As a result the city government asked that the project not impact downstream properties and conditioned its approval on postdevelopment stormwater flows (from the developed portion of the site to the conservation easement wetlands) matching predevelopment stormwater flows as closely as possible. In addition, the city had recently implemented a $1.20/ft^2 stormwater fee for newly developed impermeable surfaces. With the clinical facility and its parking areas occupying almost 7 acres, the stormwater fees could have potentially been as high as $280,000. By comparison, implementation of the hardscape stormwater management improvements totaled $160,000. Removing potential contaminants, such as petroleum products, heavy metal sediments, and excess nutrients from stormwater before it entered the restoration area was also very important to the UIHS community as they were planning to use the restored wetland and upland areas for food, fiber, and medicine.

Technologies employed in the project design included vegetated swales for stormwater detention, infiltration, and conveyance (Figure 14.6), sediment traps and flow control structures, constructed wetlands (Figure 14.7), and reduction of impervious surfaces. Constructed stormwater treatment wetland technology improves water quality through a variety of biological processes that naturally occur in wetland environments. Microorganisms that live on aquatic plants and in soils transform organic matter and nutrients present in wastewater effluent into types of nutrients that support a diverse community of aquatic pond life. Plants also play an active role in taking up

FIGURE 14.6 Vegetated swales in the parking lots at the UIHS Potawot Health Village, Arcata, California. Photo by M. Wilson (2006).

FIGURE 14.7 Stormwater treatment wetland at the UIHS Potawot Health Village, Arcata, California. Photo by E. Johnson (2004).

available nitrogen, phosphorous, and other nutrients from stormwater. By slowing down the runoff and detaining it for a period, stormwater wetlands allow particulate matter containing pollutants to settle out through the process of sedimentation.

Stormwater flows from roof and parking surfaces at the Health Village site are retained and transported through a series of swales to a treatment wetland and then released to the restored wetland area, taking advantage of the site's topography to reduce the developed area's vulnerability to flooding. This same stormwater is now a valuable supplemental water resource for the project's restored wetlands where a variety of cultural and educational activities occur. In this way, the goals of best management practice stormwater technologies and a restored, culturally relevant environment were accommodated and proved complementary. The stormwater treatment wetland at the Potawot Health Village is lined using imported clay from a nearby wetland restoration project and planted predominantly with a native wetland species, including hard stem bulrush (*Scirpus acutus*), bulrush (*Scirpus microcarpus*), slough sedge (*Carex obnupta*), and pondweed (*Potamogeton* spp.). a photograph of the recently constructed wetlands, see France 2003). Effective pest (or vector) control was also achieved by appropriately applying integrated pest management practices such as introducing mosquito fish and providing nearby swallow and bat habitat.

RESTORED WETLAND MEADOW AND UPLAND

Wetlands are important places for American Indian tribes in that they provide sources of plant and animal resources used for food, fiber, and medicine. With this in mind the wetland restoration was designed to highlight native plants traditionally used by northwest California Indians, serve as a venue for conducting activities that promote health, and convey UIHS's integrated philosophy that wetlands can have a role in community health for both local American Indian and non-Indian populations.

Initial wetland restoration activities on Ku'-wa-da-wilth Restoration Area included deepening and widening of existing wetlands. Approximately 10,000 cubic yards of soil were moved, and the topography, which had been generally leveled over the past 100 years of farming, was enhanced across the 20-acre site. Islands, stumps, and riparian vegetation were placed and designed to enhance habitat value for aquatic animals. Stormwater flows from the clinic development were

FIGURE 14.8 Yellow pond-lily (*Nupar lutea* ssp. *polysepapa*) in wetland ponds of the UIHS Potawot Health Village, Arcata, California. Seeds and roots of the yellow pond-lily are a traditionally important source of food and medicine for American Indian tribes. Photo by E. Johnson (2005).

directed into the restored wetlands. Native wetland species were planted throughout the restored wetland areas with assistance from hundreds of volunteers. Such species included hardstem bulrush (*Scirpus acutus*), cattail (*Typha latifolia*), yellow pond-lily (*Nuphar lutea ssp. polysepal*), willow (*Salix* spp.), red alder (*Alnus rubra*), slough sedge (*Carex obnupta*), monkeyflowers (*Mimulus* spp.), a variety of ferns, and many other wetland and wet soil plant species, such as twinberry (*Lanciera involecrata*), red flowering currant (*Ribes sanguineum var glutinosum*), and hardhack (*Spirea douglasii*) (Figure 14.8, Figure 14.9).

On the upland areas that surround the restored wetlands, a program to establish native wet meadow grasses, forbs, and woodland species is the focus of UIHS restoration staff time, study, and

FIGURE 14.9 Hardhack (*Spirea douglassi*) and *Salix* spp. quickly colonized the slopes of the restored seasonal wetlands, planted with *Scirpus* spp. A wetland view deck with interpretive display can be seen in background. Photo by E. Johnson (2005).

FIGURE 14.10 Hardhack (*Spirea douglassi*) in foreground with a restored meadow of tufted hairgrass (*Deschamsia cespitosa*) in the background, and the UIHS Potawot Health Village facility beyond, Arcata, California. Photo by E. Johnson (2005).

volunteer efforts. Several native grass species have been tried at the site, and the most successful have proven to be meadow barley (*Horduem brachyantherum*), tufted hairgrass (*Deschamsia cespitosa*), and Idaho fescue (*Festuca idahoensis*) (Figure 14.10). Semi-annual controlled burns are being used to foster native grass and forb communities and have helped to overcome the perennial rye (*Lolium perenne*) and velvet grass (*Holcus lanatus*) that previously dominated the site. Woodland species have been planted in clusters throughout the restored meadow environment. Species planted in these areas include primarily big-leaf maple (*Acer macrophylum*), wax myrtle (*Myrica californica*), sitka spruce, (*Picea sitchensis*), black cottonwood, (*Populus balsamifera* ssp. *trichocarpa*), and western hemlock (*Tsuga heterophylla*).

One and a half miles of walking trails hardened with pine resin wind their way around the restored wetland and upland areas. Since being established, the restored environment has allowed UIHS to host numerous native plant-related events, guided and self-guided tours, field trips, and training programs, and has provided an outdoor classroom to promote and integrate traditional land management practices such as fire management, plant identification, and gathering and processing of basketry plants into clinical and cultural programs.

TRADITIONAL RESOURCE MANAGEMENT AND "PERMACULTURE"

Traditional resource management, sustainable agriculture, and permaculture (or "permanent agriculture") practices are known to be a part of a long-standing American Indian tradition. These practices are based on a wealth of traditional ecological knowledge that rely on an understanding of seasonal winds, solar exposure, plant cycles, and availability of water to meet human needs with minimum labor and without depleting resources.

On the Potawot Health Village site grounds, activities that demonstrate permaculture include the cultivation and gathering of native plants that have traditionally been used for food, medicine, ceremonial events, and making baskets and other textiles. In order to support these activities, restoration work has included the planting of over an acre of basketry materials, including Douglas iris (*Iris douglasiana*), California grape (*Vitus californica*), hazel (*Corylus conruta sp. sericea*), and bear grass (*Xerophyllum tenax*) in the upland areas. In addition, the restoration area was designed to be an edible landscape and is planted with numerous berries including thimbleberry and salmonberry,

(*Rubus parviflorus* and *R. spectabilis*), elderberry (*Sambucus racemosa*), blueberry (*Vaccinium ovatum* and *V. parvifolium*), salal (*Gaultheria shallon*), and medicinals, including Oregon grape (*Berberis aquifolium* var. *aquifolium*), yarrow (*Achillea millefolium*), and wormword (*Artesmisia douglasiana*).

When UIHS opened and relocated to the Potawot Health Village site in 2001, the role of the community committee that worked to envision and program the landscape elements of Ku' wah-dah-with Restoration Area and Health Village site evolved into a subcommittee of the UIHS Board of Directors, now known as the Traditional Resources Advisory Committee (TRAC). Below is an excerpt from the committee description:

> *United Indian Health Services, Inc., recognizes the importance of traditional beliefs, practices, and ceremonies in the healing of the body, mind, and spirit. Traditional Health involves all things that contribute to the physical, mental, and spiritual well being of the individual, the family, and the community. It includes those traditional values, knowledge, and beliefs that exist in all of us. The primary goal of traditional health of our local Indian communities is to have balance of "all things" in our lives. UIHS encourages and recognizes a climate of respect, acceptance, and utilization in which tradition beliefs are honored.*

The TRAC committee works in coordination with other UIHS committees and staff to facilitate the numerous activities at the Potawot Health Village and Ku' wah-dah-with Restoration Area sites. A description of the Potawot Community Food Garden and UIHS's recreational, nutritional, educational, and culture programs follow.

POTAWOT COMMUNITY FOOD GARDEN/NUTRITION PROGRAM

UIHS has an enclosed 2-acre nutritional food garden and 1-acre orchard known as the Potawot Community Food Garden (Figure 14.11). The Food Garden lies within the 20-acre restoration area, produces organic fruit and vegetables for sale, and demonstrates beneficial vegetables, herbs, and flowers that can be grown in a typical back yard setting. This garden, which is encircled by native and edible, horticultural varieties of berries, and a fruit orchard provide a year-round harvest of

FIGURE 14.11 The UIHS Potawot Community Food Garden where students from Humboldt State University and "Two Feathers" youth program are learning about sustainable agriculture, Arcata, California. Photo by Eddie Tanner (2004).

FIGURE 14.12 Food stand at UIHS Potawot Health Village where food from the community garden is sold to UIHS clients, staff, and the public, Arcata, California. Photo by E. Johnson (2005).

fruits and vegetables. Interdepartmental staff coordinates activities via a Garden and Native Foods Working Group. Food produced in the garden is available to UIHS staff and clients in the following ways:

- Direct sales from the Potawot Garden vegetable stand located at the front of the Potawot clinic, open twice weekly from May to October, and occasional vegetable stands located at one of the five UIHS satellite clinics (Figure 14.12).
- "Veggie bucks" distributed to diabetes clients as incentives and redeemable at the UIHS garden vegetable stands.

Nutrition outreach and education occur in several ways. Hundreds of recipes are available free of charge from the "rola recipe" rolodex located at the vegetable stand. Regular recipe-tasting events occur adjacent to the vegetable stand, and seasonal UIHS recipe books are available at UIHS events. Cooking and canning classes are seasonally offered inside the clinic. Gardening curriculum is presented at predominantly American Indian elementary schools located in the outlying Indian communities within the UIHS service area and at the satellite clinics. A detailed gardening booklet titled the *Humboldt Gardener*, which emphasizes information and techniques unique to the local northwest climate, is being produced and will be utilized as a tool to encourage home gardening in the community. Garden internships for American Indian youth are offered throughout the growing season. Annual events such as the Harvest Festival where elementary children tour the garden and the Celebration of Food and Life luncheon occur to bring attention to healthy foods. As part of a program entitled "Food is Good Medicine," a modern food model, is being designed to mimic the historic diets of the local Indians. The model focuses on what foods are available today, and those that fall into the following categories: strength foods (good sources of protein), wellness foods (gathered plants such as leafy greens, fruits, and veggies), energy foods (starchy plants and whole grains). A circular logo has been developed and will be the basis for nutrition education to UIHS clients (Figure 14.13).

FIGURE 14.13 UIHS Food is Good Medicine logo by E. Johnson (2005).

RECREATION PROGRAMS

UIHS clients and staff are encouraged to engage in recreation and fitness activities along the walking trails of the Health Village site. A 100-mile walking club exists for clients, and is popular among clients and staff. UIHS clients and staff, as well as staff from the nearby community hospital and adjacent neighbors are observed walking during lunchtime breaks and throughout the day. Within the restoration area, the network of trails offers resting benches and interpretive signage (Figure 14.14). The UIHS Health and Wellness Committee (HAWC) hosts an annual weekend walk-a-thon and health fair at the site. Hundreds of participants attend the "HAWC Walk," participate in a variety of health activities, and conclude with a traditional lunch menu of barbequed salmon, garden fresh vegetables, and potluck items.

ENVIRONMENTAL PROGRAM

A comprehensive education and interpretation program is critical to UIHS's objective of increasing cultural and environmentally based health and wellness (Figure 14.15). Kiosks and informational signs are used to promote an awareness of valuable wetland functions, including stormwater storage, pollution control, and wildlife habitat. Interpretive materials available on-site include a self-guided walking tour booklet that addresses the traditional uses of native plants, the importance of recreation, and modern environmental management and permacultural practices. Tours of the Health Village and Ku'-wa-da-wilth Restoration Area are given to elementary, high school, and university classes, and local civic groups as well as international visitors' groups visit the facility on a regular basis. University classes are particularly active at the sites as instructors from a range of disciplines highlight the engineering, design, and social elements of the Health Village project through workshops, field trips, and class projects. UIHS also works to distribute information to other tribal communities about the project and its programs.

FIGURE 14.14 La Chompchay Kids Club utilizing the trails in the Ku' wah-dah-wilth Restoration Area, Arcata, California. The trails are paved with Resinpave™, a mix of local aggregate and nonpetroleum based emulsion made from tree resin. Photo by E. Johnson (2005).

CULTURAL PROGRAM

Cultural enhancement programs such as basket-making classes and language seminars currently occur throughout the year at the Potawot Health Village. Basket classes focus on the production of items that can be made by collecting material at the Ku' wah-dah-wilth Restoration Area. These include soaproot brush making with soaproot (*Chlorogalum pomeridianum*), eel basket making with willow (*Salix* spp.), and tule mat making with hardstem bulrush (*Scirpus acutus*) (Figure 14.16). These programs are expected to be more regularly programmed and to extend to the satellite clinics

FIGURE 14.15 Restoration education and propagation training in the UIHS greenhouse, Arcata, California. Photo by E. Johnson (2005).

FIGURE 14.16 UIHS employee harvesting bulrush (*Scirpus acutus*) from restored wetlands to make "tule mats," Arcata, California. Photo by L. Kadlecik (2004).

within the next few years. Participants in the classes include a variety of clients from the child and family counseling programs, as well as a number of other clients generally interested in such topics.

UIHS staff coordinates the annual week-long summer camp at the local state park, "La Chomp-chay" (meaning little frog in the Yurok language) Kids Club, for young children, and the Teen Advisory Group for teens. Both youth groups meet regularly and during summer camp sessions to plan and explore traditional games and sports, physical fitness activities, environmental education, and other health-related prevention activities.

UIHS also is working to implement a zero-waste initiative. Staff working on the initiative are seeking ways to reduce waste through clinic wide policies, guidelines, and outreach to medical and purchasing staff and at UIHS-sponsored events and local traditional ceremonies. Composting of lunch and coffee waste is available within the clinic, and compost is used at Potawot Community Garden.

The programs and activities offered at the UIHS Potawot Health Village site and Ku' wah-dah-wilth Restoration Area are continually expanding and increasing in usage. The vision that the original community committee and project design team members had for integrating the natural environment into the health and well being of the northwest California Indian community is continuing to grow, blossom, and fruit.

WIYOT TRIBE'S TULUWAT VILLAGE RESTORATION ON INDIAN ISLAND

BACKGROUND

The Wiyot people have inhabited the Humboldt Bay region of Northern California for thousands of years. Prior to 1860 there were over 3,000 Wiyot people living along the waterways and lowlands of Humboldt Bay where they hunted its wildlife, fished its waters, and gathered its roots for food, fiber, and medicine. In the bay is Indian Island, the center of the Wiyot world, and the location of the annual World Renewal Ceremony at the village of Tuluwat. For centuries this ceremony was a weeklong event in which the Wiyot people invited guests from other tribes to join them in asking the Creator to bless all people and the land in preparation for the New Year (ITSI, 2002).

The first recorded Euro-American settlement of Indian Island occurred in 1858 as Europeans began to inhabit the area and harvest Humboldt County's rich natural resources. On February 23, 1860, Robert Gunther purchased the island from Captain Moore (Loud, 1918). As more Europeans moved into the region, native peoples were denied access to their lands and forced into smaller and smaller areas. Conflict between the groups was common and on the night of February 26, 1860, while many of the Wiyot were gathered for the World Renewal Ceremony at Tuluwat, a group of European settlers paddled over to the island and, using hatchets and knives, massacred an estimated 180 Indian people. Most of the victims were elders, women, and children. This event was timed with several other massacres around the bay that severely decimated the Wiyot people (HWR, 2003).

After the massacre occurred on Indian Island, a series of dikes and field drains were constructed to convert the salt marsh to grazing land to raise cattle. A small shipyard was constructed on top of the Tuluwat shell mound in the 1870s, which included a drydock as well as various other structures. Archaeological fieldwork and excavation began on the Tuluwat shell mound in 1913, which left Indian Island and the Wiyot Tribe stripped of many of its ancestral and cultural archaeological relics. Indian Island changed ownership over the years, and by 1962 the City of Eureka owned much of the island. Tuluwat shell mound was designated as a National Historic Landmark in 1964 (NHLP, 1964).

Currently, there are 550 enrolled members of the Wiyot Tribe, many of whom reside on the 88-acre Table Bluff Reservation (TBR) located 16 mi south of the City of Eureka. In September 2000, the Wiyot Tribe purchased 1.5 acres on the southeast shore of Indian Island. In addition, in June 2004 an adjacent 40 acres of Indian Island were deeded to the Wiyot Tribe by the City of Eureka. These 41.5 acres of Indian Island now under the Tribe's ownership contain the remains of the sacred shell mound and Tuluwat Village site. This effort to clean up, restore, and reestablish ceremonial use of their land on Indian Island is called the Tuluwat Village Restoration Project.

PROJECT SITE DESCRIPTION

Indian Island, including tidelands, is approximately 300 acres in size, 1 mile long, and less than one half mile wide. Aside from two large Indian shell mounds and some dredge spoil fill areas, Indian Island has primarily flat topography comprised of tidal saltmarsh habitat.

The Tuluwat village site is part of a 6-acre shell mound (clamshell mound) dating back to 900 A.D, which contains remnants of the Wiyot Indian daily life, meals, tools, and ceremonies as well as many burial sites. The site also still contains remains from the 19th and 20th century European settlers who used the shell mound and surrounding areas for agricultural/cattle raising and industrial use. There are dilapidated remains and chemical contamination from a ship repair yard that operated sporadically from 1870 to 1990. Remnants include man-made dikes, a boat dry dock, and structures including the associated residence (Figure 14.17). The site also has several stands of introduced, nonnative plant species.

PROJECT GOALS

During the development of an Indian Island Cultural and Environmental Restoration Project Feasibility Study (HWR, 2003), tribal members working with others in the community developed a set of goals to help them create a vision of what the reemergence of Tuluwat would look like and how it would function. The goals of the Tuluwat Village Restoration Project are to:

- *Re-establish the Tuluwat Village World Renewal Ceremony*
  The Wiyot Tribe will restore their tradition of hosting the annual World Renewal Ceremony on Indian Island. Tuluwat Village will provide a place for tribal members and visitors to carry-on their ceremonial customs such as dancing, sharing food, singing songs, and saying of prayers.

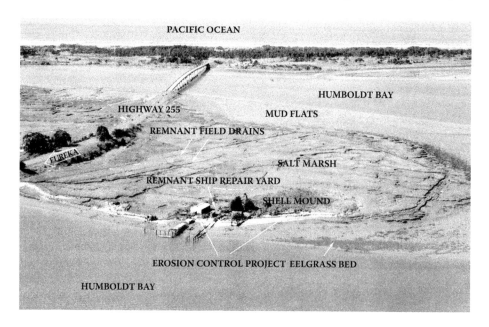

FIGURE 14.17 Aerial photo of Tuluwat Village site on Indian Island, Eureka, California. Photo by Don Tuttle (2004).

- *Protect sensitive cultural and archaeological resources*
  Over the years, archaeological digging, uncontrolled looting, and erosion have taken their toll on the site. In addition, increased visitation to Tuluwat could bring impacts to the shell mound. The Tribe's goal is to implement both physical and programmatic solutions to this problem.
- *Improve surrounding salt marsh habitat*
  The Tribe intends to restore the salt marsh habitat to a degree that it becomes self-sustaining with minimal maintenance. Their goal is to have a salt marsh habitat that thrives with healthy and diverse native plant life and wildlife. The Tribe also hopes to restore the natural tidal hydrology. Tuluwat Village will also provide a secure place for biological and hydrological studies related to Humboldt Bay, and will be a base for the Tribe's environmental monitoring programs.
- *Provide a venue for educational and cultural field trips*
  The Tribe will host educational, cultural, and research uses for tribal members and for the public on the island. The variety of uses will include field trips teaching about local Native American culture and natural history.

THE VISION

To fulfill the goals of the Tuluwat Restoration Project, the Tribe plans to redevelop a place for the World Renewal Ceremony where visitors can learn about the history of Indian Island and the culture of the Wiyot Tribe. Access to Tuluwat will be by boat or ferry. This will help to create a strong sense of transition for the visitor as they approach this special place. The dominant structural feature will be the dance house, which will be adjacent to the "Circle of Tribes" gathering area (Figure 14.18). Other structures will include a refurbished caretaker's residence, kitchen area, and a vault toilet. In order to minimize impacts from redevelopment, some of the new structures will be located on existing foundations. The look and feel of the space will be based on a pre-contact Wiyot village with architectural features that will highlight traditional Wiyot architecture. Landscaping will utilize

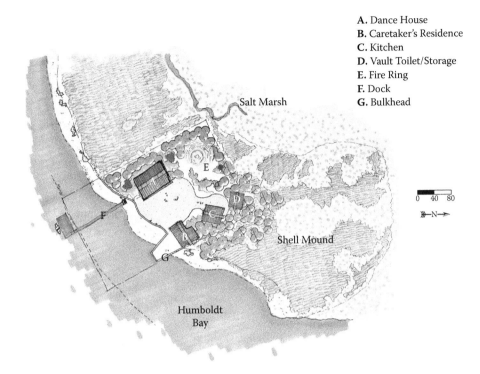

A. Dance House
B. Caretaker's Residence
C. Kitchen
D. Vault Toilet/Storage
E. Fire Ring
F. Dock
G. Bulkhead

Salt Marsh

Shell Mound

Humboldt
Bay

FIGURE 14.18 Illustrative site plan, Tuluwat Village, Eureka, California, J.A. Brennan & Associates; edited by HWR Engineering and Science (2004).

native species, be designed to facilitate outdoor gatherings, and will include tables and interpretive signage. When fully developed, the site will be designed to receive a peak of up to 150 ceremonial participants as well as twice monthly tours of up to 30 people.

The project site location and its preexisting uses as a ship repair facility have created several significant challenges to the redevelopment effort. Access to the project area is extremely challenging because there are no roads, trails, or boat landing dock. All materials must be transported to the site by barge or boat. Protection of the cultural resources, including the shell mound itself, provides a unique constraint for the development activities as well as proposed use of the site. The contamination of the site from previous shipyard activities compounds these challenges. The elements of the Tuluwat Restoration Project designed to balance the Tribe's vision with the many environmental and cultural concerns are described further below. The Tribe's Environmental Program Manager, Andrea Davis is the primary contact for the project and Mike Wilson of HWR Engineering & Science, Arcata, has assisted with many project management tasks; J.A. Brennan and Associates, Seattle, Washington provided landscape architecture services.

CULTURAL STRUCTURES AND INFRASTRUCTURE

CEREMONIAL DANCE HOUSE

The most significant element of the Tuluwat Restoration Project is the construction of a ceremonial dance house. The dance house design will emphasize traditional Wiyot Indian pole and split-plank architecture. The Tuluwat Village dance house will be approximately 40 × 50 ft in size, fully enclosed, and will have a central floor of wood or dirt. The central floor will be surrounded by rows

FIGURE 14.19 Illustrative section of proposed dock, Tuluwat Village, Eureka, California, J.A. Brennan & Associates (2003).

of elevated benches. There will also be two small dressing rooms at one end of the structure. The dance house will be constructed reusing the concrete perimeter foundation and pad and, if possible, the steel frame of an existing building.

REBUILD DOCK/DOCK ACCESS

Tribal and public access is very important in furthering the goals of the Tribe. Given the nature of Indian Island and the current conditions of the project site, access is highly limited due to the fact that there area no roads, trails, or dock. The traditional method of access to the island was by boat. This limits the number of people that can access the site and can exclude those challenged by physical mobility. It is the intent of the Tribe to provide appropriate access to the site in order to facilitate project activities. The installation of a modern dock structure is critical in this respect.

The existing dock is in extreme disrepair and requires complete replacement. Rebuilding the dock will greatly increase site accessibility by allowing vessels to easily carry visitors to the site.

The dock will be constructed of either concrete panels or aluminum. The interior width of the dock will be approximately 6 ft. The dock will have an 80-ft aluminum alloy gangway. There will be a floating dock assembly, 30 ft wide, and 20 ft deep to accommodate the berths of vessels up to 50 ft long (Figure 14.19).

REFURBISH BULKHEAD

A degraded bulkhead will be refurbished using the existing concrete footing and wall as a foundation. This structure will allow for access to the site by small barges during construction and after project completion will provide a bay overlook and a place for reflection.

LANDSCAPING AND TRAILS

The landscaping for the project will match or enhance the ecological and cultural importance of the site. It will provide privacy, and will blend/buffer site activities with the surrounding environment. The landscaping and trails will be designed to foster a park-like setting. Landscaping will include amenities such as picnic tables and benches, interpretive signs, information kiosk, and hard surface trails. A fire ring area and meadow will provide gathering areas for visitors, and will lend coherence to the overall landscape design. Nonnative vegetation will be replaced with native species appropriate to the site, such as wax myrtle (*Myrica californica*), shore pine (*Pinus contorta*), sitka spruce (*Picea sitchensis*), and twinberry (*Lanciera involecrata*).

Camp Kitchen

Food is a very important part of northern California Indian ceremonies and for this reason the project includes construction of a "camp" kitchen. This is an open, covered shelter for serving potluck-style foods during ceremonies. The kitchen will have countertop space, cabinets, shelves, and a temporary self-contained wash station.

Caretaker's Residence

For years the site has experienced ongoing problems of trespassing, vandalism, and looting. Thus, security is an important element of the proposed project. To provide security for the site, the Tribe plans to hire a caretaker/manager to watch over the area and act as a guide for visitors. In order to support this function the Tribe will restore and renovate the existing residence to be used as the new caretaker's building.

The caretaker's building will require a new roof and major remodeling to make it habitable. Total replacement may be required if the existing framing and foundation is not salvageable. It will be single-story and approximately 500 square feet, with a kitchen and a bathroom. Electricity will be supplied by solar and possibly wind generation with a gas generator to be used as a backup. Nonpotable water for the caretaker's residence will be provided via an existing well located on the project site.

Vault Toilet

For public use, the project proposes to build a 16 × 16 ft vault toilet structure with two separate toilet rooms. Typically, the tank is underground; however, in order to minimize impacts to cultural resources the tank ("vault") will be above-grade; the restrooms will be located directly above the tanks and will be accessible by stairs and/or a ramp.

Utilities

Currently, there are no utilities and none will be installed or extended to the project area. All utilities required for construction and future site use will be self-contained and/or portable. Potable water will be transported to the site, both for the caretaker's residential use and for all events/activities on site. Electricity demands will be met using solar and battery storage, or small wind systems, and/or diesel or propane generator(s). Cell phones will provide telecommunications service. All recycling and garbage will be transported off site and disposed of on the mainland.

Cleanup, Remediation, and Resource Protection

Impacts to the Tuluwat have come in many forms. A change in the way the shell mound has been managed has left it susceptible to erosion. Agricultural endeavors caused the salt marsh adjacent to the shell mound to be diked and drained, and uncontrolled, amateur explorations and looting resulted in significant soil excavations. Furthermore, the century-long usage of part of the shell mound as a ship maintenance and repair yard left the area contaminated with paints, solvents, petroleum products, and various other chemicals that are a source of environmental and human health hazards (HWR, 2003).

The cleanup, remediation of hazardous materials, and protection of the cultural resources is the first step to preparing the site for the development of access, cultural structures, and landscaping. Cleanup, remediation, and protection consists of the following tasks:

- Erosion control
- Debris removal
- Hazardous waste remediation
- Protective geotextile and fill
- Ecological restoration

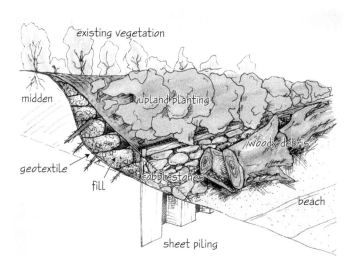

FIGURE 14.20 Illustrative cross-section of erosion control structure to protect Tuluwat Village Shellmound, Eureka, California, HWR Engineering and Science (2006).

Erosion Control

When the Tribe purchased the site in 1999 there was significant erosion impacting the shell mound. Temporary erosion control measures were implemented to protect the mound while a long-term erosion control plan was developed. In 2006 the Tribe installed an ecologically sensitive barrier to protect the shell mound that also enhanced surrounding wildlife habitat values. The 400-foot long bio-engineered revetment was installed using minimal heavy equipment and blends well with the aesthetic of the surrounding island and bay. It utilized carbon reinforced fiberglass sheet piling (Northstar Endurance CSP) as a footing to support rock and soil backfill combined with geo-grid (Stratagrid SG150) (Figure 14.20). The fiberglass sheet pile was selected for durability and aesthetics as well as minimal impact and cost. Native Hooker's willow (*Salix hookeriana*) was integrated into the structure to help bind and conceal the backfill. To provide diversity of habitat, a variety of other native plant species such as red fescue (*Festuca rubra*), red flowering currant (*Ribes sanguinium*), twinberry (*Lonicera involucrata*), wax myrtle (*Myrica californica*), yerba buena (*Satureja douglasii*), pickleweed (*Salicornia virginica*), salt rush (*Juncus lesueurii*), and saltgrass (*Distchilis spicata*) were also installed and are now flourishing (Figure 14.21 and Figure 14.22).

Debris Removal

The cleanup and remediation of contaminated materials at the Tuluwat Village site is integral to the long-term goals of the Wiyot Tribe. In 1999, the Tribe performed an initial environmental cleanup of the site in which 1,300 pounds of identifiable hazardous waste were removed. This included the removal of several dumpster loads of trash, hazardous materials such as lead acid batteries, paint, solvents, various oils, deteriorated machinery, and collapsed wooden structures (Figure 14.23). To date, over 30 tons of industrial scrap metal and wood have been removed from the project site, most of which has been recycled. Much of the debris removal has been done using volunteer labor and donated services from local contractors and businesses (Figure 14.24). Remaining debris and dilapidated structures will be removed before hazardous waste remediation begins.

FIGURE 14.21 Erosion control structure to protect the Tuluwat Village Shellmound, Eureka, California. Note: Temporary deer fencing to protect native planting and additional cobble facing still to be installed. Photo by A. Davis (2006).

Hazardous Waste Remediation

To move forward with the cleanup of the site, the Tribe requested assistance of the U.S. Environmental Protection Agency (USEPA) Brownfields program. Between 2001 and 2005 Phase I, Phase II and II-B assessments were conducted (USEPA, 2002, 2003, 2005). According to the assessments, total petroleum hydrocarbons–organics (TPH-o) and polyaromatic hydrocarbons (PAH), and metals such as copper, arsenic, nickel, and aluminum contamination were detected in near-surface soils (midden) in the former boatyard area (approximately 0.3 acres). Significant levels of pentachlorophenol (PCP) and pesticides were also detected in a limited area (approximately 0.1 acres).

FIGURE 14.22 Volunteer tending to native Hooker's willow (*Salix hookeriana*) used to reinforce the rock, soil, shell, and geogrid revetment, Tuluwat Village, Eureka, California. Photo by M. Wilson (2005).

FIGURE 14.23 Initial removal and disposal of various hazardous waste from the Tuluwat Village site, Eureka, California. Photo by M. Wilson (1999).

The Brownfields assessments indicated that groundwater beneath the area is not currently impacted with TPH or pesticides/PCBs. However, arsenic and highly chlorinated dioxin congeners (dioxin) that are generally associated with PCP contamination were detected in some groundwater samples. The source of both the dioxin and the arsenic (above background) in the groundwater appears to be from the wood treatment of ships that were worked on in the dry dock area.

After the extent of the pollution was determined, remedial activities were recommended. Due to the cultural and historical significance of the midden materials that make up the shell mound, it was determined that the remedial activities would focus on the protection of visitors to the site and, to the extent possible, removal of contaminants that may continue to threaten the surrounding environment. After a protracted negotiation process with the Tribe, the State Historic Preservation Office, National Park Service, U.S. Army Corps of Engineers, and the State Regional Water Quality Control Board, it was determined that material that contains dioxin at levels down to 50 parts per billion (ppb) would be excavated and disposed of in an appropriate landfill. This will result in the removal of up to 17 cubic yards of culturally significant midden. It is not proposed to excavate midden contaminated with metals and hydrocarbons because these constituents are considered stable

FIGURE 14.24 On numerous occasions volunteers have assisted in the removal of debris, garbage, and scrap metal from the Tuluwat Village site, Eureka, California. Photo by M. Wilson (2002).

and not significantly affecting groundwater (Table Bluff Reservation, 2006). This does not preclude the potential injection of an *in situ* treatment process to address remaining PCP, dioxin, and metals contamination, if a method is found that will not impact cultural resources.

Protective Geotextile and Fill

In addition to the excavation, a soil/geotextile cap will be installed. The fill will consist of 6 in. to 1 ft of imported soil or base rock covering the entire 0.75-acre upland zone of the former boatyard. The purpose of the cap is to protect visitors from possible contact with the residual soil contamination. A 4-acre portion of the shell mound, which does not have hazardous materials contamination, will also be capped in order to protect the cultural resources from potential visitor impacts. Existing vegetation will be removed, the surface will be lightly hand-graded, and a geowebbing will be installed. Soil will provide cover for the geoweb as well as divots and depressions left from uncontrolled archaeological digging and looting. The area will then be landscaped as a native grass meadow surrounded with red alder, shore pine, and wax myrtle, which will provide a visual buffer and wind break.

Ecological Restoration

The goal of the ecological restoration element of the Tuluwat Restoration Project is to improve the tidal salt marsh and upland environments of the project area in a way that is complementary to the cultural and historic values of the island and of the Humboldt Bay region. The objectives of the ecological restoration phase are to:

1. Redefine and/or improve the tidal channel system to allow for more natural tidalhydrology, function, and aesthetic quality.
2. Maximize native plant and wildlife habitat and diversity.
3. Create a passively managed system that minimizes maintenance needs.
4. Provide the tribal community and the general public opportunities for education and enjoyment.

The Tribe proposes to improve habitat of upland areas by removing nonnative plants and reintroducing appropriate native plants. In addition to restoring the upland areas, the Tribe plans to increase the tidal inundation to the site by piercing the remnant dike in several locations and increasing the channel sinuosity of several of the field drain/tidal creeks. The Tribe will also incorporate nesting areas and woody debris into the salt marsh area with the goal of increasing habitat structure and diversity. A detailed implementation plan will be developed before restoration activities begin.

PHASES OF THE TULUWAT RESTORATION PROJECT

The Tribe will implement the project in phases, which will allow the Tribe to efficiently manage the project and maximize funding opportunities, as listed below:

- Phase 1: Erosion Control (Completed Year 1–2)
- Phase 2: Cleanup and Remediation (Year 1–3)
  - Set-up construction mobilization and staging area
  - Remove debris
  - Clean up and remove the majority of contamination
  - Cap contaminated area
- Phase 3: Public Access (Year 3–5)
  - Cap area(s) to protect cultural resources
  - Rebuild dock

- Refurbish bulkhead
- Install landscaping and trails
- Phase 4: Cultural Development (Year 4–8)
  - Build and install:
    - Ceremonial dance house
    - Camp kitchen
    - Refurbished caretaker's residence
    - Vault toilet
    - Additional landscaping and interpretive signage
- Phase 5: Ecological Restoration (Year 4–8)
  - Redefine and/or improve the tidal channel system to allow for more natural tidal hydrology, function, and aesthetic quality.
  - Restore native plant and wildlife habitat and diversity.

CONCLUSION

Both the UIHS Potawot Health Village and the Wiyot Tribe's Tuluwat Village Restoration Project are important examples of how architecture, engineering, and environmental restoration can be important forces in efforts to support the cultural vitality of the American Indian communities of California's north coast. These projects demonstrate how traditional native values, world views, and knowledge can be integrated into modern design to achieve multiple objectives and create diverse land use projects that are of great value to both American Indian and non-Indian people.

LITERATURE CITED

Anderson, K. 2005. *Tending the Wild: Native American Knowledge and the Management of California's Natural Resources,* University of California, Berkeley, CA.

Berry, W. 1996. *Another Turn of the Crank: Essays.* Counterpoint.

Blackburn T. and Anderson K. 1993. *Before the Wilderness—Environmental Management by Native Californians.* Ballena Press, Menlo Park, CA.

Ferreira, Mariana K.L. 1996. *Sweet Tears and Bitter Pills. The Politics of Health among the Yuroks in Northern California.* Unpublished Doctoral dissertation, University of California at Berkeley and at San Francisco, CA.

France, R.L. 2003. *Wetland Design: Principles and Practices for Landscape Architects and Land-use Planners.* W.W. Norton.

Heffner, K. 1984. *"Following the Smoke" Contemporary Plant Procurement by the Indians of Northwest California.* Six Rivers National Forest, Eureka, CA.

Heizer, R.F., Elsasser, A.B. 1980. *The Natural World of the California Indians.* University of California Press, Berkeley, CA.

HWR. 1999. *UIHS Health Village Conservation Easement Management Plan,* unpublished planning document on file with the City of Arcata, CA.

HWR. 2003. *Indian Island Cultural and Environmental Restoration Project (IICERP) Feasibility Study,* unpublished planning document on file with the Wiyot Tribe, Loleta, CA.

Innovative Technical Solutions, Inc. (ITSI), 2002. *Phase I Brownfields Assessment,* prepared for U.S. Environmental Protection Agency.

Loud, L.L. 1918. *Ethnogeography and Archaeology of the Wiyot Territory.* Vol. 14, No. 3, pp. 221–436, University of California Publications in American Archeology and Ethnology, Berkeley CA.

Margolin M. 1981. *The Way We Lived—California Indian Stories, Songs, and Reminiscences.* Heyday Books and the California Historical Society, Berkeley, CA.

National Historic Landmarks Program (NHLP), 1964. National Parks Service.

Nomland, G.A. and Kroeber, A.L. 1936. *Wiyot Towns.* University of California Press, Berkeley, CA.

Roscoe, et al. 1996. *Cultural Resources Investigation of the Proposed United Indian Health Services Medical Complex.* Roscoe Archaeological Consulting. Eureka, CA.

Table Bluff Reservation-Wiyot Tribe, 2006. *Site Cleanup Plan, Tuluwat Village, Indian Island, Eureka, California.* Prepared by SHN Consulting Engineers & Geologists.

USEPA, USACE. 2002. *Targeted Brownfield Site Assessment, Indian Island, Eureka, California. Phase I Report.* Prepared by Innovative Technical Solutions.

USEPA, USACE, 2003. *Targeted Brownfield Site Assessment, Indian Island, Eureka, California. Phase II Final Report.* Prepared by Weston Solutions.

USEPA, USACE, 2005 *Targeted Brownfield Site Assessment, Indian Island, Eureka, California. Phase II B Addendum.* Prepared by Weston Solutions.

Waller, G., Kadlecik, L., Ornelas, S., and Wilson, M. 1996. *Restoration, Agriculture and Stormwater Management at Health Village.* George Waller Wetland Systems, Arcata CA.

Winzler and Kelly, Consulting Engineers. 1996. *UIHS Project: Wetland Delineation Report.* Eureka, CA.

Winzler and Kelly, Consulting Engineers. 1997. *Final Environmental Impact Report for United Indian Health Services Health Village.* Eureka, CA.

15 Preserving Cultural and Natural Resources

The Site Development Plan for the Sumpter Valley Gold Dredge State Heritage Area

Carol Mayer-Reed

CONTENTS

ABSTRACT

As the population of the western United States builds and tourism expands as an important economic industry, pressures to preserve or restore natural resources is an increasing concern. Restoration of mineral extraction sites has begun to take place. However, in Sumpter Valley, Oregon, 80 acres of an historic gold mining site has been set aside as a State Heritage Area to preserve and interpret the stark remains of mining activities that devastated tens of thousands of acres. Visitors to the site will understand the extraction processes, but also witness evidence of how nature has begun to regenerate the landscape over 50 years' time.

INTRODUCTION

Gold mining affected 26,000 acres of the Sumpter Valley in northeast Oregon from 1897 through 1954 (Figure 15.1). Several gold dredges mined the valley during this period of time, leaving vast tracts of stark, cobbled rock tailings where pristine mountain creeks, vegetated wetlands, and meadows

FIGURE 15.1 The town of Sumpter was established in close relationship to the gold mining site.

abundant with wildlife once flourished. Mining operations of the dredges left tailing patterns that ran perpendicular to the natural course of the drainage ways. Clear, cold mountain streams that flowed through the meadows scoured new channels through the massive piles of tailings. The deep, rich soils of the valley eroded. Sediments were transported by the waterways to remote places downstream. The once-rich valley floor was left nearly devoid of vegetation, except for some small islands of the landscape that had not been dredged. The environmental impacts of gold dredging in Sumpter Valley were significant.

The peak of the gold mining boom in this region was between 1897 and 1903. The town of Sumpter grew dramatically during this time, expanding to about 1,400 acres of land and a population of 30,000 people. At that time, the town was complete with a brickyard, sawmill, smelter, railroad, electric lights, and an excellent potable water system. Despite Sumpter's remote location, there were 16 saloons, several hotels, 2 banks, 4 churches, a hospital, a dairy, 2 cigar factories, and several assayers. There was even an opera house used for grand balls and traveling vaudeville shows. Quite an extensive Chinatown developed as well. In August 1917, a disastrous fire burned the town to the ground (*Gold in Sumpter Valley*, 1968). The fire, coupled with dwindling gold resources, reduced the population of Sumpter dramatically. Today, over a century later, Sumpter has a resident population of only about 175 persons. Now tourists pass through the town during the summer to reach camping and fishing destinations in the surrounding mountains.

Interestingly, time was a healing agent of the ignored and unmitigated landscape of sterile cobbled tailings on the valley floor (Figure 15.2). The environmental devastation was not remediated by the hand of man. Through nature's own restoration processes, the landscape of the Sumpter Valley has evolved during the last 50 years after cessation of gold mining operations. Within this devastated environment, new regimes of vegetation have regenerated. New wetlands have formed in low-lying, poorly drained residual waterways between tailings. Creeks that flow through the valley,

FIGURE 15.2 Riparian vegetation has colonized along the margins of the tailings over time.

despite their twisted unnatural courses, have complex margins of riparian vegetation and willow thickets. Beaver dams are present throughout the area, creating even more extensive wetlands and ponds. Populations of resident wildlife have begun to inhabit the valley and during certain seasons the area supports significant migratory waterfowl.

The healing of the landscape through natural processes is a worthy and optimistic interpretive story and focus for the Sumpter Valley Gold Dredge State Heritage area. Gaining an understanding of the gold mine industry, along with its operations and impacts, is particularly important for the increasing population of visitors to the region. Tailings are preserved rather than removed as a denial or erasure of past evidence. Preservation of the gold dredge and its landscape of raw, unmitigated tailings is key to the visitor experience. Interpretations of this stark landscape helps to portray the colorful stories and folklore that took place in northeastern Oregon environment and culture.

In 1971 the historic gold dredge, still largely in tact and positioned in its final operational location, was placed on the National Register of Historic Places. Then, 22 years later, 80 acres of the original gold mining site, including the dredge, were acquired by the Oregon Parks and Recreation Department. The main goal of the acquisition was to preserve, conserve, and interpret the historic, cultural, and natural resources of the site and establish it as a State Heritage Area.

Interestingly, the best evidence that the landscape has begun to reforest and improve aesthetically are the vacation cabins springing up in certain areas of the dredged valley. With tourism and real estate values on the rise, it is even more critical that the State Heritage Area protect the healing natural resources, wildlife habitat, and a portion of the cultural landscape that was mined.

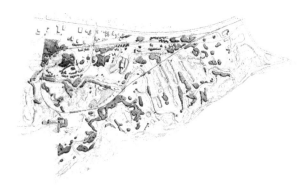

FIGURE 15.3 The Site Development Plan illustrates how visitor trails, viewpoints, and a railroad loop are incorporated into the remnant tailings.

GOALS OF THE SITE DEVELOPMENT PLAN

The Site Development Plan was commissioned by Oregon State Parks and Recreation and prepared in the year 2000 by Mayer/Reed, a Portland-based landscape architecture and environmental graphics consulting firm (Figure 15.3). This plan followed a previous overall master plan prepared by the state parks department in 1994. The 1994 plan outlined a program for facilities that would meet the needs of tourists, while the subsequent 2000 plan was developed to a greater level of detail and an expanded interpretive program (Sumpter Valley Dredge State Park Master Plan, OPRD, 1994.) The Site Development Plan includes strategies for how to plan the gold dredge site and coordinate town planning with a tourism focus.

The Site Development Plan sets forth concepts for how visitors can become immersed in the landscape and how facilities such as parking, trails, and overlooks are incorporated within the heavily impacted gold mining environment. Specific site elements are designed to preserve the unique character and authenticity of the site as well as underscore interpretive themes, rather than simply apply generic state parks standards (*Park and Recreation Structures*, 1990).

In that the Sumpter Valley Dredge State Heritage Area encompasses significant natural and cultural resources, interpretation of these resources is key to an understanding of the site changes over time. Out of these historic economic, cultural, and physical patterns caused by gold exploration and extraction operations, site visitors begin to understand the following inter-related key messages:

1. Underlying geology and water systems that produced the gold and mineral deposits in the region
2. Human discovery, motivation, and drive behind the industry of gold extraction
3. Significant regional, social, and cultural impacts of mining operations
4. Environmental impacts, technology, and machinery of the mining operations
5. Ongoing evolution of the natural environment through time since dredging was ceased in 1954

Development of the site for interpretation and use by visitors focuses on two major attractions and related landscape features. The major attractions are the historic gold dredge as an artifact of this industrial archaeology, and the narrow-gauge Sumpter Valley Railroad that transported timber for the logging industry. Themes include interpretation of the massive dredge tailings and other types of placer mining, as well as native plant communities that have established on the site and the related wildlife. An on-site visitor experience and visual comparison of the dredged site with the nondredged landscape is critical to this understanding.

In addition, the Site Development Plan describes how improved vehicular and pedestrian connections provide more direct integration with Sumpter's commercial district to improve tourism and demonstrate the close interrelatedness of gold mining operations with the town itself. The interpretive story also underscores past attitudes toward the discovery of natural resources and the economically driven choices made by the culture of that time.

SITE INTERPRETIVE OPPORTUNITIES

The 80-acre State Heritage Area is divided into zones that are determined to be most suitable for preservation, conservation, and development of visitor amenities. For purposes of the site development plan, these five zones are defined as: (1) predredge landforms, (2) cultural restoration landscapes, (3) transitional landscapes, (4) surface water systems, and (5) park activity areas.

PREDREDGE LANDFORMS

Predredge landforms are areas of the site that were not mined by dredge operations (Figure 15.4). These areas include flat meadows and original wetlands, adjacent forested slopes (even if logged), and landscapes that appear to be native or otherwise nondisturbed. Topography in these areas appears to be consistent with what would have naturally existed. Areas containing mature trees are assumed to be second-growth vegetation, but landforms are largely intact. Since the area was logged in the late 19th and early 20th centuries, original old-growth trees do not exist on site or adjacent off-site forested slopes (*An Illustrated History*, 1902).

Interpretive opportunities for pre-dredge landforms address the fundamental ecological systems of the vicinity. Visitors can imagine what Sumpter Valley must have been like in pre-dredge and pre-agricultural times before cattle ranching was prevalent in the valley. Viewpoints along trails are set up for visitors to make direct comparisons between the nondredged landscape and the dredged areas.

Wildlife habitat in current and natural states in pre-dredge landforms is protected and restored or enhanced when possible. Wetlands are preserved and the adjacent forested areas retain the character and function of pre-dredge conditions, both on- and off-site. Management recommendations for the pre-dredge landforms include removal of invasive, non-native plant species and restoration of native plant communities. Trail access must be avoided or be extremely limited in these sensitive areas.

CULTURAL RESTORATION LANDSCAPES

Intact tailings within the State Heritage Area as a cultural landscape are preserved and protected from disturbance, especially given that only 80 acres of the 26,000 acres mined will be included in the State Heritage Area (Figure 15.5). Cultural restoration landscapes are the intact dredge tailings and resultant

FIGURE 15.4 Pre-dredge landforms are unmined areas with intact native vegetation on site.

FIGURE 15.5 Cultural restoration areas are linear mounds of intact tailings.

ponds as remnants of actual dredge operations. The tailings created by the sweeping motion of the gold dredge stacker formed extensive linear mounds of stone cobbles. In some places, the tailings achieve heights of 14–16 ft. Tailings range from 170–700-plus ft long and average 95 ft wide. Tailings typically have very steep 2:1 side slopes with small detailed vertical ridges. Tailings are best observed in early morning or late afternoon when sunlight and shadows feature the texture of the ridges. The cobbled side slopes of tailings are highly unstable, unsafe, and not suitable for public access.

The interpretive value of the tailings and ponds is to show the significant alteration of the original landscape and long-term effects of gold dredge operations. In very limited areas, direct contact with the tailings allows visitors an opportunity to experience the stark, massive piles of cobbles. Devoid of soil and vegetation, the microclimate is obviously altered, particularly in the summer when the reflected heat of the tailings is acutely apparent to the visitor. After experiencing the tailings first hand, the larger picture (as conveyed through maps and aerial photos) of the environmental impact is better appreciated. The extent of tailings can be seen for several miles in Sumpter Valley; the vastness of the impact of dredging operations is truly staggering.

To enhance cultural landscape preservation, site development that allows visitors to experience the tailings is limited to the narrow gauge railroad alignment, and a few pedestrian trails and viewpoints. In the cultural landscape zones, vegetation should be removed from intact tailings in order to preserve the stark tailing landforms so that visitors better understand the true visual and environmental impacts of the mining operations. Furthermore, areas of the site adjacent to the dredge and dredge pond are key to the visitor experience. In this location, tailings are reconstructed where they were flattened or removed around the historic dredge. The reconstruction of tailings helps to complete the story of how the dredge moved through the site and created the pattern of landforms still seen.

Presently, the gold dredge is completely restored and open to the public. Visitors can tour it both inside and out to understand its mechanics, operation, and massive scale (Figure 15.6).

Transitional Landscapes

Transitional landscapes were once tailings or otherwise severely disturbed, but are overtaken with native vegetation (Figure 15.7). Between tailings, numerous linear and circular remnant ponds are filled with water, ranging in depth from a few inches to 10 ft or more. Thickets of riparian plants are established around the remnant ponds, evolving into valuable wildlife habitat for feeding and nesting. Alder, cottonwood, willow, and dogwood have taken hold in the wetlands and on creek banks, while native species of higher order vegetation such as Douglas fir and ponderosa pine are establishing on the drier stony ridges.

FIGURE 15.6 The historic gold dredge has been renovated as a visitor attraction.

The key interpretive message for the transitional areas is that in 50 years' time since the cessation of dredging, soils have collected and nature has begun to reclaim select portions of the site. New waterways have been essential to the re-establishment of wildlife over time. The landscape continues to evolve in nature despite the high degree of devastation that occurred from mining activities.

The more detailed version of the interpretive story depicts how new soil is created and accumulated a particle at a time, how seeds disperse, and how key species of native plants colonize even a highly disturbed site. The phenomenon is a demonstration of how essential microclimatic and topographic conditions, through slope orientation, moisture, and nutrients create an environment that is conducive to the most opportunistic pioneering plant communities. This is an amazing and inspiring story of regeneration (Figure 15.8).

Transitional areas of the site are undisturbed and protected to aid in the evolutionary process already underway. Visitor access is limited to several overlooks that focus on the interpretive themes.

SURFACE WATER SYSTEMS

As a result of the dredging process, the creeks and rivers of the valley were drastically altered by mining operations (Figure 15.9). Most areas of the site were dredged to a depth of about 12–14 ft in

FIGURE 15.7 Transitional landscapes are where dense thickets of native vegetation and wildlife habitat have established at the dredge.

FIGURE 15.8 Small opportunistic plants are the first to colonize among the cobbles of the tailings.

the search for gold. Original surface water systems through the site include the Powder River and tributary creeks that have remained fast flowing through the valley. On the flat original valley floor there were a number of oxbows. During and after dredging, water coursed above and below grade in channels through the massive tailings. There are a number of large remnant ponds from which the dredges operated. The tailings have very unnatural steep side slopes, yet thickets of riparian vegetation have managed to establish. A complex system of interconnected oxbows runs artificially perpendicular to the river through the tailings. Isolated ponds of slower moving water have even

FIGURE 15.9 Surface water systems include a number of small linear ponds between the tailing depressions.

FIGURE 15.10 Activity areas provide hands-on learning as children pan for gold.

greater riparian habitat and populations of waterfowl. The river, creeks, oxbows, and ponds are preserved and protected from impacts of park development and visitors.

Interpretive messages of the water systems include the Sumpter Valley watershed and its influences. The story includes the following topics:

1. An account of the sources of water on site as generated by the snowmelt in the off-site upper watersheds of the mountainous region.
2. A comparison of how water historically moved and currently moves through the site.
3. How water was the means for deposition of gold and how water was used in extraction process.
4. How water conveyed the valley soils eroded and displaced by dredge operations; and how soils have been deposited in the valleys below.

PARK ACTIVITY AREAS

Park activity areas are parts of the site that have been most heavily altered over time from grading operations, gravel extraction, and pedestrian and vehicular traffic (Figure 15.10). These areas are closest to the town of Sumpter and most suitable for development of park facilities and intensive visitor day use. Visitor use areas include parking, buildings, and general site circulation. Picnic areas are located central to the park activity area. Active recreation and/or camping are not considered to be appropriate anywhere on the site.

Interpretive opportunities for this part of the site include a tour of the historic gold dredge, the narrow-gauge railroad and train station, and a display of gold mining equipment and artifacts. A shallow beach for gold panning near the dredge is a fun, tactile, and interactive way to stimulate the imagination of children and further educate the public about this method of mining. A short interpretive trail provides an overview of site themes, with trailheads to longer trails that offer more in-depth experiences in the landscape. Interpretive materials are focused at key locations and viewpoints.

SITE CIRCULATION

The concept behind the Site Development Plan is to focus the visitor experience to activity areas, and encourage the use of designated trails and key interpretive locations in order to preserve cultural and natural resources (Figure 15.11). Two pedestrian trail systems are an important part of

FIGURE 15.11 Careful trail layout allows visitors to become immersed in the mined landscape.

the visitor experience. The primary trail system links spaces within the activity areas. Trails are barrier-free and constructed of compacted gravels and cobbles that match the surrounding color and texture of the landscape (*Standard Drawings for Construction and Maintenance of Trails*, 1996). At overlooks, visitors' attention is focused to a particular subject by integrating interpretive materials into low stone retaining walls.

The secondary trail system is extended through the larger site and links to off-site trails in the area. Secondary trails are more rustic and provide a number of interesting vantage points and ways to experience tailings, site topography, and wildlife habitat such as beaver dams, vegetation ecotypes, and waterways. Viewpoints and overlooks containing information are located in strategic places to highlight the most significant site features. Interpretive signage is kept to a minimum and appropriately scaled to reduce distraction from site features.

Adjacent to the west slope of the site, an elevated visitor viewpoint is provided in an observation tower. Designed in a heavy timber construction vocabulary similar to Forest Service fire lookout towers in the northwest, the structure affords a sweeping panorama over the site. From this vantage point, large-scale patterns of the tailing landforms, waterways, and impact of the extent of dredge operations are better observed. Interpretive information is located at this site feature.

In addition to trails, an historic narrow-gauge railroad and train station is included as part of the Site Development Plan (Figure 15.12). This train originally serviced Sumpter Valley from the turn of the 20th century. A new track alignment enters the site from the southern boundary, stops at a small visitors' station, and loops back through the State Heritage Area. The narrow-gauge train is an important tourist attraction and provides a way to experience a larger portion of Sumpter Valley both inside and outside the park. This feature is especially appealing for families and the elderly who may find the secondary trail system over tailings too much of a challenge. The train reduces the need for parking on the historic site as many visitors park at a remote off-site lot. Stories and interpretive messages are narrated for visitors during the train ride, making the experience engaging and educational.

FIGURE 15.12 The narrow-gauge train ride allows visitors to tour the site safely without causing impacts to habitat or preserved tailings.

CONCLUSION

The Sumpter Valley State Heritage Area is a valuable part of Oregon's history, preserving the cultural and natural resources of the gold dredge and site. Visitors gain an understanding of the story of gold mining, its impacts, and nature's ability to restore the altered landscape.

As a step toward greater economic stability in the future, the community of Sumpter will benefit from tourism as a sustainable industry, following decades of decline from extractive natural resource activities in the form of mining and logging. The promotion of cooperative private/public partnerships, land use planning, sound transportation planning, and the preservation of natural resources in the region are key to making the entire Sumpter Valley a successful tourist destination.

LITERATURE CITED

An Illustrated History of Baker, Grant, Malheur and Harvey Counties, with a Brief Outline of the Early History of the State of Oregon. Western Historical Publishing Company. (1902)

Gold in Sumpter Valley. Brooks Hawley, Sumpter Stage. Baker, Oregon. (Second printing—December 1968.)

Park & Recreation Structures. Good, Albert H., Graybrooks. Boulder, Colorado. (1990).

Standard Drawings for Construction and Maintenance of Trails. United States Department of Agriculture, Forest Service Engineering Staff. Washington, D.C. (December 1996)

Sumpter Valley Dredge State Park Master Plan. Oregon Parks and Recreation Department. (1994)

PHOTOGRAPHY AND ILLUSTRATION

Aerial & historic photography: Stephen Rich. Sumpter Historic Aesthetics Committee, Sumpter, Oregon.
Watercolor illustrations: Martin Kyle-Milward. Portland, Oregon. (November 2000)

Part 6

Regions

16 Extreme Projects
Ecological Restoration Needs to Address Altered Ecosystems at Larger Spatial Scales

Steven I. Apfelbaum and Neil Thomas

CONTENTS

Introduction 358
Oak Savanna Decline 361
Urban Stormwater Runoff and River Restoration 361
Kankakee Sands, Indiana 363
Littoral, Caracas, Venezuela 363
The Des Plaines River: A Case Study of the Ecological and Hydrological Changes
     and Opportunities for Watershed Restoration 366
    Introduction 366
Problem Background: Site Description 367
Results and Applications of the Findings 369
Natural Ecological System Functions and Processes Should Be Emulated 369
    Water Yield 369
    Sediment and Pollutant Management 369
    Applications 370
    Landscape-Scale Wetland and River Restoration 370
    Flood Mitigation in Restored Flood Plains 372
    Flood Mitigation throughout the Watershed 374
    Methods 374
    Results and Opportunities 376
        Stormwater Management and Flood Mitigation Benefits 376
        Hydraulic Geometry Benefits 376
        Water Quality Benefits 378
Summary 378
Acknowledgments 378
Literature Cited 378

ABSTRACT

Extreme projects represent a series of challenges not typically found in smaller projects. So called extreme projects typically cross multiple ecological, political, and regulatory zones. They also typically integrate urban, industrial, and agricultural land-uses. The midwestern

United States' oak savanna ecosystem is an example of one with hundreds of thousands of platted private parcels involved, most with owners who are entirely unaware of this ecosystem and its restoration and management needs. Currently, exotic species invasions, modified hydrology, and livestock grazing among many other threats and stressors make working with this ecosystem an extreme undertaking. Watershed and river restorations have similar socio-political and regulatory challenges. More importantly, their success depends on a larger scale of thinking. Several exemplary watershed projects are introduced in this paper. These include strategies for addressing the Venezuelan mud slides near Caracas and, in greater depth, the Des Plaines River in Wisconsin and Illinois. The results indicate that restoration of the Des Plaines River must be integrated with watershed-scale strategies. Emulating historic stormwater behavior by integrating natural upland landscape features in urban developments and agricultural lands offers stormwater management benefits such as easier maintenance, reduced cost, and improved aesthetics. Additionally, this chapter suggests that a strategic program of wetland restoration offers a substantial benefit for restoration and stabilization of such rivers as the Des Plaines by reducing flooding peaks, improving water quality, and improving fish and wildlife habitat. Some strategies and techniques in the tributary watershed are summarized later.

INTRODUCTION

Ecological changes in land, rivers, lakes, and oceans as a direct result of human intervention have created a necessity and opportunity to think big in applying ecological restoration and land management principles. Using ecological system health indicators, expansive areas of the earth, from the wilds of Brazil to the ever-expanding urban environments, present immense opportunities and needs for ecological restoration. The use of ecological system health indicators should be foundational in the design and implementation of large-scale restoration projects. Feedback loops and adaptive restoration can be simplified, monitored, and used by professionals and lay persons around the world. Key indicators of general ecological system health have the following characteristics:

1. *Biodiverse.* In general, healthy systems have the capacity to support a full range and compliment of plant and animal species (Figures 16.1 to 16.4). Studies of largely unaltered reference areas, and analysis of the trends through the process of their alteration, have provided a basis for understanding the likely trajectory and targets that may be associated with restoring biodiversity.

2. *Dynamic.* Ecological systems ebb and flow with changing climate conditions (and other conditions including biological, human, disease and insect infestation, fire, grazing ungulates, etc.), with drier site species moving down hydrological gradients during excessive dry periods and displacing or coexisting with persistent wetter site species during these times. The ability of ecological systems including component species to move across landscapes, assemble and reassemble, and maintain and recuperate from perturbations is a manifestation of the unfettered dynamic capacity of largely unaltered systems.

3. *Productive.* In general, compared to monoculture and simple polyculture agricultural systems, most native landscapes are highly productive. Healthy ecological systems produce a diversity of life and multiple levels of productivity (i.e., mammals, birds, fish, humans, vegetation, phytoplankters, etc.).

4. *Stinginess.* Healthy ecological systems generally retain and slowly release water, soils, and nutrients in antithetical contrast to the urban developed landscapes. These human created systems freely release stormwater, nutrients, contaminants, eroded soils, and life supporting energy.

FIGURE 16.1 The historic Midwestern U.S. ecosystems were largely comprised of various prairie types growing from elevated high dry soils through wetter settings. Dry site species moved down from the drier landscapes during dry periods and then moved up to these drier habitats during wetter climatic periods. The complex mosaic of ecological systems that included prairies, wetlands, forests, savannas, etc., were dynamic, productive, stingy, and diverse. These prairies formerly occupied millions of acres contributing to stormwater management functions on large landscapes. These functions have largely been lost and have not been replaced with today's agricultural and developed land uses. (Color versions of this and all subsequent figures are available from the author.)

5. *Natural capital.* Ecosystems have a storehouse of genetic, behavioral, and other self-sustaining attributes. Capital can be in the form of wood for humans or termites, clean water for waterfowl and humans, oxygen cycling, and air cleansing for all life, etc. The capacity of ecological systems to function and the latent and potential storehouse of products, energy, structure, diversity, and productivity are irreplaceable attributes. Attempts to restore these attributes are often overly costly in material and time. Thus, resulting in few ecological system functions being restored.

FIGURE 16.2 Wetlands also occupied many millions of acres in the Midwestern U.S. Their contribution to the functions of the ecosystem, including hydrological, is only now becoming appreciated.

FIGURE 16.3 The agricultural systems now present on the land do not have similar characteristics as the historic prairie and wetland ecosystems. Even with the very best conservation practices these agro-systems have opposite functions as historic ecosystems.

Applying a human health model to the above general characteristics of healthy functioning ecological systems and incorporating a testable hypothesis, simplifies designing and implementing ecological restoration plans at all scales.

Extreme ecological restoration projects have several characteristics that distinguish them. First, out of necessity the design and successful implementation always involves achieving balance

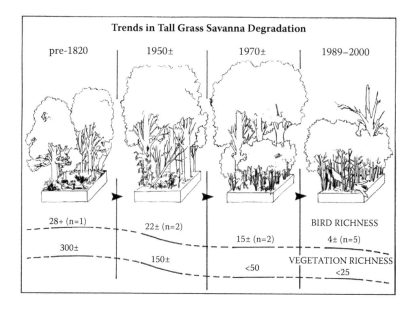

FIGURE 16.4 In the U.S., degradation of oak forests and oak savanna systems are occurring over vast landscapes. A concurrent decline of native breeding songbirds, native ground vegetation, and the oak trees has been documented. Other trends include: increased erosion and a conversion from previous ground water infiltration fed streams to systems now fed by surface runoff from stormwater. Many secondary problems are associated with these extreme, wide-ranging changes to the land.

between the economic aspirations of those who own and operate the land and water bodies, the natural resources, and often human cultural aspirations. This is very different from the small home lot prairie restoration, or wetland restoration in the town park, where in each case these three variables may play a part but, usually only to make sure people understand the project. In extreme projects, human livelihoods, regional and national economics, and cultural rethinking at many spatial and temporal scales will be required.

OAK SAVANNA DECLINE

Research conducted throughout the midwestern U.S. has suggested oak system decline is occurring in many regions (Haney and Apfelbaum 1990; Apfelbaum and Haney 1989; Apfelbaum and Chapman 1993). As depicted in Figure 16.4, this decline is directly related to a reduced frequency or absence of historic controlling wildfires, altered or absent grazing ungulates, and hydrological alterations associated primarily with dewatering from adjacent agricultural and developed land uses. Contributing factors include the invasion of exotic herbaceous and woody plant species along with fragmented ecological systems due to land development and agricultural land use. All have reduced the health of oak systems. Declining native plant diversity and productivity, decreasing abundance of birds, erosion of soils, increased water yield (runoff), and death and subsequent reduced replacement of oaks is occurring over hundreds of million acres in the midwestern United States.

In developing urban areas such as Chicago and Milwaukee, hundreds of thousands of private landholding property owners have variable sized pieces of land containing degrading savannas. The woodlot in a southern Wisconsin farm and the patch of bur oak trees in a residential yard are examples of what remains of this ecological system. This project represents an extreme project in part because of the private and public property and regulatory coordination and education that would be required to successfully manage a larger percentage of the oak savanna systems in these areas. Even in rural agricultural communities, assembling fragmented oak systems into manageable units will require an enormous effort. When coherent conservation planning is incorporated as a fundamental step in developing regional master plans, restoration strategies for this ecosystem have the greatest chance for success.

URBAN STORMWATER RUNOFF AND RIVER RESTORATION

Urban and agricultural runoff, even in communities and cities with the very best stormwater management policies and regulations, has resulted in severe impacts to downstream rivers, wetlands, and lakes. Because of the huge landscape scale where poor stormwater management practices exist, restoration of rivers is an extreme challenge. Throughout the U.S. and the world, there is extreme need to rethink our stormwater management strategies. Keeping stormwater on the land as long as possible, encouraging infiltration, and incorporating other strategies to mimic natural functioning systems are critical elements in managing stormwater. Tens of thousands of miles of rivers that have been channelized or destabilized by stormwater and land drainage policies of local, state, and federal governments. This provides an extreme restoration need, again requiring major investments in regulatory personnel and land owner coordination and education (Figure 16.5).

This process may have several points of initiation including policy, regulatory, and practical technical solutions to address existing cycles of degradation set in motion by the adverse impacts of more runoff leaving the land at higher velocities and rates.

Models of alternative ways to manage stormwater in urban developments can involve ecological restoration strategies such as through the use of Stormwater Treatment Trains (STT) where native landscapes are utilized to manage stormwater (Apfelbaum et al. 1995; Apfelbaum 1993) (Figure 16.6).

In urban developments and agricultural settings where our STT approach has been utilized such as in conservation developments using prairies, wetlands, and other native landscapes, runoff has been reduced by 50–75% and the peak discharge by 60%. The STT provides a unique opportunity for addressing stormwater management in new, retrofit, and redevelopment projects (Figure 16.6).

FIGURE 16.5 Urban stormwater management has contributed significantly to larger volumes of runoff, reduced infiltration, and base flows to streams, wetlands, lakes, and other waterways; and deteriorated water quality. Stabilizing streams can use soil bioengineering techniques. These methods use plantings with proper design and installation, and follow-up management and maintenance. They can provide significantly greater stability at a lower cost than engineered strategies.

FIGURE 16.6 We use alternative stormwater management that requires a stormwater treatment train (STT) comprised of linked series landscape features that manage the volume, rate, and quality of stormwater simultaneously. We have used this strategy in projects and reduced total volume of runoff by over 50–75%, and reduced peak discharges by over 60%. This technique can eliminate storm sewers, save money, and greatly improve water quality in lakes (Apfelbaum 2004).

KANKAKEE SANDS, INDIANA

Working with the Indiana chapter of The Nature Conservancy, Applied Ecological Services, Inc. (AES), designed and initiated the construction of the 7,300-acre project with the restoration of over 5,000 acres of wetland and 2,000 acres of prairies and savannas (Figure 16.7). This project started on a landscape used for row-cropped agricultural purposes. Miles of large drainage ditches and hundreds of miles of subterranean field tiles have been used to drain the land for several decades. The intensive use of the soils for agricultural and drainage has depleted most native species that could have been found in the seedbank. Consequently, adding native seed and grown plants has been a necessity in many areas of the project. In addition, undoing the drainage infrastructure by disabling tile lines and plugging and backfilling ditches, has also been an important part of this large scale restoration project.

By most standards, including that of The Nature Conservancy, this project is large and complex. Projects such as this are invaluable in providing baseline data for testing strategies on how to restore land at these spatial scales. In addition, these larger scale projects help to define ways to reduce costs and increase the effectiveness of the restoration.

LITTORAL, CARACAS, VENEZUELA

Working with a Harvard Design School studio, AES, Inc., was involved in assessing ecological strategies for addressing unstable slopes, eroded streams and gorges, and newly deposited areas associated with the devastating mudflows near Caracas, Venezuela, that killed up to 50,000 people in December 1999 (Figures 16.10–16.11). Twenty-three coastal communities were severely impacted by mudflows. Heavy rains caused major slope failures. Mudslides from adjacent coastal mountains,

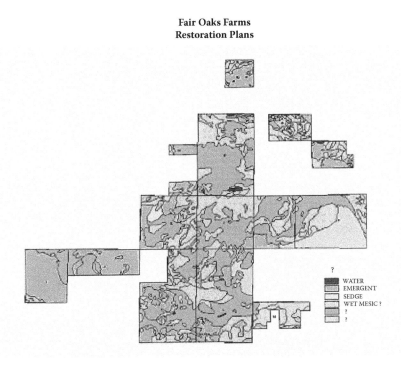

FIGURE 16.7 Larger scale restorations even of highly altered agricultural landscapes can greatly improve ecosystem functions. Seeding, soil management, exotic species management, all present opportunities for rethinking strategies for restoration on large scale projects.

FIGURE 16.8 Large landslides resulting from unusually high precipitation events, created mudflows that inundated 23 communities along the coast of Venezuela during the extreme storms of December, 1999.

and from river channels and gorges that fed down to cement-lined channels in the urban communities, contributed to the impacts (Figures 16.8 and 16.9).

We assisted in generating an ecologically based counterpoint to engineering proposals to install larger cement-lined channels from the coast to the mountain foot, rock dams, and other strategies

FIGURES 16.9 The predevelopment agricultural valley was eventually constructed. Cement lined channels that conveyed regular minor mudflows directly to the ocean became blocked by massive rocks and mud materials during the event. As a result, the mudflows diverted through the urban areas, destroying communities and killing thousands of people.

to address future mudflows. The experience during this latest event and previous events has been to continue to enlarge the cement channels. However, in response to this design change, with each mudflow large rocks and other debris clog the cement channels prompting the water and mud to depart from the channel and short-circuit through the neighboring communities. This process has occurred time and again with the recent impacts and damages being devastating (Figure 16.10).

FIGURE 16.10 The plan for preventing future major mudslide events included the installation of slope and grade controls within upper reaches of the drainage ways, rock dams, and larger cement lined channels to convey materials to the ocean. (Such as those used in Caracas, Venezuela). This strategy has proven to be ineffective.

FIGURE 16.11 Cement lined channels have constrained mudflows, and instilled a false sense of security which has resulted in urban development adjacent to the channel environment. (Color version available at www.gsd.harvard.edu/restore.)

Several restoration strategies to prevent future disasters included: replanting the steep mountain slopes, remeandering and removing the straightened cement channels, creating a cross section for the new channels that included a substantial floodplain environment, and creation of green corridors as parks and open spaces within the urban core areas to reduce the risk of flooding and mudflows in the future (Figures 16.11 and 16.12).

THE DES PLAINES RIVER: A CASE STUDY OF THE ECOLOGICAL AND HYDROLOGICAL CHANGES AND OPPORTUNITIES FOR WATERSHED RESTORATION

INTRODUCTION

This example project is provided in more detail than the previous examples of extreme projects. This is done to illustrate often encountered complexity on such projects.

Diverse and productive prairies, wetlands, savannas, and other ecological systems once occupied hundreds of millions of acres in presettlement North America. Today, however, these ecological systems have been replaced by vast tracts of tilled and developed lands. Land-use changes have modified the capability of upland systems, including former depressional wetlands in the uplands, to retain water and assimilate nutrients and other materials that now flow from the land into aquatic systems. The historic native prairie plant communities that were dominated by deep-rooted, long-lived, perennial species have been primarily replaced by annual crop species or shallow-rooted, non-native species found in lawns and brome grass fields. The native vegetation was efficient at using water and nutrients, and consequently maintained very high levels of carbon fixation and primary productivity. Agricultural plant communities are also productive but their production is primarily above ground. In contrast, prairie ecosystems typically produce 70% of the biomass below ground in highly developed root systems. These changes in the vegetated landscape, coupled with intentional, conventionally engineered stormwater drainage systems, have increased the rate and volume of stormwater runoff from the uplands. As water leaving the landscape has increased, the hydraulic geometry of streams such as the Des Plaines River has been drastically altered.

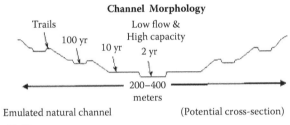

Channel Morphology

Trails   Low flow &
                         High capacity
        100 yr
               10 yr   2 yr

                    200–400
                    meters

Emulated natural channel   (Potential cross-section)

-Multipurpose
-Fosters reinvestment
-Creates community opportunities
 (Los Chorros)
-Recreational opportunity
-Increased channel capacity
-Reduced risk of failure

FIGURE 16.12 Redevelopment of the inundated areas could begin with a redefinition of a flood and mudflow prone area, establishment of parks and open space features to accommodate these areas, and redirecting redevelopments to areas that are safe. Use of natural channel cross sections instead of the cement channels, where terrace environments provide an increased capacity to convey water and mudflows, may prove to be more effective for ensuring safer communities.

This example was prepared to:

1. Provide an historic perspective on landscape-scale ecological changes and resulting hydrologic changes of an example stream, the Des Plaines River.
2. Summarize methods and alternative land development techniques that contribute to the restoration of the hydrology of the Des Plaines River.
3. Summarize a recent study of large wetland restoration that could further restore the Des Plaines River.

PROBLEM BACKGROUND: SITE DESCRIPTION

Comparing historic and recent stream discharge data illustrates trends and changes that have occurred on the uplands and how these changes have affected the hydrology of wetlands and aquatic systems. The Des Plaines River was chosen as a study watershed because of the availability of historic data and traceable changes in watershed land uses (Apfelbaum 1993).

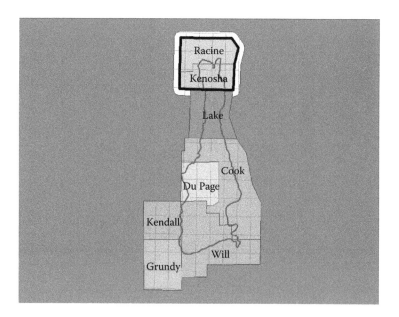

FIGURE 16.13 The Des Plaines River watershed includes 420,000 acres (620 square miles), starting in southeastern Wisconsin and northeastern Illinois. Many jurisdictions including two states, multiple counties, and over 150 municipalities are present in this watershed.

The Des Plaines River originates southeast of Burlington in southeastern Wisconsin (Figure 16.13). It flows for over 90 river miles through agricultural, urban, and suburban landscapes in northeastern Illinois and the Chicago region. It then flows west and south before joining with the Kankakee River to become the Illinois River. The historic data presented are from a case before the Illinois Supreme Court (U.S. Department of War vs. Economy Power and Light, 1904) that dealt with the navigability of the Des Plaines River. The data were derived from a gauge station installed and operated at present-day Riverside, Illinois, from 1886 to 1904. The U.S. Geological Survey has maintained this same station since 1943. Historic data from 1886 to 1904 included a single-stage measurement per day and weekly discharge measurements (rating curves). For our study, duration flow curves were created for the years 1886 to 1904 and 1943 to 1990. The data were compared using median values of discharge (50 percent) and also using low and high levels of discharge as indicated by the 75% and 10% values derived from the annual duration flow curves in the periods from 1886 to 1904 and from 1943 to 1990. The watershed area gauged at Riverside (Figure 16.15) is approximately 620 square miles (400,000 acres).

In the late 1800s, about 40% of the watershed had either been tilled or developed. In contrast, approximately 70 to 80% of the watershed is now developed or tilled annually for agricultural production. Annual duration flow curve values based on linear regression analysis suggests very significant increases in discharge since 1886, perhaps by as much as 250 to 400 times (Figure 16.15). In 1886, the median discharge was 4 ft³/s. Trends in low, medium, and high flow values for the Des Plaines River have undergone very significant increases.

Watershed hydrologic modeling suggests that the watershed and stream discharges from 1886 to 1904 had already been modified by development and agricultural land uses. The Des Plaines River watershed was settled in the late 1830s. Thus, agricultural and urban development had occurred for 50 years before the 1886 data were collected. Other data resulting from the litigated court case suggests very clearly that the discharge of water from the Des Plaines River was significantly less between 1886 and 1904 compared with present day discharges. Because the litigation contested navigability, evidence was presented using daily stage, discharge, and water depth data to illustrate the opportunity for commercial navigation on the river. The data suggested that between 1886 and 1904, for an average of 92 d per year, the river had no measurable discharge. For an additional 117 d per year, the river had 60 ft³/sec. or less discharge, which was equal to a depth of less than 3 in. at Riverside. Based on these statistics, it is clear that the 400,000-acre watershed yielded no water or such low flows that navigation was not possible or reliable for more than 60% of each year. During another 10 to 25% of each year the river was covered with ice.

Additional supporting evidence comes from original land surveys which were contracted by the U.S. Government Land Office to document vegetation types and to identify, where possible, the widths and depths of streams when they were encountered during the process of laying out the section lines. The original land survey records for parts of the Des Plaines River where section lines were surveyed identified that many stretches of the river had no discernable channels. Where channels now occur, surveyors in the 1830s found wet prairies, swamps, and swales but usually no conspicuous or measurable channel widths. Channels and "pools" were identified in some locations, particularly with greater frequency downstream in the watershed.

RESULTS AND APPLICATIONS OF THE FINDINGS

These data very clearly suggest that significant changes in the hydrology, hydraulics, and water yield from the Des Plaines River watershed have occurred since settlement. Other studies of major river and watershed systems have yielded similar results, suggesting that the concepts and conclusions reached from the studies of the Des Plaines River may generally be transferable across the North American landscape. These findings and their applications are discussed below.

NATURAL ECOLOGICAL SYSTEM FUNCTIONS AND PROCESSES SHOULD BE EMULATED

WATER YIELD

Historically, natural landscapes "managed" stormwater very differently than it is managed by human strategies. Historic data clearly indicate that a relatively small percentage of the precipitation in a watershed actually resulted in measurable runoff of water leaving the watershed. In fact, preliminary analysis suggests that an average of 60 to 70% of the precipitation in the watershed did not leave the watershed through the Des Plaines River; this water was lost through evaporation and evapotransporation. Analysis predicts that approximately 20 to 30% infiltrated and may have contributed indirectly to base flow in the streams and directly to base flow in wetlands in the watershed. During a full year, the balance of the water directly contributed to flow in the "river."

Present-day water management strategies involve collection, concentration, and managed release of water. These activities are generally performed in developed parcels in the lower topographic positions. Historically, a greater percentage of water was returned to the atmosphere through evaporation and evapotransporation in upland systems. In these situations, micro-depressional storage and dispersed storage occurred, rather than concentrated storage. Weaver (1968) documented the ability of the foliage of native perennial grassland vegetation to intercept over an inch of rain with no runoff generated, which suggests that depressional (e.g., wetland) storage may have been less important historically for stormwater management.

SEDIMENT AND POLLUTANT MANAGEMENT

Because many pollutants in stormwater require water to dislodge and translocate the suspended solids to which they are adsorbed, there is a great opportunity to emulate historical functions by using upland systems to perform biofiltration functions, increase lag time, and reduce total volume and rate of runoff.

Increased discharge and velocity of water moving through channels has been documented to greatly affect in-stream water quality. In Illinois, as much as 70% of current in-stream sediment loads come from channel and bank destabilization associated with the higher velocity waters, and with solufluction and mass wasting of banks after floodwaters recede. Stabilizing or reducing hydraulic pulsing in streams can best be accomplished by reduction of tributary stormwater volumes and runoff rates from uplands. This can be accomplished in many areas by integrating substantial upland perennial vegetation buffers throughout developments and agricultural lands.

Buffers are designed not only to convey water and minimize erosion (i.e., grassy waterways) but also to attenuate hydraulic pulsing, settle solids and adsorbed nutrients, and reduce and diffuse the velocity, energy and quantity of water entering rivers, wetlands, and other lowland habitats. The use of upland micro-depressional storage, perhaps in the form of ephemeral wetland systems and swales in the uplands, would also emulate historical landscape conditions and functions.

APPLICATIONS

Several model projects of "conservation developments" are now being completed, which integrate up to 30% or greater of the urban development as open space planted to perennial native prairie and other upland vegetation communities, wetlands, and wet swales. These not only serve to emulate the stormwater management functions of the historic landscape, but they are appreciated by residents as aesthetically pleasing site amenities.

Prairie Crossing is a 677-acre residential project in Grayslake, Illinois, designed to offer comprehensive on-site stormwater management in uplands, wetlands, and created lake systems. The development was designed around a "stormwater treatment train" which slowly moves water away from developed areas through a series of natural landscape features. Extensive upland prairie and wet swale systems provide runoff biofiltration and enhance water quality while reducing the quantity of water reaching wetlands and lakes in the development (Apfelbaum et al. 1995).

In these types of projects, upland vegetation takes several years to fully offer stormwater management benefits. In created prairies, surface soil structure develops a three-dimensional aspect in 3 to 5 years. The development of this structure seems to have an important role both in offering micro-depressional storage and further retaining water in the upland systems.

Restoration and native species plantings have also provided benefits where ecological system degradation has led to increased water and sediment yields. Where ecological degradation is occurring indirectly (where human activities on the landscape have reduced or eliminated major processes such as natural wildfires), restoration can provide vegetation and stormwater management benefits. Wildfires have been all but eliminated since human settlement has occurred, especially in areas that contain forests or savannas. In oak woods and savannas, where fire has been eliminated, dense shading is caused by the increased tree canopy and uninhibited growth of a dense shrub layer. Where this occurs, the growth of ground cover and soil-stabilizing vegetation is suppressed due to the low-light conditions. Consequently, on these degraded savanna sites, highly erodible topsoil's are carried away by the increased volumes and rates of storm water runoff. Re-establishment of ground cover vegetation is key to reducing runoff, improving water quality, and re-establishing an infiltration component in degraded timbered systems (Haney and Apfelbaum, 1990, 1993; Apfelbaum 1993).

LANDSCAPE-SCALE WETLAND AND RIVER RESTORATION

In addition to the degradation of water quality and ecological systems caused by the loss of historic stormwater functions on the landscape, another damaging result has been the increased frequency and severity of flooding in our riparian corridors. Across the United States, conventionally engineered flood control projects have cost billions of dollars, and still have not been able to adequately replace the natural functions of the historic landscape. AES conducted two additional studies of methods to use natural systems for flood mitigation. The first was a study of the potential benefits of restoring the 12,000 acres of currently drained agricultural lands in the Des Plaines floodplain along the river's main stem. The second study reviewed wetland restoration potential in all drained hydric soils in the Wisconsin reach of the watershed over an area of 125 square miles (Figures 16.14 and 16.16). An overview of both is presented below:

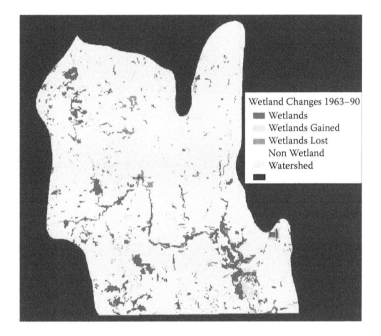

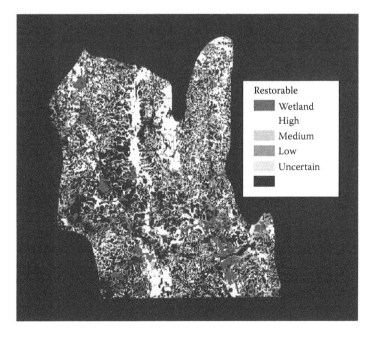

FIGURE 16.14 A comparison of the wetlands that were drained or filled and wetlands that reappeared between 1963 and 1990 suggested substantial ecological restoration opportunities in the watershed.

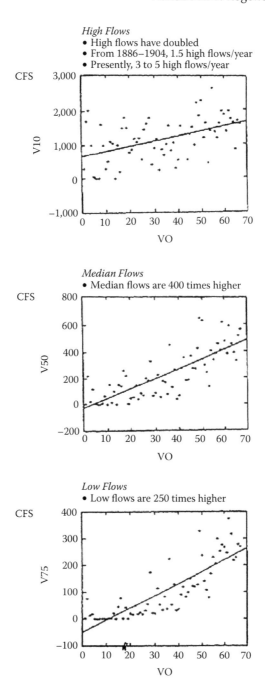

FIGURE 16.15 Linear regression analysis was used to evaluate the hydrological changes associated with the loss of wetlands from the late 1800s through present day. This analysis suggested low (V75%), median (V50%), and high flow flood events (V10%) have changed significantly over time (Apfelbaum 1993).

FLOOD MITIGATION IN RESTORED FLOOD PLAINS

To determine whether the restoration of natural landscapes, and wetlands in particular, could successfully address the issue of flood abatement, AES conducted an exploratory study of an ecological approach as an alternative to a traditional flood control project often proposed by the U.S. Army Corps of Engineers.

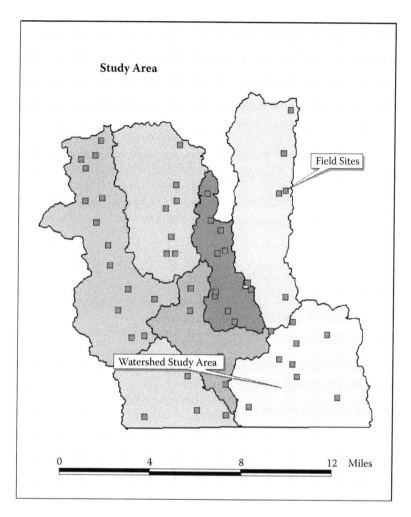

FIGURE 16.16 Representative, existing, and historic wetlands were examined and measured to determine their restoration potential and the watershed hydrological benefits that could be achieved.

This agency's flood mitigation program reportedly proposed to spend approximately $150 million dollars to address 43% of documented flood control needs. Much of the project would have been located on existing public lands, and would have contributed to ecological system deterioration along the main stem of the Des Plaines River.

Alternatively, an evaluation was performed of purchase (fee title or easements) of floodplain lands along the river to restore wetlands. In addition, the purchase of drainage/flood easements in other areas was also explored. This was predicted to reduce all but a few localized flooding problem areas, or over 70% of the flood mitigation need. This method decentralizes floodwater management and seeks to restore a base flow hydrology between wetlands and the river, thus reducing the overland surface water flows and potential flood water contribution to the river (Apfelbaum, unpublished). In this alternative, the restoration of upland buffers along the river, tributaries, and wetlands was estimated to reduce stormwater input to the river through infiltration. Additional benefits included increasing runoff lag time and increasing public open space opportunities for trails and wildlife habitat.

Traditionally, engineers view the lowland environment (i.e., rivers, wetlands) as a perfect location for treating or physically removing water contaminants and addressing other hydrological problems created in the uplands. Conversely, this study suggests that stormwater, sediment loads, and varied contaminants may be best managed on upland systems. Although land costs for using upland environments for stormwater management are higher, efficiencies in flood abatements and reductions in potential contaminant problems may be greater. A landscape with many upland micro-depressional storage opportunities and a large buffering capacity offers more efficient processing of stormwater than a single biofiltration wetland or floodwater detention reservoir downstream. Each buffer or depressional wetland treats a smaller volume of water and contaminants. Also, dispersed stormwater treatment facilities in the uplands have significantly reduced long-term maintenance costs vs. conventionally engineered storage systems, and represent a more sustainable approach to the management of stormwater. Centralized biofiltration wetlands, on the other hand, have high maintenance requirements and problems, including reduced removal efficiencies for some materials in both the short-term and long-term. The results from this exercise stimulated a larger study of the opportunity to manage stormwater in wetlands throughout the watershed, not just along the flood-plain of the Des Plaines River's main stem.

FLOOD MITIGATION THROUGHOUT THE WATERSHED

A larger study was undertaken to investigate flood mitigation benefits of wetland restoration throughout the watershed, not just along the river's floodplain.

The study had three key components:

1. Field analysis to map and document the extent, type, costs and benefits of using wetland restoration to address flood control in the basin.
2. GIS mapping and modeling to view the wetland restoration in the basin spatially. Generation of statistics for hydrologic and hydraulic modeling based on GIS modeling.
3. Hydrologic and hydraulic modeling to test the flood mitigation benefits of landscape-scale wetland restoration.

METHODS

1. Field analysis
   GIS base maps were prepared using mosaics of aerial photographs, topographic maps and digital maps of all hydric soils (Figures 16.14 and 16.16) and existing and impacted wetlands. These maps were used in the field to conduct a general field review of the wetland restoration opportunities in approximately 125 square miles of the water-shed in Wisconsin.
   Prior to a second round of field work, we randomly chose and mapped approximately 70 drained hydric soil locations representing a mix of drained wetlands that were isolated in headwater locations on main stem and side tributaries to the Des Plaines River (Figure 16.16). Following this mapping, each site was visited to confirm the accuracy of the mapped boundaries, existing soil conditions, and land uses. In addition, the existing vegetation systems and estimated or measured topographic depth and dimensions of any depressional basins were described. Measurements taken in the 70 locations were extrapolated to similar settings throughout the Des Plaines River watershed in Wisconsin.
2. GIS methods
   A GIS system was used for all mapping and integration of the data layers used in the field analysis. GIS was used to determine acreages of depressional storage, landscapes features, and hydric soil locations. These data served as input into hydraulic/hydro-logic modeling.

3. Hydraulic/hydrological modeling (HMS)

Riverine wetlands have different characteristics than upland wetlands for affecting run-off rates, and different modeling processes are applicable to each (Krause, 1999).

Riverine wetlands should be modeled as a function of the river channel (or sub-basin channel) reach using the Modified Pulse routing (Krause, 1999). This method accounts for the effects of additional storage provided by the riverine wetland and for releases of water from wetland storage as an uncontrolled reservoir. For the Des Plaines River HMS analysis, a base storage vs. discharge rate was established in a Southeast Wisconsin Regional Planning Commission (SEWRPC) study for the Wisconsin sub-basins of the Des Plaines River. A similar base relation was used by Krause to model wetland effectiveness in riverine floodplains.

Krause recommended that the SCS curve number method be used to quantify the precipitation loss, and resultant runoff, from the drainage areas. The Clark method was recommended to transform this runoff from the drainage area into runoff at the channel reach. The Clark method uses a constant (K) to calculate this runoff transformation:

$$K = R/(T_c + R)$$

where R is the storage attenuation coefficient and T_c is the time of concentration. The effect of wetland storage increases the time of concentration as the amount of time that the wetland takes to fill with the runoff before the runoff is released to the reach. If the storage attenuation coefficient is held constant, then the K will increase as wetland volume is added to the system.

Work done in the Red River watershed in northwestern Minnesota (Eppich, Apfelbaum, and Lewis, 1998) quantified the effectiveness of upland wetlands in terms of a curve number adjustment within the SCS methodology for depressional wetlands which have water elevation control only (uncontrolled runoff occurs after the wetland is filled). As a result, the tributary constant (Cn value) used to define runoff potential was modified based on land use, soil type, and general land slope within the tributary area. Appendix A provides a derivation and example of this methodology. The adjusted Cn value can be applied to the SCS lag equation to define a wetland adjusted SCS lag period (an input needed to quantify runoff in the HMS model). This method replaces an analyst's judgment factor required to define the K value in the Clark method (and the consequent storage attenuation coefficient) with a mathematically derived process using the quantifiable wetland volumes to calculate the storage attenuation with the SCS lag equation.

This study was not as concerned with defining actual flows in the Des Plaines River during the hydrologic modeling analysis as it was with providing comparison values between present land use conditions and with potential conditions with varying amounts of wetland restoration. Consequently, only minor attention was given to flow calibration, and the Des Plaines River watershed was divided into six fairly large sub-watershed areas for analysis. Existing land use and soils data were used to calculate a value for Cn for each of the subwatersheds. The topographic data and Cn values were used to compute a SCS lag period and initial rainfall abstraction value for each subwatershed. These numbers, together with the watershed area, were used as input into the HEC-HMS model with assumed base flow of 0.2 cfs/square mile of watershed area (SEWRPC Planning Report No. 44). The restored wetlands were assumed to have an outlet control, which would provide positive drainage at a very low rate so that the wetland would hydrologically function similar to a retention basin but which would have stormwater storage available 2–3 weeks after a major storm event.

Stormwater runoff was modeled using rainfall information published in the Rainfall Frequency Atlas of the Midwest, (Huff and Angel, 1992). Rainfall events of 24-h and 7-d duration were modeled, and recurrence intervals of 5, 10, and 100 years were used.

Field investigations and analysis of the watershed characteristics showed that approximately 27,800 acres of nonwetland area within the Des Plaines River watershed in Wisconsin had high potential for restoration as wetland. This area is approximately 36% of the river's Wisconsin watershed area. However, limited area was available in the Brighton Creek and Jerome Creek sub-watersheds and somewhat limited area was available in the Kilborne Creek subwatershed for wetland restoration. Topographic analysis indicated that these depressional areas typically had depths of at least 18 to 24 in. (in many cases considerably greater depths—up to six ft) available for stormwater runoff storage. Wetland restoration options that were evaluated included the following:

Maximum practical watershed area restoration as wetland

15% watershed area restoration as wetland

5% watershed area restoration as wetland

2% watershed area restoration as wetland

No wetland restoration

The wetland depths used for storage were modeled as 18-in and the maximum practical watershed area restoration alternative used the following wetland restoration percentages for the subwatersheds:

Center—30%

Kilborne—20%

Brighton—15%

Upper—30%

Lower—30%

Jerome—15%

Composite—21%

RESULTS AND OPPORTUNITIES

Stormwater Management and Flood Mitigation Benefits

Results from the hydrologic model analysis showed that the restoration of upland wetlands would significantly reduce both the runoff volumes conveyed during the high flow periods after major storm events and the peak runoff rates in the Des Plaines River. Figure 16.17 shows the peak runoff rates quantified at the Wisconsin state line from several design storm scenarios that were modeled as part of the analysis.

Wetlands sited in existing floodplains could also reduce runoff rates and volumes similar to those shown by Figures 16.17 and 16.18 in the Des Plaines River. These riverine wetlands would require a means of detaining the stored water long enough to avoid discharging stormwater into the river system during the period of elevated flow in the river. If this control were not provided, release from the wetlands would be at a considerably higher rate and at an earlier time, negating much of the flood control potential of the wetland.

The floodplain wetlands would provide little or no benefit to the streams and creeks in the upper areas of the watershed compared to upland restored wetlands.

Hydraulic Geometry Benefits

Implementation of a restored wetland system would provide hydraulic geometry benefits to downstream channels by requiring these channels to convey less water at lower flow rates (Figures 16.17

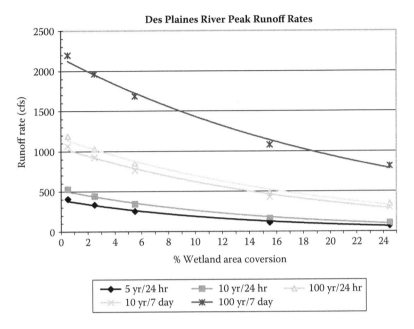

FIGURE 16.17 Des Plaines River peak runoff rates and relationships with the percent of wetland area restored. Several design storm scenarios modeled using data collected at the Wisconsin state line.

and 16.18). Greater overall benefit to the watershed would be provided by siting the wetlands in the upper areas as these locations would produce a more naturalized flow pattern in the smaller upper area streams. The benefit to the Des Plaines River, however, would be similar if the wetlands were located in the floodplain areas assuming adequate outlet controls, if the wetlands were spread throughout the river's length.

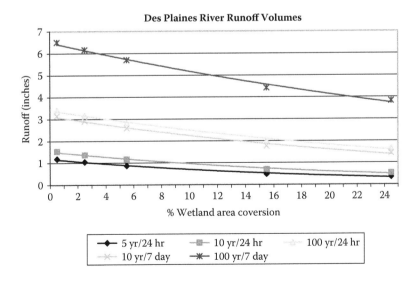

FIGURE 16.18 Des Plaines River peak runoff rates and relationships with the percent of wetland area restored. Several design storm scenarios modeled using data collected at the Wisconsin state line.

Water Quality Benefits

Upland wetland restoration areas will capture sediment generated by the areas that are tributary to the wetland, and they will attenuate the runoff flow rates in the smaller streams that flow into the major Des Plaines tributary creeks. The attenuated flow rates will reduce the sediment generated by bed and bank erosion during high flow periods. The attenuated flow rates will stabilize the stream banks by decreasing the saturated soil depth in the banks, making the banks less prone to slope failure. Location of the restored wetlands in the upland areas of the watershed would reduce the production of sediment compared to a floodplain wetland location. However, wetland restorations in either location would provide a water quality benefit to the Des Plaines River. Base flow contributions to the Des Plaines River could be achieved by restoring wetlands in uplands or those in floodplain environments. In some specific locations in the tributary basins, where surface and subsoil types are more conducive to infiltration, greater opportunities exist to support base flows.

SUMMARY

Extreme projects represent unique challenges. The ecological challenges are often significant, as are addressing political, landownership, and regulatory impediments. New tools are needed to address these cross jurisdictional ecosystem restoration needs, if the vital services and functions from healthy systems are to be maintained. This paper provides a few examples of the many challenging projects where a direct link exists between the well being of people (e.g., ecosystem provides safety and sustenance) and the health of ecosystems. When humans begin to understand their dependency on healthy functioning ecosystems, socioeconomic, political, and regulatory barriers to maintaining these systems seem to diminish, also. Symbolic and simple ways to instill the trust in these links will be an important step to decrease the need for future extreme restoration projects.

ACKNOWLEDGMENTS

Funding for a series of ongoing studies summarized in this paper was provided by the Wisconsin Chapter of The Nature Conservancy and the Liberty Prairie Foundation in Grayslake, Illinois. Assistance in parts of the original study was provided by Dr. Luna B. Leopold and staff at Applied Ecological Services, Inc., in Brodhead, Wisconsin.

LITERATURE CITED

Apfelbaum, S.I. and A. Haney. 1989. Management of degraded oak savanna remnants in the upper Midwest: Preliminary results from three years of study. pp. 280–291 in the Proceedings of the First Annual Meeting of the Society for Ecological Restoration and Management, Oakland, California. Society of Ecological Restoration, Madison, Wisconsin.

Apfelbaum, Steven S.I. 1993. The Role of Landscapes in Stormwater Management. National Conference on Urban Runoff Management: Enhancing Urban Watershed Management the Local, County and State Levels. USEPA/625/R-95/003, April 1995, pp. 165–169.

Apfelbaum, S.I., et al. 1995. The Prairie Crossing Project: Attaining Water Quality and Stormwater Management Goals in a Conservation Development. Proceeding of Conference Using Ecological Restoration to meet Clean Water Act Goals, pp. 33–38, USEPA conference.

Boyce, M. and A. Haney 1997 *Ecosystem Management, Applications for Sustainable Forest and Wildlife Resources.* Chapter 16: S. Apfelbaum, and K. A. Chapman Ecological Restoration: A Practical Approach, pp. 301–321.

Haney, A. and S.I. Apfelbaum 1990. Structure and dynamics of Midwest oak savannas. pp. 19–30 in J. Sweeny, ed. *Management of Dynamic Ecosystems: North Central Section of the Wildlife Society*, West Lafayette, Indiana. 180 pp.

Huff, F.A. and J.A. Angel, 1992. *Rainfall Frequency Atlas of the Midwest*, Bulletin 71. Midwestern Climate Center and Illinois State Water Survey. Champaign, IL.

Krause, A. 1999. Modeling the Flood Hydrology of Wetlands using HEC-HMS. Master's thesis, University of California, Davis.

Schneider, S., and R. Campbell. 1991. Cause-effect linkages II. Abstract presented at the Michigan Audubon Society Symposium, Traverse City, MI, September 27–28.

Weaver, J.E. 1968. *Prairie Plants and their Environment: A 50-year Study in the Midwest*. Lincoln, MB: University of Nebraska Press.

17 Sudbury, Canada
From Pollution Record Holder to Award Winning Restoration Site

John M. Gunn, Peter J. Beckett, William E. Lautenbach, and Stephen Monet

CONTENTS

ABSTRACT

The case history of Sudbury is a remarkable story of restoration of a severely damaged-landscape near a very large mining and smelting complex in Northern Ontario, Canada. Given enough time, the natural recovery of this landscape was expected. After all, this area has recovered time and time again from natural disasters, including being repeatedly scraped clean by advancing glaciers in past millennium. However, the surprise in this particular story is the scale and nature of human involvement, and the benefits that a community achieved from active participation in environmental restoration.

INTRODUCTION

People are usually both horrified and fascinated by scenes of death and destruction, whether it is an image of the aftermath of Hurricane Katrina, the Asian tsunami, the collapse of the World Trade Center buildings, or simply the sight of a car in the ditch along a highway. Thankfully, images of

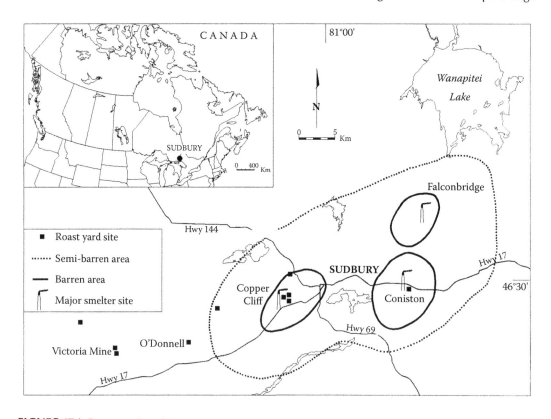

FIGURE 17.1 Barren and semi-barren areas surrounding the Ni and Cu smelters Sudbury, Canada.

renewal and restoration also have the same powerful appeal. For example, we are drawn instinctively to the sight of a new baby, the sound of melting ice after a long winter, and the appearance of green shoots after a brush fire. The story of Sudbury, Canada, should appeal to both of those extremes in human nature.

The mining and smelting region of Sudbury (Figure 17.1) has variously been described as a "dead zone," "moonscape," or "major ecological damage zone," and few environmental textbooks are complete without a historic picture of Sudbury with its blackened and barren landscape (Figure 17.2)

FIGURE 17.2 A historic picture (circa 1970) of the devastated land near the smelter in Sudbury, Canada (photo by K. Winterhalder).

FIGURE 17.3 The stunted white birch forest in the semi-barren areas of Sudbury, Canada (photo by K. Winterhalder).

and the belching smoke stacks. The superlatives and statistics about Sudbury are hard to resist: home of the world's largest metal smelting complex; the world's tallest smoke stack, tailings ponds that contain about a half a billion tons of acid-generating waste, 20,000 ha (49,400 acres) of barren land surrounding the smelters and old roast beds; an additional 80,000 ha (197,000 acres) of semi-barrens, where the original mixed coniferous and deciduous forest was reduced to a near monoculture of stunted white birch (*Betula papyrifera*) (Figure 17.3). At its pollution peak in the 1960s, Sudbury was the largest point source of acid-forming emissions on Earth, with a maximum annual release of approximately 2.5 million tons of sulfur dioxide (Potvin and Negusanti 1995). By contrast the whole industrial nation of Japan today releases about 1 million tons. A century of acid emissions (over 100 million tons) also led to widespread acidification of area lakes (more than 7000 lakes damaged). In addition, the smelters released tens of thousand of tons of metal particulates (Cu, Ni, Cd, Co, etc.) into the atmosphere, severely contaminating surface soils and lakes sediments in a zone extending out approximately 30 km (19 mi) away from the smelters.

With such an environmental history you can imagine the fun that the media had when Apollo 16 astronauts Charles Duke and John Young visited Sudbury in July 1971, just a few months before walking on the moon. Although they were there as geologists touring the area, the lunar jokes about this city were soon all the rage. Perhaps worse still for the image of the city was the occasional practice in early acid rain literature of referring to a "Sudbury" as a unit of pollution (Glass and Brydges 1982).

Well, where is the good news? How did this city of 160,000 people turn itself around, and in April 1991 become identified by a prominent Canadian magazine as one of the ten best Canadian cities in which to live? Grander still was the recognition in June 1992 when Sudbury received one of the 12 Local Government Honours Awards at the United Nations Earth Summit in Rio de Janeiro for its municipal Land Reclamation Program. How did this happen, what more needs to happen to maintain the progress in Sudbury, and what can others learn from this experience?

SOURCE OF THE PROBLEM

Both the environmental problems and the great wealth of Sudbury have an ancient common origin—the vast sulfide ore deposits that lie along the 150 km rim of the Sudbury Basin. The basin is thought to be a crater that formed when the earth was struck by a huge meteorite 1.8 billion years ago (Figure 17.4). This violent event left in its wake one of the world's great ore bodies and, in particular, the largest known concentration of nickel on earth. The ore also has a large

FIGURE 17.4 Radar image of the 30-km-diameter Sudbury Basin, the result of a meteorite impact 1.8 billion years ago (photo courtesy of the Canadian Centre for Remote Sensing).

amount of copper as well as significant levels of other valuable metals (including platinum, palladium, gold, silver, and many others), but it was the high sulfur content of the ore that was both a gift and a curse. The gift was the high sulfur content that made the ore readily combustible. It could be roasted in the open in vast heaps by simply igniting it with cordwood (Figure 17.5a,b).

(a)

(b)

FIGURE 17.5 A 1920s view of the 2286-m-long O'Donnell roast yard, before (a) and after (b) ignition (archive photos by Inco, Ltd.).

FIGURE 17.6 The 381-m superstack replaced the much smaller smokestacks in 1972 (photo by D. Pearson).

This high sulfur content, therefore, eliminated the need for expensive external fuels. However, the resulting clouds of sulfur released from the roast yards (1883–1929) and the early smelters rained down as sulfuric acid, wreaking havoc on local lakes, forests, and crop lands. Lifting the pollutants into the atmosphere with taller, and taller smokestacks only spread the damage further and further afield (Figure 17.6).

OBVIOUS SOLUTIONS

The extensive destruction of land and water by industrial emissions were not intentional. In fact, it can be argued that over the years "best practices" were used to prevent their occurrence. But, unfortunately, severe damage did occur. A landscape that once supported a rich variety of natural resources—forest, fish, wildlife, etc.—was reduced to a barren wasteland in a few decades of mining and smelting.

There is no particular magic or uniqueness in the solutions to environmental damages in Sudbury. The same solutions apply everywhere: (1) reduce or eliminate the contamination and (2) repair the damage.

CLEANING THE AIR

The damaging effects of the smelter fumes were well known at the turn of the century but the first control orders by the Ontario government were not issued until 1969 and 1970. These regulations were solely directed at improving local air quality and led to the construction of the famous 381 m (1250 ft) "super stack" at the Copper Cliff smelter of Inco, Ltd. (Figure 17.7). It is ironic that the commissioning of the world's tallest smokestack occurred in 1972, the same year that the United Nations held its first environment conference, the Stockholm Conference. It was probably just about the same time that company officials were organizing the celebrations for the new smoke stack to solve Sudbury's local problems that delegates at the Stockholm conference were arguing vehemently that "acid rain" (a term not yet in popular use) from countries like Britain and Germany were causing severe damage to sensitive lakes and forests in far off Scandinavian countries. The damage was caused because industries were releasing pollutants high in the atmosphere through the

FIGURE 17.7 The world's tallest smokestack (photo by *Sudbury Star*).

use of tall stacks. Sudbury and its super stack would not go unnoticed in the debate that followed. Our mammoth "solution" to our local pollution problem soon became an international icon that was emblazoned (with the accompanying bright red X) on tee shirts of protestors around the world. Green Peace and other international NGOs helped lead the campaign.

Somewhat lost in the initial protests about Sudbury's air pollution was the fact that not only had pollutant dispersal increased with the changes in 1972, but that total emissions from the Sudbury smelter were also greatly reduced. However, these reductions were not sufficient to prevent severe ground-level fumigation events from occurring, leading to frequent vegetation damage. More and more regulations and industry pollution control investments were still required. This reduction continued over the decades until total emission of both sulfur dioxide and metal particulate had been reduced by nearly 90% by 2005 (Figure 17.8).

It should be noted that even after the emission reductions, Sudbury smelters were still very large point sources of SO_2 in the world, with combined total annual emissions of approximate 200,000 tons of SO_2 today. From a financial perspective it should also be noted that to date more than a billion dollars were spent on pollution control abatement efforts, but the local mining companies have not gone broke as a result of these regulatory changes. Production of metals from Sudbury still remains high, and profits are now close to an all time high. In fact, Inco, Ltd.'s own analysis showed that the massive infrastructure costs (Can. $600 million) for the renovations of the smelters in the mid-1980s (reputed to be one of the largest corporate investments in the environment) paid for itself in energy efficiency and worker productivity in about 10 years (Bouillon 1995)—proof yet again that a clean company is a more competitive and efficient company.

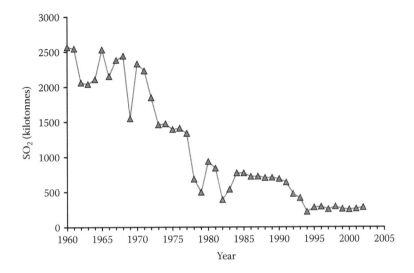

FIGURE 17.8 Sulfur dioxide emissions have reduced by 90% since 1960.

REPAIRING THE DAMAGE

Planning for a land reclamation program in Sudbury began to take shape in 1973, shortly after the Regional Municipality of Sudbury was incorporated (Gunn 1996). One of the earliest priorities identified by the newly established Planning Department was to improve the image of the region. In addition, the area was experiencing high unemployment in the late 1970s and early 1980s. The principal goals for any future regional land reclamation program were therefore to (1) improve the appearance of the area, especially along the principal travel corridors and near neighborhoods and (2) provide short-term jobs for unemployed workers and students. The catalyst that initiated the Sudbury Regional Land Reclamation Program, was the announcement in September 1977, that Inco, the region's largest employer, would be laying off 3,500 Sudbury employees and would not be hiring summer student laborers. Regional government staff and community leaders responded rapidly to this crisis by securing funds from federal and provincial agencies to provide employment through a large-scale, land-reclamation program. In May, 1978, the program was officially launched. Thus began one of the largest community efforts in reclamation of industrially damaged lands ever undertaken.

From the beginning, the regional government provided the administrative support and structure and acted as the general contractor for the work. Technical expertise, advice, and support for the program was provided by the Planning Department and a public advisory committee: VETAC (Vegetation Enhancement Technical Advisory Committee). VETAC provides not only expert technical advice on treatment procedures, but assists with the selection of sites for treatment, conducts biological and chemical monitoring, and helps with fund-raising and communication. One of the first tasks of the advisory committee was to help design and test appropriate soil-treatment procedures (Figure 17.9). These first trials depended heavily on the expertise of local university, government, industry, and horticultural experts.

A simple treatment procedure was developed for use in direct application without tillage, on the stony slopes along the roadways (Lautenbach et al. 1995). The procedures adopted involved manual surface-application of lime (10 tons/ha of crushed dolomitic limestone), fertilizer (N_6-P_{24}-K_{24} at 400kg/ha), seeds (five grasses: redtop [*Agrostis gigantea*], red fescue [*Festuca rubra*], timothy [*Phleum pratense*], Canada bluegrass [*Poa compressa*], and Kentucky bluegrass [*P. pratensis*] and two legumes: bird's foot trefoil [*Lotus corniculatus*] and Alsike clover [*Trifolium hybridum*]; at a rate of 40 kg/ha for all species combined) (Figure 17.10a,b,c,d). Trees were also planted including

FIGURE 17.9 Test plots to develop simple methods of land reclamation (photo by K. Winterhalder).

(a)

(b)

FIGURE 17.10 Land reclamation activities (a,b) manual soil liming, (c) seeding and fertilizing, (d) tree planting (photos by City of Greater Sudbury).

(c)

(d)

FIGURE 17.10 (Continued)

jack pine (*Pinus banksiana*), red pine (*Pinus resinosa*), white pine (*Pinus strobus*), spruce (*Picea glauca*), larch (*Larix laricina*), northern red oak (*Quercus rubra*), and black locust (*Robinia pseudoacacia*). These procedures were designed to trigger natural recovery processes that would lead to self-sustaining ecosystems. The use of minimal treatment rates was important in this approach. For example, seed application rates were light and tree planting density low to encourage increased species diversity through natural in-filling. Tree species were matched to habitat features (soil moisture, exposure, slope, etc.) and planted in naturally-appearing groupings. Species selected were tolerant of acidic soils and low nutrients, and preference was given where possible to native species. Non-native legumes were used to rapidly restore nitrogen reserves or, in the case of black locust, as a "nurse plant" to establish a cover at extremely dry and exposed sites.

ACCOMPLISHMENTS TO DATE

Thus far, more than 3,367 ha (8,320 acres) of land along approximately 100 km (62 mi) of municipal roads and throughout many of the affected neighborhoods (Figure 17.11a,b,c,d) have been limed and seeded (Figure 17.12a,b). In addition, almost 8.5 million trees have also been planted. The appearance of another 1,000 ha (2,470 acres) of land along the highways has been improved by removing dead trees, stumps, and other unsightly debris. The treated area represents about 50% of the most severely damaged land, including almost all the highly visible sites that are accessible for manual application.

(a)

(b)

FIGURE 17.11 Changing view of Sudbury neighborhood: (a) 1981, (b) 1987, (c) 1994, and (d) 2000 (photos by K. Winterhalder).

(c)

(d)

FIGURE 17.11 (Continued)

With conifers making up about 95% of the planted trees, the reclaimed sites provide a welcome view of greenery both summer and winter. Monitoring studies also reveal that survival and growth rates of planted trees have been very good. Other key findings include: (1) pH values in soils has increased from 3.5–4.0 before treatment to 4.6–6.5 after treatment, (2) metal uptake by plants has declined, (3) insects, birds, and small mammal populations have increased at reclaimed sites, (4) spontaneous colonization of herbaceous and woody species has occurred at treated sites, and (5) land treatment procedures in drainage areas have improved water quality in some Sudbury lakes.

In addition to the ecological and visual effects, over 4,400 students and unemployed individuals have received temporary employment through the land restoration program. To date, the total cost of the municipal program has been approximately $23 million, with most (90%) of the funds coming as grants from national, provincial, and municipal governments. The local mining companies, Inco and Falconbridge, Ltd., have also contributed funds, trees, and technical assistance to the municipal program.

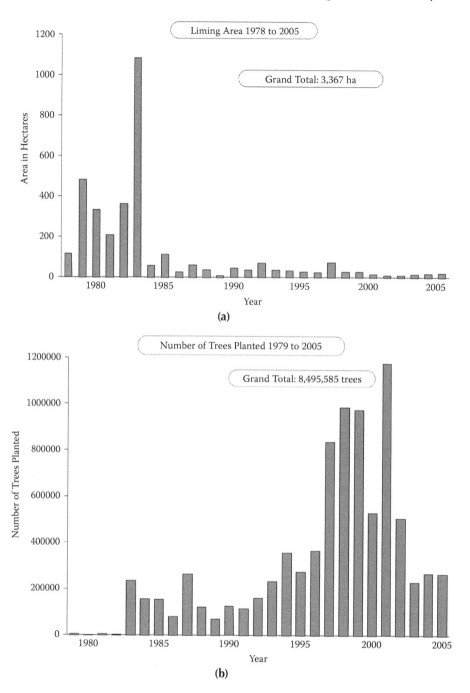

FIGURE 17.12 Annual activities of municipal land reclamation program: (a) soil treatment, (b) tree planting.

CHANGING EMPHASIS IN THE PROGRAM

Three distinct trends in the program have developed. In the first years (1978–1983) the emphasis was on soil improvement and job creation. During this period 76% of the land liming occurred (Figure 17.12a) and 61% of the jobs were created. Because wages make up approximately 80% of the cost of this program, the early period was also the period of most of the financial expenditures.

The year 1983 became a turning point in the program. That year massive layoffs of workers from the local mines led to creation of a federal/provincial employment program to help provide temporary jobs for unemployed workers. A total of 1,281 individuals were hired to work on the Land Reclamation Program in 1983. More than 1,000 ha of land were treated and 235,000 trees planted in that year at a cost of over $7 million.

VOLUNTEERS

The second trend has been the shift during the past 10 years to a lower-cost program that employs fewer paid worker, involves more volunteers, and concentrates on tree planting instead of soil treatments. During the period 1984–2005, about 97% of the trees were planted (Figure 17.12b). Trees were planted at barren sites that had previously been limed and seeded, as well as on some semi-barren areas where survival was possible without soil treatments. The past two decades also saw the spontaneous rise of a much expanded volunteer effort to assist with the tree planting. To date more than 7,500 volunteers have planted over 270,000 trees on damaged lands. In addition, over 400,000 trees were given out to individuals to plant on their own property.

EXPANDED INDUSTRY ROLE

The expanded role of industry in land reclamation was the third trend. Well over a billion dollars has been spent by Inco and Falconbridge on emission control technologies and environmental improvement projects in the Sudbury area. By far the greatest expenditures have been directed at rebuilding and modernizing smelters to reduce smelter emissions. Nevertheless, both companies have a long history of land reclamation work, particularly of tailings areas (Peters 1984, Michelutti and Wiseman 1995). To date, they have revegetated a total of approximately 2,000 ha (5,000 acres) of tailings and sand pits using highly mechanized and intensive treatment procedures (Figure 17.13). More than three million trees have been planted and several former industrial sites have been converted into valuable parks, recreational areas, and wildlife conservation areas (Figure 17.14a, b).

Beginning in the summer of 1990, the mining companies increased their effort to treat the more visible damaged lands on their properties. The initiation of an aerial liming (Figure 17.15)

FIGURE 17.13 Agricultural-type seed bed preparation of tailings areas (photo by Inco, Ltd.).

(a)

(b)

FIGURE 17.14 Copper Cliff tailings area with windblown dust (November 1973) prior to the establishment of a surface cover of trees (a) and after (b) vegetation cover was established (July 1994) (photo by Inco, Ltd.).

FIGURE 17.15 Aerial liming of watersheds to improve drainage water quality (photo by P. Beckett).

and seeding program by Inco and Falconbridge was particularly important, with more than 1,200 ha treated between 1990 and 2005.

APPEALING TO THE SACRED

Sustaining a land reclamation and community involvement program for decades is not easy. Volunteers come and go, funding priorities change, and new uses for the same land arise. To continue to expand the restoration efforts—or even to hold the hard-won ground of "restored green space" against the ever mounting pressures for properties for housing, transportation routes, or industrial development—restoration groups often have to appeal to things considered "sacred." For example, we have developed a Tree Trust and "memorial forests" where trees are planted in memory of the deceased. However, rather than being treated as a cemetery, the restored area is considered a "ribbon of life" around a lake or other valuable natural asset. Such areas become quite immune from political pressures for development.

Sacred, too, in a world of ever-shrinking supplies of clean freshwater are water resources. Sudbury is fortunate in this regard, with more than 300 lakes within the city borders (Figure 17.16). Many of these lakes were badly damaged by smelter emissions; many are still damaged, but a large number have recovered naturally (without direct or watershed liming) and recreational fishing is now popular in many lakes (Figure 17.17a,b). Dozens of lake stewardship groups have sprung up around this "city of lakes" to advocate for water quality protection. One of the most successful groups has been a volunteer group working to improve a 20-km urban stream, Junction Creek, which flows through the center of the city. Environmental "celebrities" such as Dr. Jane Goodall have helped with their restoration effort (Figure 17.18a). Having family groups, politicians (Figure 17.18b), schools, and the media participate in the fish stocking and other "adopt a creek " activities in this project has also infused the sense of the sacred into a clean up effort that would otherwise be quite mundane (Figure 17.18c,d).

Providing public access can also be considered a sacred activity. The Trans-Canada Trail now weaves through the reclaimed areas of Sudbury. How unlikely the prospect that walking trails would exist in the once barren land? Many other trails have been built or being planned in restored areas to increase the value of restored areas to the community (Figure 17.19a,b).

FIGURE 17.16 With over 300 lakes, Sudbury is a "city of lakes" (photo by E. Snucins).

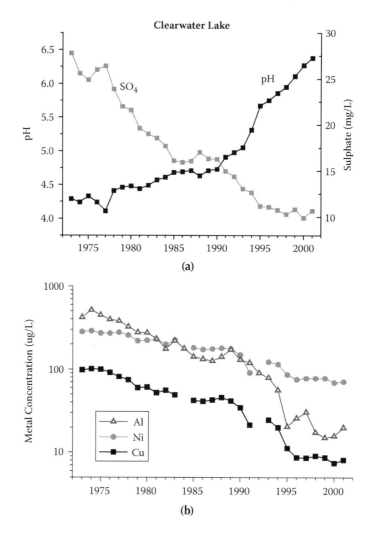

FIGURE 17.17 Water quality improvements in Clearwater Lake, a 76-ha headwater lake in Sudbury, Canada, located 11 km from the Copper Cliff smelter. (a) pH, SO_4, (b) metal concentrations. This lake is the longest running monitoring site in Canada for assessing the effects of air pollutants on lake ecosystems. The lake was fishless for more than 50 years. With improved water quality there are now four fish species present. (Data from W. Keller, Ontario Ministry of the Environment).

BECOMING A SUSTAINABLE COMMUNITY

It may seem odd to use a case history of a hard-rock mining area to discuss ecosystem "sustainability." High quality ore deposits are nonrenewable resources that can be rapidly depleted with current technologies. Mining has traditionally been a transient industry involving short-term use of the land without regard to future uses of that land. The presence of vast areas of derelict land is the legacy of mining in most countries (Moore and Luoma 1990, Young 1992).

The size of the ore body in Sudbury has delayed the inevitable, but no doubt the end of mining will come. What then for this community? There have been a number of planning initiatives in this regard (e.g., Pearson et al. 1992). One of the most recent is that a large number of community partners have come together under the banner of Earth Care Sudbury (www.earthcarecanada.com) to develop a local action plan to address this future. The future they are planning for makes extensive

(a)

(b)

FIGURE 17.18 Exciting community events: (a) participation of Dr. Jane Goodall in fish release on Junction Creek, (b) Sudbury's Mayor D. Courtemanche and his family release brook trout, (c) "bug search" by local school children, (d) creek clean-up day.

(c)

(d)

FIGURE 17.18 (Continued)

use of the restored nature to attract a new knowledge-based industry that will be drawn by the quality of life that this once devastated area now offers. Planning for the confounding effects of climate change and the need for food security is part of the group's mandate.

It is noteworthy that Sudbury's newly drafted municipal plan (i.e., official plan) now incorporates the idea of watersheds as municipal planning units. A decade earlier the land reclamation groups had begun to organize their efforts on a watershed basis for soil treatment and tree planting, so that both the land and the downstream drainage lakes would be improved (i.e., a win-win situation), but municipal planning was slower at recognizing these ecological boundaries as significant. Now that they have, the quality of both groundwater and surface water resources should be better protected, and the ecological "services" provided by restored green spaces and wetlands within the watershed will be more highly valued.

SOCIO-ECONOMIC EFFECTS

Sudbury is a much different place now from the way it was in 1972 when the first major emission control program was initiated, or in 1978, when the municipal Land Reclamation Program began. The air quality is much better—in fact, it is equal to or better than conditions in many Ontario cities—and the view of the landscape is far more pleasant (Figure 17.20). At the same time Sudbury is also no longer a single-industry town. Creative community planning and communication initiatives, and aggressive political lobbying have brought about not only the Land Reclamation

(a)

(b)

FIGURE 17.19 Encouraging public access: (a) boardwalk along lakeshore, (b) trail through reforested area (photos from the City of Greater Sudbury).

FIGURE 17.20 Sudbury's new image (photo by E. Snucins).

FIGURE 17.21 Science North, Sudbury, Canada. The science center is an important part of the revitalization of this mining town (photo by the City of Greater Sudbury).

Program but economic diversification through an expansion of health services, education, and communication and hospitality industries, and by attracting several government offices to the area (Smith 1993). Sudbury now has a symphony orchestra, professional English and French theatres, an annual international film festival, open-air farmers' markets, a new science center (Figure 17.21), and much-expanded urban sport fishery—all developed within the last 30 years.

It is not possible to relate the socio-economic and cultural improvements in Sudbury to environmental changes alone, but it is obvious that environmental improvements have removed a serious constraint on economic diversification. It is inconceivable that the tourism industry could rise to revenues of more than 140 million dollars per year in an area considered a "moonscape," or that the landscaping and florist business would increase by over 400% between 1970 and 1986 without a profound change occurring.

NEED FOR PARTNERSHIPS

The recipe for success in large-scale and long-term reclamation always starts with the involvement of dedicated people. You need strong leadership and champions, both in government and among private citizens. Positive change can also happen very quickly if passionate environmental champions exists within industry (e.g., Aitkens 1991, Peters 1984). Sudbury has been fortunate to have had such leadership, both individual and group leadership for more than 30 years. For example many of the original technical advisory committee members are still leading the VETAC activities today.

It is an overused word, but "partnership" is really the only way forward. Strong partnerships between industry, government agencies, universities, and community organizations were essential to Sudbury's transformation. From our experience, guiding principles of successful partnerships include:

1. Understanding of each partner's needs
2. Avoiding blame and spreading the credit
3. Support for a common vision
4. Frequent and effective communications
5. Flexibility (i.e., many goals) to take advantage of opportunities
6. Consistent monitoring and reporting of progress, including the small steps
7. Celebration of success through awards, certificates, and media attention (Figure 17.22a,b)

(a)

(b)

FIGURE 17.22 (a) Selection of national and international awards; (b) commemorative stone recognizing tree planting accomplishments (photos by the City of Greater Sudbury).

There are immediate and tangible benefits for partners involved in restoration projects, such as tax credits and positive publicity of sponsoring industries. However, there has been little objective analysis of the factors such as the number of partners, level of financial contribution, project duration, and administrative structure that contribute to success or failure, particularly for partnerships in large-scale ecological restoration projects like the one in Sudbury. Given the environmental challenges we still face here and throughout the world, such analysis should be encouraged so that more effective partnerships can be created to address these challenges.

INTERNATIONAL OUTREACH

The Sudbury story started out with a focus on a local problem, but we soon learned that dealing with this problem in isolation was not possible (Table 17.1). Even with the tall stacks, the reduced emissions, and the closure of obsolete plants, acid rain remained a serious problem in the Sudbury area. For example, in the early 1980s, scientists in the Sudbury area were surprised to discover that during an extended (8-month) shutdown of the Sudbury smelters because of a labor dispute, acidic deposition did not improve significantly in the local area (Lusis et al. 1986). This finding indicated that the background pollution from the rest of North America was still so large that the closure of the largest point sources of sulfur dioxide in the continent was seemingly unimportant to the acidifying effects of air pollution in the city. National and international cooperation in achieving regional emission reductions were therefore needed, even to deal with our "local" problem.

In recent decades considerable progress has been made in both Canada and the United States in reducing air pollutions (Jeffries et al. 2003). We in Sudbury were proud that our data on recovery of acid damaged lakes (Keller and Pitblado 1986, Gunn and Keller 1990) was well used in the transboundary negotiations between Canada and the United States. In particular, the Sudbury data was very important in the debate leading up to the revisions of the 1990 Clear Air Act in the U.S.

TABLE 17.1
Time Line of the Emergence of Sudbury, Canada, as a Model Community for Restoration

Period I—Sudbury Considered a Large but "Local" Pollution Site

1888–1929	Open roast beds (largest 2.3 km long) used local cordwood as fuel to ignite ore
1957	First experimentation with lake liming as a remedial measure for acidified lakes. Lake neutralization remained experimental in Sudbury; only eight lakes were limed in 1970s and 1980s
1960	Sudbury is the largest point source of SO_2 globally (2.5 M tons per annum). Gorham, E. and A.G. Gordon publish some of the first papers on acid rain effects on lakes (*Can. J. Bot.* 38: 307–312) and vegetation (*Can. J. Bot.* 38: 477–487)
1972	Commissioning of world's tallest "superstack" in Sudbury. Concurrent reductions in emissions, increased dispersal and closure of obsolete plants leads to 50% drop in SO_2 concentrations in air. Beamish and Harvey (1972) publish first paper documenting regional fish losses (*J. Fish. Res. Board Can.* 29: 1131–1143)
1978	Sudbury's land reclamation program begins to revegetate barren land, enhance city image, and provide needed employment for Sudbury residents

Period II—Recognition as Internationally Significant Recovery and Reclamation Site

1986	Keller and Pitblado (*WASP* 29: 285–296) publish first paper documenting beginning of regional recovery of damaged lakes
1985–1990	Clean air negotiations, U.S. Congress; Sudbury case history counters "irreversibility argument"
1990	Gunn and Keller (*Nature* 345: 431–433) document biological recovery including resumption of reproduction in damaged fish populations
1978–2005	City of Greater Sudbury revegetates 3,328 ha of land, plants 8.0 million trees, and provides employment for more than 4,400 people through its reclamation program

Period III—Emergence as a Useful Model System for International Exchange

1990	First Healthy Cities Conference in Sudbury
1992	Sudbury receives U.N. "Local Honours Award" at Earth Summit Conference in Rio
1995	Hosts first Sudbury Mining and the Environment Conference
2001–2002	Sudbury scientists conduct courses and hold restoration science workshops in mining regions of Russia
2005	Sudbury brings its experience and expertise to the Sustainable Cities Initiative cities of Reynosa and Matamoros, Mexico

Congress, because it showed that investments by industry in pollution control would pay good dividends in terms of recovery of damaged resources. It wasn't all just "water under the bridge."

Transferring environmental problems to other cities, countries, or other continents is not right. Atmospheric pollution can be directly transferred but, more commonly, the problems are indirect, related to globalization, where the push for cheaper and cheaper goods for consumers in rich countries often leads to environment destruction and waste accumulation in poor countries. This is a tragedy in the recipient country, but has scary consequences for all, because of the anger and misery this inequality creates.

What can a relatively small mining city in North America contribute in this regard? Sudbury, through its university, industry, and community partnerships, is trying to meet some of the international responsibilities that came with the 1992 U.N. recognition for its restoration program. For example, the Sudbury Mining and the Environment Conference has been established (1995, 1999, 2003) that attracts a broad international group of scientists and mining officials to encourage better management of environmental problems related to mining throughout the world. Individual scientists have also traveled extensively conducting courses and seminars related to ecosystems rehabilitation and urban renewal in Sudbury—always emphasizing the importance of partnerships in achieving success (Figure 17.23a,b,c). This international outreach work has not yet become

(a)

(b)

FIGURE 17.23 Participation of Sudbury scientists in CAES Ph.D. course "Industrial Impact on Natural and Social Environment," Kola Peninsula, N.W. Russia, September 2001. (a) Nikel smelter near the Norwegian border. Smelter is similar in design to the Sudbury's smelters in 1960. (b) Touring a devastated landscape near Pechenganickel in Monchegorsk, Russia. (c) Russian scientist showing results from test plots using Sudbury's land reclamation techniques in Monchegorsk, Russia (photos by J. Gunn).

(c)

FIGURE 17.23 (Continued)

widely accepted as a mandate or a focus for our community, but it is hoped that in time this outreach work will grow to meet the many needs elsewhere.

ACKNOWLEDGMENT

We appreciate the assistance of Tina McCaffery, Karen Oman, and Dorothy Robb in preparing this manuscript.

REFERENCES

Beamish, R. J., and H. H. Harvey. 1972. Acidification of the La Cloche Mountain lakes, Ontario, and resulting fish mortalities. *J. Fish. Res. Board Can.* 29: 1131–1143.

Bouillon, D. F. 1995. Developments in Emission Control Technologies/Strategies: A Case Study. pp. 275–285. In J. M. Gunn (ed.). *Restoration and Recovery of an Industrial Region*, Springer-Verlag, New York.

City of Greater Sudbury. Economic Development and Planning Services. 2004. Land Reclamation Annual Report. 15 p.

Glass, G. E. and T. G. Brydges. 1982. Problem Complexity in Predicting Impacts from Altered Precipitation Chemistry. pp. 265–286. In Johnson, R. E. (ed.) *Acid Rain/Fisheries*. American Fisheries Society Publication. 357 p.

Gorham, E., and A. G. Gordon. 1960. Some effects of smelter pollution northeast of Falconbridge, Ontario, Canada. *Can. J. Bot.* 38: 307–312.

Gorham, E., and A. G. Gordon. 1960. The influence of smelter fumes upon the chemical composition of lake waters near Sudbury, Ontario, and upon the surrounding vegetation. *Can. J. Bot.* 38: 477–487.

Gunn, J. M. 1996. Restoring the smelter—damaged landscape near Sudbury, Canada. *Restoration and Management Notes.* 14(2): 129–135.

Gunn, J. M., and W. Keller. 1990. Biological recovery of an acid lake after reductions in industrial emissions of sulphur. *Nature* (London) 345: 431–433.

Jeffries, D. S., T. A. Clair, S. Couture, P. J. Dillon, J. Dupont, W. Keller, D. K. McNicol, M. A. Turner, P. Vet and R. Weeber. 2003. Assessing recovery of lakes in the southwestern Canada from the effects of acidic deposition. *Ambio.* 32: 176–182.

Keller, W., and J. R. Pitblado. 1986. Water quality changes in Sudbury area lakes: a comparison of synoptic surveys in 1974–76 and 1981–83. *Water Air Soil Pollut.* 29: 285–296.

Lautenbach, W. E., J. Miller, P. J. Beckett, J. J. Negusanti and E. K. Winterhalder 1995. Municipal Land Restoration Program: The Greening Process. pp. 109–122. In Gunn, J. M. (ed.) *Restoration and Recovery of an Industrial Region*, Springer-Verlag, New York.

Lusis, M. A., A. J. S. Tang, W. H. Chan, D. Yap, J. Kurtz, P. K. Misra and G. Ellenton. 1986. Sudbury impact on atmospheric deposition of acidic substances in Ontario. *Water Air Soil Pollution*. 30: 897–908.

Michelutti, B. and M. Wiseman. 1995. Engineered Wetlands as a Tailings Rehabilitation Strategy. pp. 135–141. In J. M. Gunn (ed.) *Restoration and Recovery of an Industrial Region*, Springer-Verlag, New York.

Moore, J. N., and S. M. Luoma. 1990. Hazardous waste from large-scale metal extraction. *Environ. Sci. Technol.* 24: 1278–1289.

Pearson, D., J. Blanco, I. Filion, F. Hess, P. O'Sullivan, B. Rogers and P. Smith. 1992. Tomorrow Together: Towards Sustainable Development in the Regional Municipality of Sudbury. Unpublished Report.

Peters, T. H. 1984. Rehabilitation of Mine Tailings: A Case of Complete Reconstruction and Revegetation of Industrially Stressed Lands in the Sudbury Area, Ontario, Canada. pp. 403–421. In P. J. Sheehan et al. (eds.) *Effects of Pollutants at the Ecosystem Level*. John Wiley & Sons, New York

Potvin, R. R., and J. J. Negusanti. 1995. Declining Industrial Emissions, Improving Air Quality, and Reduced Damage to Vegetation, pp. 51–62. In J. M. Gunn (ed.) *Restoration and Recovery of an Industrial Region*, Springer-Verlag, New York.

Smith, P. L. 1993. Sudbury: A Case History of Environmental and Economic Transformation. Report for Office of Training in the Environment, Industry, Science and Technology, Ottawa, Canada. 32 p.

Young, J. E. 1992. Mining the Earth, pp. 99–118. In L. R. Brown et al. (eds.). *State of the World 1992*. W. W. Norton, New York.

Passerelle: Bridging Concerns, Contentions, and Conflicts in the Sociology of Restoring Serenity to Venice[1]

> Rather than pity, this city inspires rage.
>
> —**Brondi, C.** *in* **Pertot, G.,** *Venice: Extraordinary Maintenance: A History of the Restoration, Conservation, Destruction and Adulteration of the Fabric of the City from the Fall of the Republic to the Present* (2004)

> Human behavior, which was at the heart of the build up of ground levels in Venice over the centuries and which is also one of the main reasons for the *aqua alta* today, will be at the centre of the city's future.
>
> —**Ammerman, A.** *in* **Fletcher, C.A. and T. Spencer (Eds.)** *Flooding and Environmental Challenges for Venice and Its Lagoon: State of Knowledge* (2005)

INTRODUCTION

Venice is beginning to be looked upon in environmental circles with the same gravitas as it has been in architectural circles since Ruskin offered it up as the ultimate archetype of form and sentinel of (dis)function. Musu (2001, *in* Musu 2001) believed that Venice is the most relevant example of the relation between environment and development and the tradeoffs existing between ecosystem preservation and economic resourcism. And Rinaldo (*in* Musu 2001) went even further in his assertion that the Venetian challenge is the central episode of the crisis of modern civilization.

Although originally referred to as the *Serenissima Republica* or the Most Serene Republic, Venice's environmental history has been filled with anything but serenity in terms of the tone of the debates that have raged over the years. Quite simply put, "Venice excites huge emotions" (Fletcher and Da Mosto 2004). The 17th century, for example, was filled with written polemics and soapbox speeches about the causes, consequences, and corrective measures in relation to sedimentation in the lagoon (Crouzet-Pavan 2002). Today's debates concern a wide variety of structural and non-structural solutions that have been posited to save or at least salvage Venice from the rising tides of water and tourists that threaten to engulf her.

BUILDING RESTORATION/RENOVATION/RECONSTRUCTION DILEMMAS

There is an extremely checkered history of building restoration in Venice (Lauritzen 1986). Much of the early debate dealt with the role of conservation vs. restoration, and how the latter may or may not differ from reconstruction, refurbishment, and renovation. Ruskin in his immensely influential *The Seven Lamps of Architecture* certainly did not hold back any punches when he admonished his

---

[1] An expanded version of this essay will occur in the book *Waterlogged: Environmental Reflections on Venice.*

409

contemporaries: "Do not let us talk then of restoration. The thing is a lie from beginning to end." Ruskin believed that the stones of Venice, which he considered to be our "most precious of inheritances," were deliberately being vandalized by restorers whose demolitions carried out under the guise of restoration ruined the very buildings they were trying to preserve in their misplaced zeal to improve upon things. Restoration in this light therefore could be equated with mutilation (Pemble 1995). And there were indeed egregious examples of this, perhaps none more infamous than the 19th century work on St. Marks's described by William Morris as "criminal" when he learned that the restorer had replaced apostles names with his own in several of the mosaics there (Pertot 2004). The restoration debate was certainly not settled in Victorian times.

Throughout much of the 20th century the concept of restoration continued to be envisioned by the modernists as rescuing buildings from time's ravages through making the end-products appear to be somehow "better" than the original ones. Others, who might be referred to as the abolitionists, however, championed the cause of letting time take its toll by preserving the picturesque decay that the city was so well known for. In this regard, it was believed that the odor of death must be preserved and the process of relentless disintegration sustained (Pemble 1995). For the conservationists, the big question was how desirable was the need for a return to some imagined pristine state (before the Baroque accretions) and in terms of history, just how important was fidelity to the past or should it be denied altogether like a painting a corpse with rouge.

Eventually Ruskin's opinions about the validity of restoration carried the day. A 1984 document warned of the dangers of "recreating instead of repairing" and a 1988 report was titled *Venezia "restaurata,"* the latter in inverted commas to suggest the futility of turning the clock back (Pertot 2004). Consequently, by likening architectural recreation to waxworks, many churches have actually been "rescued" from previous heavy-handed restorations through a process of essentially "de-restoration" (Pertot 2004).

This is not to say that mistakes have not nor will continue to be made. For example, removing the seemingly unsightly but actually protective black layer from the Istrian marble by well meaning but naïve restorationists actually serves to destabilize building foundations. Use of steel and concrete may have enabled cheap repairs but unlike brick and wood these modern materials have no elasticity to absorb the shifting forces of the essentially unstable Venetian subsoil and thus will contribute to future problems. Modern critics such as Pertot (2004) are particularly bothered by the restorationists' shift from using natural colors to harsher and garishly bright colors of plaster, transforming the city into an alien entity from its past, whose buildings are decked out like tarted-up beachside cottages rather than the venerable structures that they are. And finally, let us not forget the incredible irony involved with moving the historic gilded bronze horses from the open outside balcony to a confined corner inside St. Mark's in order to "protect" them from the polluted air of Marghera without ever considering that the damage wrought by the excited breaths of the tens of thousands of gaping-mouthed tourists are probably much more serious an insult. Certainly, this decision could not in any way have been related to that fact that it is now possible to charge admission to see the statues.

MOSE—BIG SOLUTION OR BIG WASTE OF MONEY?

The question of whether Modulo Sperimentale Elettromeccanico (MOSE) will protect or damage Venice and its lagoon has divided the citizens from the politicians and become the central defining environmental debate about Venice's future (the fact that the majority of the metropolitan city's voters live on the mainland in Mestre, an area that is not threatened by floods, further complicates matters). Cocks (*in* Fletcher and Spencer 2005) believed that it is almost impossible to judge the MOSE project because most of the information is unpublished or difficult to find and is not in English, therefore excluding the opinions of the international scientific community. These limitations have not, however, in any way hampered the generation of strong opinions about the efficacy and feasibility of building the mobile barriers.

Rinaldo (*in* Musu 2001) is representative of many of MOSE's proponents in believing that the highly artificial nature of the lagoon necessitates major engineered interventions, arguing that no matter how many water-absorbing wetlands are built this will never be enough to solve Venice's flooding problems. And he is especially critical of polemical arguments "tainted by political agendas" that have polluted reasonable debates over the years. It is certainly true that the heat of the debate about MOSE has drowned out more complex and subtle arguments about Venice's fate (Fletcher and Da Mosto 2004).

Alternatively, Lauritzen (1986) echoed the view held by many opponents of MOSE when he stated: "The arguments in favor of the barriers have been presented with such conviction, and backed with such an authoritative array of pseudo-scientific data, that this strategy has assumed the character of a foregone conclusion." These opponents are also exasperated by what they see as the sneaky way in which the MOSE supporters have circumvented or marginalized the voicing of credible, alternative viewpoints. In an article in the June 15, 2006, *Guardian* the director of Italy's World Wildlife Foundation (WWF) agreed that the close links between government and big business in Italy has corrupted decision making in Venice (for a particularly telling example of this see Berendt's (2005) interpretation of the restoration of the Venice opera house). Many are, therefore, harshly critical of the Consortium for—as they believe—its acceptance of bribes from its member contactors. What big engineering firm participating in the decision-making process would not want to build the gates and therefore do everything in its power to bring about their construction?

Amazingly, it wasn't until 1995 that an environmental impact assessment was ordered see how the mobile gates would affect the city and lagoon. This, of course, angered many, turning those previous "gate sitters" who may have been ambivalent to MOSE into vocal opponents. As Keahey (2002) stated: "It is inconceivable that this step, common elsewhere in the developed Western world during the decade of the 1990s, took so long in Italy." In his mind, and those of many others, this is an inexcusable omission that severely compromises the validity of the entire project.

The so-called "independent review" by, of all people and in all places, engineers from MIT, is a ludicrous farce, MOSE's opponents cry, similar to asking a fox if there is a perceived problem in the open door to the henhouse. And opposition to the mobile barriers certainly did not diminish when in 2000 the regional administration authority issued a decree annulling the previous decree of 1998 which had been negative to the project (Pertot 2004). The fact that experts in the Ministries of the Environment and of Cultural Heritage had both come down against the project despite the blessing from the "independent" review panel (to be fair, composed of accomplished individuals from several institutions, in addition to MIT) seems to have been conveniently ignored by the decision makers in charge.

Environmentalists are the biggest opponents to MOSE, worrying that the gates will be closed too often and thereby damage the lagoon ecosystem (Nova 2002). They regard the project as needless tampering, the "disaster of doing." But engineers argue that because everything one sees in the lagoon is the result of massive interference by humans and not the product of natural evolution, MOSE is merely part of a grand tradition of Venetian environmental management. Keahey (2002) counters this when he states that "the recent history of Venice and its region is a history of failed engineering," citing the creation of Porto Marghera within sight of sublime historic center as "strike one," the failure of a nearby dam in the mountains that resulted in the deaths of thousands in the 1960s as "strike two," and predicting that MOSE if built would be "strike 3—you're out."

And the debates continue. Some models argue that the gates could lower flooding tides by 8 to 12 inches whereas others say MOSE will only work for rapid in-and-out tides and not those that linger for more than a full day, going as far as to suggest that the flooding in 1966 would still have happened had MOSE been operational due to high river discharge and watershed runoff at that time (Plant 2002). The WWF is opposed to not only the mobile barriers but also to the accompanying permanent earthwork structures erected at either side of the channel mouths, believing that they will definitely change the flow of tides and currents, "provoking consequences of unpredictable proportions" (Lauritzen 1986). This is countered by proponents of MOSE who intriguingly (and perhaps teasingly) state that it is possible that the strategic opening and closing of the three sets of gates in relation to tidal cycles could be used to induce a circulation pattern to flush out pollution and thereby actually *improve* water quality (Harleman *in* Fletcher and Spencer 2005). Given all these uncertainties, critics say the MOSE project is simply bad wisdom, putting all the eggs in a single basket of unproven effectiveness (Nova 2002). Nonsense, say MOSE's supporters, stating that it is just one element in a much vaster plan of interventions to be undertaken throughout the entire lagoon and city (Scotti 2005). In this light, Venice cannot be saved without investments in both the barriers *and* the lagoon and to ask people to choose between them is to set up a false dichotomy (Cooks *in* Fletcher and Da Moso 2004). Many concede that, although the gates meet present problems, they might not adequately address future concerns (Keahey 2002). Consortium spokespeople themselves state that the "barriers are never the final solution in a changing world but would buy a century of usefulness" (private lecture and tour, Venice, July 2006).

The key question in deliberating about the effectiveness of MOSE really comes down to the projected frequency and duration of barrier closings in relation to sea level changes. Predictions from official sources are for about seven closures a year, an amount that has led most experts to agree that this will not be a significant problem in terms of a buildup of contaminants within the lagoon. Although the closures are predicted to be relatively short in duration, questions persist about the effects of repeated closures over the entire season. Ammerman (*in* Fletcher and Spencer 2005) dismisses impact studies that have failed to consider the seasonality of flooding events, whereas MOSE proponents argue that gate closings in winter are not worth worrying about due to low biological productivity in the lagoon at that time. But remember that Venice is sinking, and the sea is rising, and gate closings will consequently have to increase.

Given that data on Venice's flooding is compromised due to the short-term views in both analyses of the problem and its proposed solutions, Ammerman (*in* Fletcher and Spencer 2005) considered the hubris with which the Consortium engineers involved with MOSE make their predictions to be unjustified. His own archeological work predicts a relative sea level rise of 30 cm over next century, a value that is seven times larger than the low scenario used in three official and overly optimistic impact reports (Nova 2002). Given such a sea level rise, Harleman (*in* Fletcher and Spencer 2005) predicts that the gates could be closed for 10% of the time or 20% of the time in winter months, an obviously not unsubstantial duration whose implications on the consequent buildup of contaminants cannot be ignored. Independent analysis suggests that by 2050 the gates may be up for much of the wet season (i.e., not the seven closures per year that proponents say, but more like 24 times each winter month, for about 150 closings day after day for 4 months).

At best, it seems that MOSE will buy time. But how many years will it take before the gates become obsolete? Based on a set of simulations which (unlike some official scenarios) include leakage due to flip-flopping gates resulting from oscillations due to waves hitting them obliquely, some work suggests that MOSE will become obsolete within only a few decades at which time the gates will have to be demolished to allow for the construction of a more effective way to separate lagoon and sea. It is no surprise then that even among those who agree in principle about the predicted efficacy of MOSE, some would conclude that, like Lauritzen (1986), the whole project might very well be "a terrible waste of taxpayers' money."

Surprisingly, the MOSE project has not really been seriously examined from the triple bottom-line perspective of sustainable development which considers and gives equal weight to the social and economic factors in addition to environmental ones. Recently, several groups of Harvard and Ca'Fosari University students, working under the supervision of professors Peter Rogers and Stefano Soriani, undertook just such an investigation. One group determined that in the end it made more economic sense to take the money otherwise spent on building (2½ billion Euros) and maintaining (12 million Euros a year) MOSE and invest it in the stock market, which over a projected 60-year lifespan of the gates, would generate 40 billion Euros, an amount that is more than adequate to pay off the estimated 4 million Euros a year necessary to compensate individuals and institutions for flooding damage (this, of course, ignores the nonquantifiable discomfort involved in living with repeated flooding). And the other group concluded that the lack of, in their minds, an adequate competitive bidding process, cost–benefit analysis, climate change accountability, and alternative comprehensive solutions, made the project a risky endeavor. MOSE, in this case, the group concluded, was really an acronym for "Major Obstacle to a Sustainable Environment."

And the debate continues. Proponents emphasize their steadfast conviction that there will be no perceived operational nor construction problems to the lagoon environment associated with MOSE. To blame the gates for the deterioration of water quality in the lagoon is not valid, they argue. Environmentalists should address the causes of this deterioration in terms of more effectively managing agricultural runoff and industrial discharge. And just when it seemed that the debate was over once the construction had begun, the future of MOSE may once again be in doubt. The head of the Insula (a supposed partner with MOSE) complains that no money is going to his group's efforts to repair infrastructure, all of it instead being sucked into MOSE. "The problems of sewers in the city are just as important as high tides," he states. With the recent elections, the new mayor, supporting the majority of the new city council who are opposed to the project, backed a council proposal to examine alternatives such as inflatable rubber "sausage" barriers. And the European Union is finally entering into the fray, stating that the money should be spent on building maintenance, and threatening to take Italy to court unless a correct environmental impact statement can be carried out (Clarke 2006). And so it goes on....

TOURISM MANAGEMENT

In *Watermark*, his wonderfully moving homage to the city he so loved, Nobel Laureate Joseph Brodsky came down in favor of gates as being the only viable solution to Venice's serious flooding problem. This poet differed from the scientists and engineers, however, in where he would place these protective barriers. Rather than in the lagoon between the outer islands and against the Adriatic, Brodsky believed that the gates should be erected at the exit from the train station and against the influx of tourists. He is not alone in holding such an opinion. At a recent debate in London on whether Venice should be allowed to die, economist John Kay concurred that "the sea of tourists may be a lot more threatening than the Adriatic." Given that the economic returns to Venice from day tourists is minimal in relation to their contribution to the decline of the city, some have suggested since the early 1980s that admission be charged (Plant 2002). Kay (2006) believes that Venice should be managed as a theme park with strict controls: "Today 12 million people a year pay 50 Euros to visit Eurodisney. It is clear ... that if the Disney Corporation owned Venice, Venice would not be in peril." Because the unmanaged sea of tourists is drowning Venice, at the very least, many would now agree that the time has come to find ways to manage tourist flows (van der Borg and Russo *in* Musu 2001).

All tourist destinations have a carrying capacity at which point the pressures consume site resources and irreparable damage ensues such that visits begin to decline. Analysis of life-cycle models of tourism development have suggested that Venice can withstand a critical limit of 25,000 tourists per day of which 14,000 are daytrippers, a level which is now, however, being exceeded for two-thirds of the year (van der Borg and Russo *in* Musu 2001). Venice may therefore soon reach the point where the classic Yogi Berra remark finally makes sense: "No one goes there any more, it's too crowded." On the other hand, people will probably always travel to Venice, for as one tourist wrote in the *Times* letters of June 5, 2006: "Venice is a congested, over-priced, kitsch-ridden tourist trap. It is also the most imaginative, awe inspiring place I have ever traveled to."

There is a desperate need for strategic, integrated urban planning and design to prevent the danger of Venice becoming little more than a tourist theme park (Magnani and Pelzel *in* Musu 2001). Many would agree that it is the type of tourist that needs to be changed before this can happen. The onslaught of partying tourists to the annual Carnival and to big draw events like the infamous 1989 Pink Floyd concert in Piazza San Marco should to be curtailed in favor of cultural tourism.

Sustainable tourism development needs to transform tourist activities into an economic base which contributes to financing the maintenance of Venice's infrastructure, monumental, artistic, and environmental capital (Musu, *in* Musu 2001).

SINKING INTO VENICE'S INSTITUTIONAL QUAGMIRE

If there is a single phrase that describes the inertia of governmental decision making in Venice it is "bureaucratic lethargy" (Keahey 2002). Despite passage of a series of special laws over the last three decades making protection of Venice a national priority, really little has been accomplished. The reason for this, Pertot (2004) believes, is due to the inoperable and paralyzing bureaucracy, production of grandiose plans with no actions taken, fatalistic acceptance of corruption as a way of life, and exodus of sensible Venetians and invasions of insensitive tourist hordes, that together spell a recipe for disaster. Since the early 1970s, for example, plans have existed to stem emigration to the mainland by providing restored buildings at low rent as well as decent schools and hospitals, none of which has really occurred in any widespread and effective manner. Such procrastination, a form of tortuous death by a thousand plans, or what Pertot (2004) refers to "the production of plans by means of plans," follows a long established Venetian tradition. One doesn't have to look farther than the famous Accademia Bridge which was built of wood in 1933 as a temporary replacement for the original Austrian iron bridge. By 1984, before a final decision had been made on the permanent design, the wooden structure had deteriorated so much that it had to be torn down—only to be replaced by another such "temporary" structure (Lauritzen 1986).

Restoration plans have fared no better. In the 2002 Nova documentary on the "sinking" of Venice there is a segment showing a woman sitting at a desk with a telephone receiver in each hand and accompanied by the narrator's voice saying "Since 1966 Italians have been talking ,,,, and talking … and talking … ." This is followed by the statement that the MOSE gates have been a political hot potato tossed from one administration to the next, which in Venice means an incredible 35 times. And at another point in same documentary a disgruntled local refers to politicians with the statement "blah … blah … blah … ." However, it is important to put this in context: the 30 years of debating MOSE gates is still pretty minor compared to the 150 years of discussion it took before action was taken in relation to diverting the Piave River from entering the lagoon (Keahey 2002).

As Italians bickered throughout the 1970s, 1980s, and 1990s, increasing numbers of foreigners became more and more frustrated at Italy's inability to save its primary cultural jewel (Keahey 2002). If the consequences of the repeated failures of the various Italian governmental bodies to adequately address the problems of Venice were not so dire, the entire fiasco would have the making of a farce in which the incompetence of the players seems scarcely believable. Both Musu (Musu 2001) and Vellinga and Lasage (*in* Fletcher and Spencer 2005) conclude in the two most important scholarly books on Venice's predicament with incitements that the plans for solving problems almost never become operational in a satisfactory way due to the structure of governance in Venice and the inherent lack of transparency, accountability, and legitimacy of different groups. The bottom line is that many common practices that would be unthinkable and unacceptable by any other Western country are commonplace in Italy on the whole, and in Venice, in particular.

The foreign criticism is insulting, almost vitriolic, and quite possibly meritorious. For example, from Kay (2006):

> The problems of Venice are not pollution, technology or finance; they are problems of politics, of organization and of management.

> A sad series of accidents has placed so many of the best of Western Europe culture and civilization in the hands of Europe's most dysfunctional political system.

And from Fay and Knightly (1976):

> The very nature of government in Italy, its inherent instability, its system of political favors, and its crushing bureaucracy, made it unsuitable to handle a problem like Venice, and it is the basic reason for its spectacular lack of success.

> The fact is that politics in Italy *are* different, and politics in Venice ... are unusual even by Italian standards.

> There is an illness in government today and Italy has it worse than most. When a politician takes power then he also must take responsibility. The pleasures of power are among the rewards for taking that responsibility. But in Italy the people who take power decline responsibility.

De Mosto et al. (*in* Fletcher and Spencer 2005) identify one of the major reasons for the glaring lack of progress in solving Venice's environmental problems as being a system of fragmented and over-lapping institutional responsibilities between various administrations (and even among departments within a single administration) that leads to a piecemeal and ineffective institutional governance of the city and lagoon. Again, one should perhaps not be surprised at this jumbled morass of governance as it simply follows a long Venetian tradition of making the complex complicated. Take for example, the process of electing the *doge* or leader of the Republic: "In a nutshell: 30 men were selected by lot from the *Maggior Consiglio*; they reduced themselves by lot to 9 members; they elected 40, who reduced themselves to 12, who elected 25, who reduced themselves to 9, who elected 45, who reduced themselves to 11, who elected 41, who finally elected the doge—25 votes was the winning number. This rigmarole could last quite a while" (Buckley 2004, paraphrasing McCarthy 1963).

This "woeful Italian tradition" (Pertot 2004) has long been recognized and has born the brunt of many prejudicial jokes concerning the frequency with which Italians change their governments (58 governments in 55 years). The sad state of Venice's environment is, however, no laughing matter. By 1979, UNESCO's reference to the blunders and negligence that were "characteristic" of the Venetian and Italian governments led to calls for a foreign doge to be called in to run the city perhaps as a private city-state again, such as the Vatican, and save it from Venetians (Plant 2002). The suggestion that UNESCO take over the responsibility of Venice was, of course, a slap in the face to Italian governmental independence and resulted in the agency being temporarily run out of town as in a bad Western movie.

And if governmental incompetence was not enough, there was the messy issue of widespread corruption which continues to interfere with building restoration works, as colorfully described by Berendt (2005) with reference to the rebuilding the Venice opera house. On smaller scales, it is well known that unless one resorts to bribes it is very difficult to modernize any building (Lauritzen 1986). The same problems plague the environmental restoration of Venice wherein promised loans were never given and donations never received, instead disappearing into the morass of the corrupt system (Keahey 2002). This sorry history of financial mismanagement and graft led Fay and Knightly (1976) to be harshly critical of the Italians' ability to save Venice, summarizing the entire affair as "a tragedy within a farce." The greed of the Italian government is quite unlike anywhere else in Europe, causing them to actually tax (at 12%) all international and national donations given to restoration charities in Venice. This final insult to common sense and serious impediment to restoration activities is what finally caused many to suggest that Venice needed to be saved not for, but *from*, the Venetians and Italians. The founder of the Central Institute for Restoration, an Italian, was one obviously unable to curb his frustration and anger at the blight upon the landscape caused by shortsighted industrial greed: "Marghera is neither a city, nor a community ... but a jumble of factories, pipes, and chimneys, facing Venice, which looks on appalled at the monster it has pupped. Therefore Venice is dying by its own hand, and too many of its citizens deserve the fate of Martin Fabiero," who happened to be executed (Pertot 2004).

In summary, the city government in Venice often acts against the wishes of the populace and contributes (by either incompetence or greed) to "the widespread rape of the area" (Pertot 2004).

As a result, there is a great need to reach out to "the actors who, until now, have remained somewhat excluded from the political debate (van der Borg and Russo *in* Musu 2001).

PUBLIC PARTICIPATION, THE KEY TO A SUSTAINABLE VENICE?

Vocal opposition to installation of the flood barriers continued to grow due to the public's perception of increased pollution resulting from creating an embayment sealed off from the sea. Construction was therefore temporarily halted in order to acquire more environmental impact data. Gerritsen et al. (*in* Fletcher and Spencer 2005) describe the ensuing public consultation process and acknowledge that it helped considerably to raise the final effectiveness of the barriers: "Our overall conclusion is that the key factor in the success of the public information process has been the open character of the process, the systematic analyses and the free and full availability of the analyses whereby its results and conclusions have been available to all." "All" here means local municipalities, city governments, NGOs, the scientific community, interested individuals, and the press; and the "open process" means making the environmental impact assessment documents available for commentary on the web with hardcopies placed in libraries as well as production of summary brochures, news articles, and meeting report minutes. The problem with all this in the present context is that it is referring to the flood barriers of St. Petersburg, Russia, *not* those of Venice! Far from being open and egalitarian, Venice's process for reviewing and finally deciding upon their own flood barriers could be described as being secretive, technical, exclusionary, and quite possibly corrupt, in short, and not without a bit of irony, the same adjectives long used by the rest of the world to describe machinations within the old Soviet Union.

In 1391, a special commission of patricians was sent by the Senate in Rome to decide how to strengthen Venice's embankments, based on the advice of the local common people, the *proborum hominum* (Caniato *in* Fletcher and Spencer 2005). Today, if there is a single problem in relation to determining the environmental future of Venice, it concerns the need to engender a shared vision of the city as a social as well as a material system (Rullani and Micelli *in* Musu 2001), which can only be brought about through finding ways in which to actively engage people in the process.

The seminal book *Flooding and Environmental Challenges for Venice and its Lagoon: State of Knowledge* (Fletcher and Spencer 2005) is based on a series of workshops held at the University of Cambridge on such topics as urban flooding, engineered solutions to storm surges, physical-chemical processes in the lagoon, hydrodynamic modeling, lagoon morphology, and water quality. But where in all this is the public?

Because the emphasis of the book was directed toward exploring scientific issues, only a few contributors discussed sociological concerns. Ammerman (*in* Fletcher and Spencer 2005) believed that studies by scientists and engineers are inherently weakened by ignoring the humanity in Venice. Carrera (*in* Fletcher and Spencer 2005) dealt with altering human behavior and materials management and transport in the city in order to lessen the effects of boat traffic and called for creation of a central GIS-based data bank that would be transparent and open to all in an effort to build a city-wide knowledge library and thus aid in decision making. Scotti (*in* Fletcher and Spencer 2005) considered that engineering solutions alone cannot solve the long-lasting problems of the lagoon. In addition to the natural scientific factors of biology, chemistry, and morphology, human elements such as culture, history, and a whole suite of socio-economic concerns need to be given equal weight in any decision making. And in the same vein, De Mosto et al. (*in* Fletcher and Spencer 2005) concluded: "And with regard to policy making, the 'polymorphism' of the city and its lagoon, given its long history of man/nature interaction, and the vital underlying natural dynamics that maintain the system, demand the wider participation of a well-informed public in decision making."

Following decades of polarizing debate about the MOSE barriers one inescapable truth that emerges is that the Venetian public and the concerned international community need to become much better informed about the multitude of factors determining the health of lagoon and preservation of the city. Environmental communication to foster ecological literacy is one of the cornerstones of facilitating watershed management (France 2005). One major step forward in this regard was the production of the companion book generated from the Cambridge meetings. *The Science of Saving Venice* (Fletcher and De Mosto 2004), published in both English and Italian editions, could very well serve as a model for communicating complex socio-ecological issues anywhere in the world. Highly illustrated and with a text pitched at about the *National Geographic* level of detail, it is designed to fill the gap between the debates among experts and the concerns and interests of the general public (De Mosto et al. *in* Fletcher and Spencer 2005), a job it does admirably. Likewise, the information brochure and poster produced by the Consortium engaged in the MOSE project (MITVWACVN 2006a,b) and their open-door information center in central Venice, as well the information boards erected around the city near ongoing *insulae* projects, are all positive steps in the direction of technology transfer.

But is education alone enough to help guide Venice on its path toward an uncertain future? After all, in order to be truly effective, watershed management must be really based on direct public participation (France 2005). Below, I briefly explore four approaches to civic engagement—the Local Agenda 21 Protocol, the Vision for Venice strategy, alternative futures scenario planning, and the emerging paradigm of restoration design—that if applied could provide the guidance needed for Venice to become one the signature examples of regenerative landscape design.

1. LOCAL AGENDA 21 PROTOCOL

In Vellinga and Lasage's (*in* Fletcher and Spencer 2005) chapter at the end of the Cambridge book they consider Venice's future in terms of the three pillars of sustainability: social, economic, and ecological. For the first, they conclude that Venice needs low cost or subsidized housing to restore demographic balance to the city. For the second, they believe that a system of tourist user fees needs to be implemented. And for the third, they favor the MOSE barriers. Another surprising (if not shocking) omission in the nearly 700-page long Cambridge book (in addition to the relative dearth of material about public engagement) is that outside of this last of the 68 chapters, sustainability seems to have been mentioned only eight times, with little or no citation on the subject of Musu's edited 2001 book, *Sustainable Venice: Suggestions for the Future*!

Venice's future very much depends on applying principles of sustainability based on the preservation of the essential features of a lagoon ecosystem as the basis for formulating any solutions (Musu, *in* Musu 2001). In this regard, the emergence of social consensus on how Venice's solutions can evolve must involve all of the actors, both the local community and the national and international stakeholders, in other words, all who have an interest and an opinion to express on the problem of Venice (Beierle and Layford 2002).

Believing that strategic planning is desperately needed in order to save Venice from the sea and from itself, Dente et al. (*in* Musu 2001) explored the Local Agenda 21 movement that had its origin in the 1992 Rio conference on sustainability (R. Abbott, pers. comm., 2007). The Local Agenda 21 Protocol is a program for environmental action with concrete plans at the local level based on consensus and dialogue between citizens, local associations, businesses, and public authorities (Dente et al., *in* Musu 2001). Its process aims include combining development and environmental protection in a long-term perspective, change in population lifestyles through a participation strategy, dissemination of knowledge on environmental matters to make citizens more aware, and improvement of environmental services by the municipality. For the cities of Hamilton–Wentworth, Canada, and Seattle, Washington, for example, a shared vision of sustainability in the long-term was considered an indispensable precondition and the starting point for the entire Agenda 21 process. On the international scene, the Agenda 21 process is often established by local authorities under pressure from environmental groups. What these individual processes share in common is the "the absolute centrality of the environmental concern." For Venice, this would radically change the nature of the debate about Venice's future (Dente et al., *in* Musu 2001).

Dente et al. (*in* Musu 2001) studied 10 cities that have participated in the Local Agenda 21 process and which could offer guidance for how Venice should approach its own sustainable future:

Hamilton–Wentworth, Canada; Seattle, Washington; London; Lancashire County Council, U.K.; Cheshire County Council, U.K.; Leicester City Council, U.K.; The Hague, Netherlands; Kanagawa Prefecture, Japan; and Troyan, Bulgaria. Environmental and social issues that were dealt with included natural areas, air, water, rubbish, energy consumption, land-use, transport, public health, agriculture, economy, pollution, unemployment, civic commitment, economic viability, mobility and accessibility, noise, open spaces, environmental education, territory, water consumption, naturalistic issues, participation, houses, culture and spare time, building construction, information transfer, lifestyle, sustainable cities, socio-economic and environmental systems, and environmental cooperation among towns; the instruments for participation included civic forums, focus groups, specialist working groups, workshops, conferences, press and television coverage, technical committees, distribution of questionnaires, discussion groups, experts panels, open meetings, surveys, information leaflet distribution, and information in schools.

Dente et al. (*in* Musu 2001) then surveyed the number of organizations involved in a myriad of restoration projects currently underway in Venice, coming up with the following tabulation: political bureaucracy—30, technical experts—31, economic—21, and social—15. They concluded that the "most significant fact that emerges remains the extreme under-representation of economic and social interests." Also, the fact that only 15 local associations in Venice had participated in any of the restoration projects was in stark contrast to the group of international communities adopting the Local Agenda 21 process which had successfully engaged between eight and ten times that number of local groups! Clearly, an essential element that is missing in Venice is an involved populace in the decision-making process.

The Local Agenda 21 process brings the local community into strategic planning through satisfying two requirements: first, increased knowledge such as the identification of problems, and accurate analysis of their causes by way of creating a set of sustainability indicators and coherent strategies that combine environment and development, and, second, increased involvement such as building collaboration between various actors and developing a shared community-driven creation of sustainable development. It could be argued that with the recent publication of *The Science of Saving Venice* by the Cambridge study group, as well as the open-door reference library and distributed information posters by the Consortium, that Venice is finally moving toward satisfying the first requirement of the Local Agenda 21 strategy. The absence, however, of a shared vision for Venice's sustainability seems to be a requirement that remains unfulfilled.

2. Vision for Venice

In 1998, the Forum per la Laguna group brought together local residents, politicians, entrepreneurs, and decision makers to begin a process of trying to establish a common view of Venice. With a goal of attempting to develop an economy from within (DEW) as opposed to one dependent on mass tourism, the group formulated a multifaceted and holistic approach in which all suggested actions would be implemented in an integrated fashion based on listening to the concerns of hitherto socially excluded groups of people. Today, a DEW project funded by the EU and based at the International Institute for the Urban Environment in The Netherlands (IIUE 2007) is actively engaged in exploring these issues (Boele et al. 2007). This group produced the following vision statement: "By the year 2020, Venice and the Laguna will have a high quality environment and lifestyle. The traditions and culture of Venice will be alive and contribute to the economy that draws upon the knowledge and abilities of local people and a sustainable transport system that supports the community of islands."

To reach these lofty goals, the group developed a series of action plans based on encouraging activities structured under five strategic objectives (Boele et al. 2007). The Citizenship Strategic Objective explored the following initiatives, activities, and measures: equality of access and opportunity, participatory democracy, belonging and identity, and heritage. The Natural Environment Strategic Objectives explored the following initiatives, activities, and measures: water pollution, waste and energy, and wildlife. The Economic and Business Development Strategic Objective explored the following initiatives, activities and measures: business development, business start-up, and inward investment. The Training and Employment Strategic Objective explored the following initiatives, activities, and measures: employability, cyberskills, and capacity building. The Transport and Accessibility Strategic Objective explored the following initiatives, activities, and measures: sustainable transport and a forum for transport improvement. And the Built-Environment Strategic Objective explored the following initiatives, activities, and measures: building restoration and infrastructure. Each of these initiatives, activities, and measures were examined in terms of the actors who would participate, potential sources of funding, types of communication facilitation, and what the implications of implementation would be in terms of social, economic, and environmental impacts.

Some of the exciting possibilities that emerged from these workshops and planning exercises included: development of a citizen charter and neighborhood forums, creation of a floating environmental education center for both residents and tourists, implementation of energy conservation programs, development of a micro-business center to provide small loans for local ventures in new sectors, programs for training local youth in restoring buildings and life-long learning, promotion of pollution-free transport, and a variety of approaches to help foster a sense of community spirit and to encourage native Venetians to remain in their city.

The group particularly targeted tourism (Boele et al. 2007), developing an action plan as an example of how this sector of the economy, planned and managed with the input of local knowledge, could be used to develop a sustainable economy for a revitalized Venice and its lagoon. In this perspective, tourism was investigated through the lens of the strategic objectives established and summarized above. Some of the interesting ideas that were considered included: creation of heritage and science/technology tours, micro-financial backing for imaginative start-up initiatives, promotion of wedding and scientific/technology conference-affiliated tourism, encouragement of the development of educational facilities, creation of a floating market for local organic produce, and development of ecotourism throughout the lagoon.

The Vision for Venice group also developed a framework to establish financial support for its suggested activities and how these resources would be managed during implementation of its strategic objectives. The important issue of communication and marketing was also addressed in terms of creating a newsletter and a promotional video, and operating a mobile information kiosk in the form of a boat that would circulate around the city (Boele et al. 2007).

The comprehensive nature of this envisioning process, addressing and exploring the interfaces among social, economic, and environmental concerns, stands in stark contrast to the obstinate fundamentalism and much more narrowly focused way in which the MOSE project was discussed, "deliberated," and then acted upon. One absence in the Vision for Venice project is a realistic vision about how these mainly social, nonstructural "activities, initiatives and measures" might unite to change the future physical landscape of the city and how this might differ from the city and lagoon that would develop based on its current trajectory.

3. ALTERNATIVE FUTURES SCENARIOS

UNESCO (1979) believed that Venice is a victim of development, not of decadence. There are many problems and questions relating to the existing development of Venice's mainland suburbs, Mestre and Porto Marghera, as well as the existing development of its tourism industry in the old city center. In terms of the former, Plant (2002) criticized modernist planners who as late as the 1980s still regarded water with ambiguity and treated it as a constraint, and Pertot (2004) raised the difficult question that given that the factory workforce of 30,000 in the 1970s had decreased by more than half by the 1990s, had the time come to close down the industrial operations all together? And in terms of the latter, Dente et al. (*in* Musu 2001) reflected that the extreme tourist priorities of Venice over other cities could mean that the Local Agenda 21 process might not be an ideal fit and would need to be retooled or fused with a different model.

Gaps often exist in the production of scientific information and the delivery of those insights in a useful form to act upon. Furthermore, there is need to develop place-based relationships between environmental, social and economic aspirations, and in a form that not only supports but actually encourages objective discussion about what the future might look like on the ground. One approach that is beginning to receive widespread interest among land-use planners, particularly those dealing with large-scale, highly contentious watershed development issues, is alternative futures scenario modeling (France 2006). This technique enables predicting impacts of land-use alterations on ecological processes, integrating human dimensions into effective planning, and developing an understanding of the uncertainty of impacts and associated risks of various development options, including those related to tourism (Steinitz et al. 2001). A geographic information system (GIS) are used to generate explicit comparisons of present land use with past land uses, as well as with various alternative future predicted land uses to help provide a perspective (both spatial and temporal) for decision makers. In so doing, these methods can have the capability to catalyze change in cultural attitudes and even inspire actions in terms of how and where land development occurs and what might be possible corrective measures. Such a futures approach takes abstract goals such as enhancing water quality or restoring biodiversity or creating alternative tourism opportunities, and translates these into specific land-use practices. One very important part of the alternative-futures model building is the production of the actual change scenarios through extensive use of a citizen- or expert-driven approach or some combination of the two, each having its own set of benefits and detriments (Hulse et al. 2004).

An example of an alternative-futures study that is particularly germane to Venice is Steinitz et al.'s study (2005) of the nearby city of Padova. The parallels between the two sister cities is striking: both have important historic centers and are major tourist attractions, and both are abutted by enormous industrial developments. In Padova's case, the ZIP industrial zone along its eastern edge, one of the largest such areas in Italy, is almost the same size as the entire city itself. In addition to the industrial zone, the alternative-futures analysis also dealt with diverse landforms that included parks and gardens, canals and rivers, and many farms. The context of the study included addressing green corridors and transportation networks, green spaces and linkages, demographic trends, water systems for flood management and waste water treatment, as well as the potential expansion of the industrial zone. Study goals included exploring various solutions to the sewage-polluted river, sediment contamination of soils and canals, a dangerous transportation network, poor access to existing public parks, a shortage of green space, a lack of neighborhood identity, and a possible expansion of the industrial zone.

Following a site visit and interviews with key players, management teams were formed to research the following issues (Steinitz et al. 2005): techniques of waste water management through use of constructed wetlands, levees and flood protection, lake construction, transportation structures, possible planting palettes, green industries, green eco-roofs and parking lots, water collection and reuse opportunities, and analysis of existing features of the ZIP industrial area. Assortments of these variables as well as a set of common constraints were then used to generate and comparatively evaluate (in terms of function and design) three spatially-explicit scenarios of what the site could become in the future and how it might fit into the larger region.

Though initiated as a student project, local authorities were so impressed with the outcomes in terms of the innovative possibilities suggested to transform the landscape that they have decided to move ahead by selecting a subset of the various interventions to implement in the near future. A Padovan Landscape Alliance was formed to operate as a nonprofit group to facilitate the development of the redesigned area with responsibilities for fundraising, marketing and public outreach, programming, advocacy, organization of local volunteers, and planning a maintenance program for the public spaces.

Such planning exercises often rely upon input from the public through a type of charette process in order to help formulate the various scenarios to be critically analyzed and later adjudicated by experts. It is these experts who will, in the end, based on their own accumulative knowledge, develop a set of alternative solutions that will be put forward to the community decision-makers for adoption and (re)action (Steinitz, C., pers. comm., 2007). In addition to the conceptual participation of citizens early on in such a process, there is a need to actively engage the public through direct physical participation in rebuilding their future landscape in terms of helping to implement the proposed corrective actions.

4. RESTORATION DESIGN

Humans exist in a mindset of disseverance from nature and desperately need to find positive ways in which to reconnect to their environment. One method to do this is to physically engage in the actual process of ecological restoration (e.g., Light 2000; Jordan 2005). The act of restoring, remediating—in other words, healing—degraded landscapes is an act of reciprocity, important not only for improving the quality of the outside environment of nature, but also that of the internal environment of the psyche, or human nature (France 2003b). "Restoration design" is the process by which participants creatively develop physical and conceptual relationships to engage repaired nature through the architectural transformation of their inhabited ecological space as well as their internal consciousness (France 2007b).

In Venice, environmental restoration is most frequently preoccupied with attempting to return degraded nature to its imagined original state. Restoration design, however, is very much its own entity, separate in spirit and purpose from, and more holistic in both its conception and execution than, restoration ecology (France 2007b). Indeed, though elements of the latter are subsumed within the former, restoration designers are also much more honest in their acknowledgment of the role of humans in shaping the natural world, such as the artificial state of the Venice Lagoon. As a form of landscape architecture, restoration design is one of the most integrative of all environmental disciplines and has a rich history that extends back to the 19th-century pioneering work of Fredrick Law Olmsted in Boston, that other historic city built atop a coastal saltmarsh.

Environmental restoration as conducted in Venice is undertaken by scientists and engineers with little or no role played by environmental artists. Restoration design, in contrast, is neither pure science not pure art, but a creative blending of the two, an odd hybrid that at times is seemingly straightforward but may often be a perplexing paradox of intent and execution (France 2007b). Restoration designers are those individuals who effortlessly either integrate these two important spheres of human creativity themselves or recognize the importance of assembling interdisciplinary teams that can collectively provide such integration, something which, given Venice's own rich artistic heritage, would seem to be an ideal fit to develop there. Restoration design is in the end about thinking with the heart, and feeling with the mind, a soulful union of respect and remembrance with that of hopeful anticipation.

Environmental restoration in Venice is a nuts-and-bolts physical activity focused on achieving products, paying little or no attention to the conceptual processes involved in generating those end results. A restoration designer, in contrast, has an inherent process-minded, Gandhian sensibility, coupled with a physical imperative to act upon that sensibility, knowing that what really matters the most is not the ends but the means used toward achieving those ends (France 2007b). The long-term success of environmental restoration projects in Venice will best be achieved through the public acting as stewards by tending the healed landscapes so that the latter do not become orphaned after the reparation is completed and once the visiting experts, necessary for planning the process and guiding the implementation of the products, have moved to work on other projects located elsewhere.

In Venice, environmental restoration remains largely the purview of specialists. Restoration design, in contrast, engages lay participants in landscapes of memory where their thoughts and actions are anchored to and guided by the accumulated wealth and wisdom of history, both cultural and ecological (France 2007b). By instilling a sense of community developed through shared actions, concentrating on culture as much as on nature, restoration design has much to offer Venice, a city whose sociology is just as damaged as is its ecology.

Environmental restoration as currently practiced in Venice is a doom-and-gloom scientific/engineering enterprise filled with Baconian rationalism in which little or no room exists for mystery, mythology, or metaphysics. Restoration design, in contrast, includes a richness of celebratory approaches involving ritual and performance in which to practice community-based design (France 2007b). The strongest piece of commonality that characterizes all those involved with restoration design is an overwhelming feeling of unbridled hope brought about through participation, something which Venice, given its need to uplift its spirit just as much as its pavements, must adopt if it is to have any chance of a sustainable future.

CONCLUSIONS

It has long been recognized, as Fletcher and Da Mosto (2004) summarize, that the future health and survival of Venice and its lagoon depends on creating some form of balance between the needs of the environment, industry, agriculture, tourism, and the Venetians. The critical issue in all this is just exactly what is that point of balance and exactly just for whom is the city being saved for: the 60,000 residents as a living and working landscape set in a sustainable environment or the 19 million tourists as a preserved fossil, a bauble set in amber.

Venice's future continues to be hotly debated (e.g., VIP 2007). Kay (2006), for example, considered that too much money has been wasted on restoring Venice and that the managed theme-park option is the only hope of salvation. Berendt (2006) believed that the salvation of Venice was worth any price. And Campbell-Johnston (2006) pessimistically concluded that Venice should be allowed to drift slowly to a stately death, "to sink back to the mirage upon which it has always floated" such that it could "live forever there in the fairytale land of the imagination" rather than suffer the fate-worse-than-death of being turned into a modern Disney-fied fairyland exhibit.

If Venice is to survive into the next century it needs to do a much better job of constructing methodological *passerelle* or pedestrian bridges that would allow its citizens to have a greater opportunity to walk above the rising waters of discontent and be able to participate in and thus help to set the course ahead. Otherwise, unless some timely action in this regard is taken, the future of this most remarkable of the world's cities, if we are to believe Plant's (2002) dire warning, looks bleak indeed: "Meanwhile the seas are rising. In the city of the apocalypse, the four golden horses are at the ready, pawing at the porch of San Marco, waiting to haul the city out of the waters and into the sky."

LITERATURE CITED

Albano, C., P. Frank and A. Giacomelli. 2007. Innovative design approach for the Fusina treatment wetland. Abstract, IWA Internat. Conf. 2007, Padova.

Berendt, J. 2005. *City of Falling Angels*. The Penguin Press.

Beierle, T.C. and J. Layford. 2002. Democracy in Practice: Public Participation Decisions, Island Press.

Berendt, J. 2006. Venice; A City Beyond Price. *The Times*, June 10.

Boele, N., et al. 2007. Strategic Objectives for Venice. Part A and B. www.urban.ul.

Brodsky, J. 1992. *Watermark*. Farrar, Straus & Giroux.

Brown, P.F. 1996. *Venice and Antiquity: The Venetian Sense of the Past*. Yale Univ. Press.

Buckley, J. 2004. *The Rough Guide to Venice and the Veneto*. Rough Guides.

Campbell-Johnston, R. 2006. If You Love Venice, Let Her Die. *The Times*, June 5.

Chang, C.Y. 2006. A Local Emergence of Cultural Ecotourism: Westergasfabiek and Lazzaretto Nuovo. Student essay. Harvard Design School.

CELI (Collegio di Esperti di Livello Internazionale). 1998. Report on the Mobile Gates Project for the Tidal Flow Regulation at the Venice Lagoon Inlets. Regional Press Venice.

Clarke, H. 2006. Tide of Opinion Turning against Venice Dam. *The Daily Telegraph*, January 29.

Crouzet-Pavan, E. 2002. *Venice Triumphant: The Horizons of a Myth*. Yale Univ. Press.

Debray, R. 1999. *Against Venice*. North Atlantic Books.

Fay, S. and P. K. Knightly. 1976. *The Death of Venice*. Andre Deutsch Publ.

Fletcher, C. and J. Da Mosto. 2004. *The Science of Saving Venice*. Umberto Allemandi & C.

Fletcher, C.A. and T. Spencer. 2005. *Flooding and Environmental Challenges for Venice and its Lagoon; State of Knowledge.* Cambridge Univ. Press.

Forum per la Laguna. 2007. Urban Forum for Sustainable Development. www.forumlagunavenezia.org.

France, R.L. 2003a. *Wetland Design: Principles and Practices for Landscape Architects and Land-Use Planners.* W.W. Norton.

France, R.L. 2003b. *Deep Immersion: The Experience of Water.* Green Frigate Books.

France, R.L. 2005. *Facilitating Watershed Management: Fostering Awareness and Stewardship.* Rowman & Littlefield.

France, R.L. 2006. *Introduction to Watershed Development: Understanding and Managing the Impacts of Sprawl.* Rowman & Littlefield.

France, R.L. 2007a. *Wetlands of Mass Destruction: Ancient Presage for Contemporary Ecocide in Southern Iraq.* Green Frigate Books.

France, R.L. 2007b. *Healing Natures, Repairing Relationships: New Perspectives on Restoring Ecological Spaces and Consciousness.* Green Frigate Books.

France, R.L. 2008a. *Regenerating Devastated Cultural Landscapes: Lessons for Iraq and Elsewhere.* Routledge.

France, R.L. 2008b. *Restoring the Iraqi Marshlands: Potentials, Perspectives, Practices.* Sussex Acad. Press.

Hooper, J. 2006. Population Decline Set to Turn Venice into Italy's Disneyland. *The Guardian,* August 26.

Hulse, D., et al. 2004. Envisioning alternatives: Using citizen guidance to map future land and water use. *Ecol. Appl.* 14: 325–341.

IIUE. 2007. Developing the economy from within. www.urban.nl.P_Dew.

Jordan, W.R. 2005. *The Sunflower Forest; Ecological Restoration and the New Communion with Nature.* Univ. Calif. Press.

Kay, J. 2006a. Venice's Real Problem is Organization and Management. *The Art Newspaper,* July.

Kay, J. 2006b. The Magic Kingdom Could Save Venice. *The Financial Times,* June 13.

Keahey, J. 2002. *Venice Against the Sea: A City Besieged.* St. Martin's Press.

Lasserre, P. and A. Marzollo. (Eds.) 2000. *The Venice Lagoon Ecosystem: Impacts and Interactions between Land and Sea.* The Partheonn Publ. Group.

Lauritzen, P. 1986. *Venice Preserved.* Michael Joseph Ltd.

Light, A. 2000. Restoration and the value of participation, and the risks of professionalism. *In* Gobster, P. and B. Hull (Eds.) *Restoring Nature: Perspectives from the Social Sciences And Humanities.* Island Press.

McCarty, M. 1963. *Venice Observed.* Harcourt inc.

MIPIM. 2006. Venice at the MIPIM 2006 Project: Aims, Local Operators Involved, and Promoted Projects. www.forumlagunavenezia.org.

MITVWACVN (Ministry of Transport, Venice Water Authority, Consorzio Venezia Nuova). 2006a. Measures for Safeguarding of Venice and the Lagoon. Information poster.

MITVWACVN (Ministry of Transport, Venice Water Authority, Consorzio Venezia Nuova). 2006b. Venice. Mobile Barriers at the Inlets to Regulate tides in the Lagoon. Information brochure.

Musu, I. (Ed.) 2001. *Sustainable Venice: Suggestions for the Future.* Kluwer Academic Publ.

Nova. 2002. Sinking City of Venice. Nova Documentaries. PBS.

Pemble, J. 1995. *Venice Rediscovered.* Clarendon Press.

Pertot, G. 2004. *Venice: Extraordinary Maintenance.* Paul Holberton Publ.

Plant, M. 2002. *Venice: Fragile City 1797–1997.* Yale Univ. Press.

Resini, D. 2006. *Venice: The Grand Canal.* Vianello Libri.

Ruskin, J. 1960. *The Stones of Venice.* Da Capo Press.

Sabaeadv, J. 1987. *Twenty Years of Restoration in Venice.* Arch. Di Venezia.

Steinitz, C., et al. 2001. Nature and Humanity in Harmony: Alternative Futures for the West Lake, Hangzhou. Harvard Design School.

Steinitz, C., et al. 2005. Padova and the Landscape: Alternative Futures for the Roncajette Park and the Industrial Zone. Harvard Design School.

UNESCO. 1979. *Venice Restored.* UNESCO Publ.

VIP (Venice in Peril). 2007. News articles of 2006. www.veniceinperil.org/news_articles.

Finale

Conclusion

Reparative Paradigms: Sociological Lessons for Venice from Regenerative Landscape Design

REPARATIVE PARADIGMS

Regenerative landscape design, while incorporating elements from a family of related professional disciplines such as environmental engineering, ecological restoration, low impact development (LID), and sustainable development, is unique in terms of the relative emphasis it places upon these different attributes (Table 1). Also, regenerative landscape design, in building upon these previous reparative strategies, is often more holistic in its overall outlook, balancing the needs of both nature and people. It is important to distinguish regenerative landscape design from its sibling disciplines.

Environmental engineering advances technological solutions to repair environmental problems. Often, however, these solutions are put forward with little or no regard given to a cost-benefit analysis of the long-term economic sense and sustainability of undertaking such activities. Environmental engineering is a professional discipline of engineers and other technocrats with no input from the artistic community. And due to the high degree of expertise required to implement the techno-fix solutions, environmental engineering is very much a "top-down" process that often excludes input from the public. Finally, environmental engineering is a practical, hands-on field that operates without a substantial corpus of background theory.

A representative case study of environmental engineering is that of the water supply situation in Sydney, Australia. Rapid urban development in Sydney has led to a looming water supply crisis. In a process with Venetian echoes, a consortium of environmental engineering companies, meeting with little or no public participation, made decisions about large-scale infrastructure such as constructing a major desalinization plant rather than opting for a decentralized alternative of water micro-management such as rainwater harvesting. All economic analyses were therefore based on the single "solution" of large-scale engineering.

Ecological restoration is a scientific discipline primarily focused on repairing the broken bits and pieces of nature with generally little regard paid toward the human inhabitants of damaged or soon-to-be repaired landscapes. Much consideration and effort is given to using, whenever possible, a tool-box of "soft" bioengineering approaches rather than relying exclusively on highly mechanized technological solutions. Because of this, it is occasionally possible to use public participation in the implementation of ecological restoration projects. Ecological restoration remains, however, most often the purview of professional biologists with little or no input from environmental artists. This is a discipline that is primarily concerned with nonurban locations in which economic considerations play only a minor role. Finally, ecological restoration is rich in theory and has been the subject of a series of ongoing and lively exchanges as to its overall role in environmental management.

A representative case study of ecological restoration is that of the Florida Everglades. As a result of rampant agricultural and housing development, as well as egregiously short-sighted, earlier water management manipulations, the Everglades, the famous "river of grass," has suffered incredible stress. Today, the restoration of this signature ecosystem has become the largest and most expensive such undertaking in the world. Almost all the restoration efforts are being led by

TABLE 1

Reparative Paradigms of Environmental Engineering, Ecological Restoration, Low Impact Development (LID), Sustainable Development, and Regenerative Landscape Design (RLD)

	Environmental Engineering	Ecological Restoration	LID	Sustainable Development	RLD
Environmental	* *	* * *	**	* *	* *
Technical	* * *	*	* * *	* *	* *
Economic	—	—	*	* * *	*
Artistic	—	—	*	—	* *
Public Participation	—	*	*	*	* * *
Theory	—	* *	—	* *	*
Case Study	Sydney Water	Everglades	Alewife	Dockside	Vegas Wash

Note: The number of stars denotes the relative weight placed on each descriptive attribute (see text). Representative case studies of each paradigm are listed at the bottom.

professional groups of engineers and ecologists with little public participation in the hands-on process. And much of the emphasis has been directed toward rebuilding the natural areas in order to foster wetland biodiversity.

Low impact development or LID employs a combination of technological solutions to restore predevelopment hydrological conditions. Although the rationale for engaging in such small-scale, techno-fix reparations is primarily environmentally based, some attention is paid to the economics of implementing these decentralized solutions in contrast to centralized approaches characteristic of conventional environmental engineering. In LID, the interests of humans with respect to development can supercede those of nonhumans in terms of natural habitat. However, because many of these hydrological solutions require only minor technological expertise for construction, it is possible (in contrast to environmental engineering) to enable homeowners and sometimes environmental artists to participate in LID. However, like environmental engineering, LID is a practical discipline operating without a major theoretical base.

A representative case study of LID is that of the Alewife Reservation and Quadrangle in Cambridge, Massachusetts. The Alewife region contains the largest urban wild in the metropolitan Boston area and is one of the last bits of underdeveloped and marginalized real estate in the city of Cambridge. At the same time, the area suffers from the worst flooding and combined sewer overflows in the eastern part of the state. Decades of planning and dozens of documents have been produced about how to best integrate environmental and human interests in the developing area, culminating in the recent generation of a master plan for the park and an integrated development plan for the hard-space. Included in both is a suite of approaches advanced for addressing the hydrological problems in the area through a program of direct LID interventions such as raingardens, bioretention swales, and stormwater treatment wetlands. To date this has largely been a process run by professional consultants with technical solutions being implemented without the participation of either the public or the artistic community, though plans are to become more inclusive in the future.

Sustainable development is a multifaceted program of societal strategy and change based on preserving the environment in a healthy and largely unaltered state for future generations. Emphasis is placed on the economic implications to designing and developing human communities in balance with sustaining existing or restoring past environmental conditions. Sustainable development is largely a technical endeavor with limited or no participation by environmental artists or the general public. As an important emerging discipline with wide societal implications, sustainable development is rich in theory.

A representative case study of sustainable development is that of Dockside Green, Victoria, Canada. Dockside Green is an award-winning redevelopment of a 15-acre, former contaminated brownfield site that is being transformed into one of the world's first LEED (leadership in energy and environmental design)-rated communities. The project is firmly based on the triple bottom-line mindset of sustainable development with strong emphasis placed on economic, social, and environmental benefits. In addition to incorporating many LID techniques such as planned for the Alewife area redevelopment described above, Dockside Green moves the bar higher in terms of treating water as natural capital in truly innovative designs of its buildings and their infrastructure, particularly in relation to energy supply and use. Not only will the development be comprised of green architecture, but there will be many nonstructural elements such as provision of shared "smart" cars and environmentally beneficial regulatory bylaws. Though much effort was spent on cleaning the on-site contamination, there does not appear to be as strong a motivation for reconstructing or "restoring" a natural habitat as is the case, for example, for the regenerative landscape design project in Las Vegas described next.

Regenerative landscape design is a multifaceted strategy based on recognition that the wish of humans to reside in livable, environmentally healthy communities can only come about through direct public engagement in many phases of the overall environmental reparation process. Like its sibling disciplines discussed above, regenerative landscape design is focused on environmental improvements though not as monolithically so as is ecological restoration. Regenerative landscape design can often be a technological pursuit similar to sustainable development (but not as much so in this regard as either LID or environmental engineering). More than the other disciplines, regenerative landscape design often utilizes the efforts of environmental artists and landscape designers in its successful implementation. Where regenerative landscape design is most distinct from its predecessors is in its explicit endorsement of, and frequent reliance on, a "bottom-up" process of civic engagement. And as a constituent subdiscipline of the emerging paradigm of restoration design, regenerative landscape design has a developing theoretical basis.

A representative case study of regenerative landscape design is that of the Las Vegas Wash restoration. Las Vegas is the fastest growing city in America and represents an interesting comparison to Venice in that it is second only to that city in terms of the number of worldwide tourists it receives every year. As Las Vegas has grown, its runoff has increased to such a degree that it now severely erodes the channel of the downstream Wash, thereby leaving riparian zones perched and desiccated, and thus ruining the valuable wildlife habitat of this rare desert wetland ecosystem. The project to restore the wetlands is part of an integrative master plan to transform the area into the Clark County Wetlands Park. Like the ecological restoration work in the Everglades, the restoration of the Wash does rely upon highly engineered installations. Where it differs, however, is in an active program of directly engaging the public in all stages of the process, from initial hearings to actually encouraging them to participate in building and managing the new park. This, as well as an explicit understanding that the park is as much or more for the benefit of people as it is for wildlife—with considerable attention being paid to a comprehensive watershed management program focused on improving the quality of human developments—indicates that the Las Vegas example is a true regenerative landscape design project.

SOCIOLOGICAL ISSUES OF REGENERATIVE LANDSCAPE DESIGN (AND LESSONS FOR VENICE)

The chapters in this volume address a range of environmental concerns present in a diversity of landscape types. In consequence, the technological solutions proposed or implemented are themselves diverse as a result of being occasionally idiosyncratic to the particular location or problem at hand. The important point to emphasize is that, just as for integrated watershed management, there is no uniform or standard "one size fits all" cookbook approach to successful regenerative landscape design. That said, there is at least one common element shared by all the case studies in this volume and that is the central role of humans, either as participants in the process of restoration or as design targets for the final products. In this respect, regenerative landscape design really shares much in common to what I have referred to elsewhere as "the sociology of watershed stewardship and management" (France, R.L. 2005. *Facilitating Watershed Management: Fostering Awareness and Stewardship*). In the following section, I have extracted a series of sociological issues from the chapters in order to create a template of regenerative landscape design from which to examine how Venice has approached its own restoration challenges, which I have summarized in italics.

CHAPTER 1—NAGLE

- Waste products are as much part of our collective heritage as are the buildings or processes that generated the refuse in the first place and in this regard the end receptacles of our waste are also landscapes of memory. (*Venice treats its waste as garbage.*)
- Because postindustrial landscapes represent our common history, it makes eminent sense to move toward redeveloping them into end-use common spaces for the public. (*Venice's plans for the new wastewater treatment park in Fusina will reclaim a postindustrial site.*)
- When controversies develop in the regenerative landscape design process it may be necessary to put the project on hold in order to listen to and possibly accommodate differing visions for the intended end-use and recuperative strategies being suggested or employed for the landscape. (*Planning for MOSE in Venice continued with little revision despite the vociferous complaints of many.*)

CHAPTER 2—CRAUL AND ROWE

- The regeneration of drastically disturbed sites necessitates inclusion of a high amount of scientific acumen and experience which should not be undertaken in a vacuum without frequent opportunities for educating the public about the technologies employed. (*Venice has recently undertaken a technology transfer program to better inform the public about MOSE.*)
- The final product in regenerative landscape design needn't be an attempt to mimic past site conditions but rather can be a move toward a new future in terms of creating public spaces that incorporate recreation and wildlife habitat enhancement together in a park. (*The Fusina wastewater treatment park will become an important civic green space in greater Venice.*)
- Small-scale regenerative landscape design efforts should be integrated into larger recreational networks through landscape architecture perspectives. (*This seems to be an objective in the Fusina wastewater project and appears to be part of the mandate of the Forum di Laguna group.*)

CHAPTER 3—CONRADIN AND BUCHLI

- National laws can be used as the impetus to initiate localized actions which are best implemented following creation of a city-wide revitalization plan. (*Many laws have been passed to facilitate restoration activities in Venice but there is certainly no egalitarian, city-wide agreement or plan on what should be done.*)
- The end goals in undertaking regenerative landscape design projects are as much about place-making in revitalized neighborhoods with ensuing economic benefits as they are about restoring ecological or hydrological benefits. (*These are issues that have not been given the weight they deserve in Venice.*)

- As a result, both site aesthetics and concerns about human habitability play major roles in regenerative landscape design. *(Restoration activities in Venice are concerned with these issues.)*
- Because of this, there is a need for interdisciplinary cooperation across both the design teams and the governmental departments. *(Restoration in Venice has developed into an antagonistic contact sport.)*
- Project success comes about through direct communication with the public to ensure enough momentum to move beyond the frequent governmental institutional resistance to change. *(Open and respectful communication is not a characteristic that can be used to describe the restoration activities in Venice.)*

CHAPTER 4—CLARK AND MIDDLETON

- Concepts of a restored ecology need to be linked to recreation and social values; i.e., restoration coupled with redevelopment that will produce parks as places for people and nature. *(Natural area restoration in Venice is not linked to creating parks.)*
- Community grassroots organizations can often lead in envisioning, initiating, and completing regenerative landscape design projects. *(These groups have been largely excluded from the MOSE deliberations in Venice.)*
- Fostering stewardship is essential with volunteers being mobilized for routine maintenance as well as in engaging local residents in appropriate aspects of the regeneration work itself. *(Environmental restoration in Venice is almost exclusively a professional activity with little participation from the lay public.)*

CHAPTER 5—O'NEILL AND GAYNOR

- In the absence of government will, regenerative landscape design projects need to rely upon grassroots, "bottom-up" efforts of concerned citizens. *(The activities of the Forum di Laguna represent an important vehicle for sociological change in Venice.)*
- There should be local ownership of the project at all development stages rather than simply bringing in the public at later stages after most of the important decisions have been made. *(The public seems to have had little impact on determining the restoration projects in Venice.)*
- Regenerative landscape design entails collaboration with academics and professionals in addition to citizen advocates and may require a large degree of perseverance to circumvent government bureaucracy and city administration intransigence to innovation. *(It is almost impossible to get around the bureaucracy in Venice.)*
- Regenerative landscape design projects deal with establishing social reconnections to repaired landscapes in addition to ecological improvement, both of which together, neither one more important than the other, being essential for creating livable, healthy communities. *(Restoration in Venice is mostly an environmental engineering activity.)*
- It may be difficult to find financial support for regenerative landscape design projects from traditional "green" sources as the latter still operate under a false belief in a dichotomy of nature being apart from culture, and in ecological preservation being more important than environmental/sociological regeneration. *(There is a legacy of absconded funds intended for restoration projects in Venice that scares away present-day potential donors.)*
- Shifts in the political landscape can greatly influence whether regenerative landscape design projects can be implemented. *(This is certainly characteristic of the situation in Venice with respect to MOSE.)*

CHAPTER 6—FOSTER

- Regarding regenerative landscape design as a philosophical exercise allows examination of existing principles rather than an acceptance of them at face value and enables deliberation among various proposed contentious reparative options. *(Such philosophical discourse appears to have played a minor role in the MOSE saga.)*
- Regenerative landscape design, regardless of science and policy, cannot succeed unless it is firmly rooted in the lives of the people and approached from a viewpoint of long-term objectives. *(Decisions in Venice seem to give little credence to the wishes of the constituents.)*
- As a result, regenerative landscape design projects should inspire people through community processes and serve to engage the public in the physicality of the act of regeneration as well as in the conceptual appreciation of its implications. *(This does not occur in Venice.)*
- Advocacy groups are instrumental for creating community regenerative landscape design initiatives because unless the public is involved in direct and meaningful ways the many barriers to implementing these projects are unlikely to be circumvented. *(These voices have been largely ignored in Venice.)*
- Ensuring public access to regenerative landscape design sites and developing programs of instilling environmental literacy through education are key elements in the eventual success of these projects. *(Movements have been made toward environmental education in Venice and the future wastewater treatment park is an important new direction in creating civic space.)*
- Regenerative landscape design should embrace concepts of adaptive management by continually reexamining prescribed reparative methodologies and if necessary alter them; i.e., this approach is based on suppressing scientific/engineering egos to prevent projects from falling into fundamentalism. *("Fundamentalism" is a very apt word to describe the attitudes of technocrats and decision makers in Venice.)*

CHAPTER 7—GOREAU AND HILBERTZ

- Modern environmental management often suffers from being a largely "top-down" imposition by outside agencies which tends to override or ignore real-world expertise. *(Local opinions are little regarded in Venice's restoration planning and interventions.)*
- Frequent lack of government support means that innovative regenerative landscape design often has to rely upon mobilizing communities to undertake the work through such "bottom-up" processes that are based on local wisdom rather than "expert" knowledge imported from distant sources. *(This is the reserve of the situation in Venice.)*

CHAPTER 8 —RASMUSSEN AND HURLEY

- Community support is essential for the success of regenerative landscape design projects and is garnered through strong emphases being placed on education outreach (including use of models), as well as an inclusionary process (meetings and workshops) in which the public feels its concerns are being seriously addressed. *(Again, concerns of the public have not played a major role in Venice.)*
- Engagement of volunteers in the physical acts of restoring damaged landscapes builds feelings of environmental stewardship in the community. *(Such engaged stewardship appears little evident in Venice.)*
- Green infrastructure provides an aesthetic solution to environmental problems and if deemed too provocative to the established engineering community, can be approached as a nonthreatening pilot or demonstration project. *(The several new wastewater treatment projects in Venice represent important new strides in this direction.)*

CHAPTER 9—VOGEL

- Following economic boom times, cities can collapse inward with crumbling infrastructure and populations fleeing to the suburbs and in such cases, therefore, regenerative landscape design is really about undertaking urban revitalization (dealing with transportation, economic development, political and neighborhood institutions, and the hopes, dreams, and aspirations of sometimes downtrodden people) in concert with a policy of returning nature. *(Lack of serious efforts along these lines has meant that Venice continues to hemorrhage away its citizens.)*
- The creation of "middle landscapes" characterized by renewed human spirits and restored surrounding ecologies can be facilitated through the use of art installations and exhibitions, workshops and design charettes, and other forms of public presentations to educate about and to satisfy the hunger of urban dwellers for nature. *(A few such preliminary activities have been underway in Venice but to limited effectiveness.)*

CHAPTER 10—SENOS

- Traditional environmental restoration is scientifically and politically based, frequently ignoring broader social needs and benefits and as a result needs to become more holistic in scope and execution. *(Restoration in Venice is not holistic in this regard.)*
- It is critical for regenerative landscape design to be based on a strategy that addresses both social and environmental complexities in urban settings where natural systems are frequently marginalized and city dwellers disconnected from them. *(Lack of public space in Venice's old city has meant that nature is little thought about by residents and tourists, all the more reason to be critical of the limited opportunities available for getting them out into the surrounding lagoon on nature trips.)*
- Participatory regenerative landscape design heals both the physical and social environments by building community and instilling responsibility, fostering, therefore, a new partnership with nature that is based on ethical and not merely tinkering relationships with nature. *(Restoration activities in Venice are not participatory endeavors.)*
- Social goals of education outreach, site aesthetics, improved local economic development, and increased neighborhood spirit are important objectives in regenerative landscape design. *(Venice is beginning to make progress in these directions.)*
- Regenerative landscape design is an inclusive process characterized by a high degree of transparency in decision making and a great flexibility in proposed solutions. *(Restoration decisions in Venice are anything but inclusive or transparent.)*

CHAPTER 11—WESSELS AND HOOGEVEEN

- Both a recognition of the importance of site history and an appreciation of the idiosyncrasies of local politics are essential to informing decisions about regenerative landscape design. *(Restoration in Venice is very much about this.)*
- Regenerative landscape design is a strategy embedded in an understanding that cultural heritage preservation needn't be about preserving museum pieces but instead about developing functional components of restored working landscapes. *(Venice is on the cusp of evolving into being a museum rather than a living city.)*
- Social as well as physical scientists should be involved in establishing feelings of trust with local communities through frequent consultations. *(The rancor of the restoration debate in Venice has meant that little trust or respect exists among the various players.)*

CHAPTER 12—CORMIER

- Regenerative landscape designers can be comfortable in describing their projects through personal perspectives just as through technical, professional documents. *(The restoration literature in Venice remains very technical.)*
- Creation of multi-, not uni-, functional landscapes is a key element in regenerative landscape design, which leads to the concept of green infrastructure: the idea of building parks with both form and function rather than highly engineered structures with only the latter attribute. *(Such projects are beginning to be planned for Venice.)*
- Citizens need to take ownership of the design concepts in order to develop personal involvement with the project. *(Citizens are for the most part excluded from restoration projects in Venice.)*
- Education outreach is essential for community building and for empowering citizen groups to mount effective challenges against unsympathetic developers or politicians. *(Until recently there has been a muzzling of information in Venice about restoration initiatives being planned or undertaken.)*

CHAPTER 13—HURLEY AND STROMBERG

- Unglamorous, mundane locations are ideal targets for regenerative landscape design with goals of enhanced human livability being as important as those of improved environmental health. *(Actions of the Insula authority in Venice are working toward repairing street infrastructure and livability.)*
- Aesthetic considerations are a major part of regenerative landscape design as is revealing the working mechanisms of ecological processes, i.e., successful projects blend both art and science. *(Restoration projects in Venice remain engineering techno-fix reparations with little or no attempt made to foster ecological literacy or environmental art.)*
- Education is therefore extremely important for instructing about urban ecology. *(Venice is making progress towards environmental education but this is still largely an indoor classroom process.)*
- Community participation serves to reduce the isolation of urban dwellers from nature and from their human neighbors and can lead to motivating environmental activism. *(Few serious opportunities exist for Venetian citizens to become physically engaged in the process of restoring their lagoon.)*

CHAPTER 14—KADLECIK AND WILSON

- Regenerative landscape design projects are often bottom-up grassroots initiatives. *(Again, restoration in Venice is very much a centralized, top-down initiative.)*
- Linking environmental with individual and cultural healing is important in the rebuilding of sustainable, socially-conscious communities. *("Healing" is a word rarely voiced in the restoration rhetoric of Venice.)*
- Engineering principles blended with design aesthetics can create multipurpose landscapes of high civic quality in terms of providing space for recreation and reflection. *(Restoration projects in the Venetian lagoon are justified for their scientific, not civic, benefits.)*
- In regenerative landscape design, ecological restoration supports historical preservation. *(This has become the new modus operandi in the restoration of Venice and its lagoon.)*

CHAPTER 15—MAYER-REED

- Regenerative landscape design acknowledges and celebrates postindustrial heritage rather than attempting to wipe the palimpsest landscape clean of human history by setting the clock back in time through returning the site to some previous "natural" state. *(Due to the widespread recognition that there is nothing about Venice's lagoon that could be termed "natural," historical fidelity is not a preoccupying issue there.)*
- Nature has an incredible, inherent regenerative capacity to heal itself and sometimes the best strategy is to simply let it do so on its own, especially if that can be incorporated into a didactic lesson about cultural landscape evolution. *(Undeveloped opportunities exist for educating about the nature of culture and the culture of nature through restoration projects in Venice.)*
- Tourism in terms of creating parks for both people and nature can be a motivation in the recuperation of derelict landscapes for purposes of regional economic revitalization. *(Only a few fledgling attempts have been made in Venice in this direction.)*
- Such projects rely upon careful attention paid to site accessibility and historical and natural interpretation. *(Tours and programs focusing on the ecocultural history of the Venetian lagoon have been little realized.)*

CHAPTER 16—APFELBAUM AND THOMAS

- Larger spatial-scale projects embrace political as well as ecological complexity. *(This is certainly the case in Venice.)*
- Regenerative landscape design at the watershed scale recognizes concepts of culturally-storied landscapes and attempts to restore natural processes through creating new human living spaces. *(Restoration in Venice is focused on either repairing human inhabitations or damaged natural areas rather than in developing new sites that integrate both culture and nature.)*
- In this respect, the well-being or health of human inhabitants of a landscape are intimately linked with the same for their nonhuman neighbors or co-inhabitants of those spaces. *(Perceptions of the nature–culture dichotomy remain very prevalent in Venice.)*
- It is necessary, therefore, to educate about the dependency of humans on the "natural capital" of ecosystem goods and services, and once this is understood, socio-economic, political, and regulatory barriers to maintaining these essential ecosystems will be lessened. *(Venice has long recognized the importance of its lagoon to determining its economy.)*

CHAPTER 17—GUNN, BECKETT, LAUTENBACH, AND MONET

- Regenerative landscape design strongly believes that even the most tortured of landscapes are worthy of our restorative efforts. *(In Venice, toxic waste landfills have been contained with many opportunities remaining to develop these into postindustrial use civic spaces.)*
- Undertaking such projects can revitalize regional economies through creating jobs. *(The concept of a "restoration economy" has yet to take root in Venice.)*
- Successful regenerative landscape design is often tied to innovative community planning, communicative initiatives, and aggressive political lobbying. *(Restoration in Venice continues to be largely an autocratic process uncoupled from sustainable development planning.)*
- Creation of a mosaic of local, neighborhood stewardship groups is an important element in overall project success. *(Though such groups do exist in Venice they do not have the visibility they deserve in terms of setting and guiding the restoration agenda.)*

Robert L. France
Venice, Italy

Index

Milton Keynes UK
Ingram Content Group UK Ltd.
UKHW051941071024
449327UK00026B/2113